Human Chromosomes

Fourth Edition

D0885119

Springer
*New York
Berlin
Heidelberg
Barcelona
Hong Kong
London
Milan
Paris
Singapore
Tokyo*

Human Chromosomes

Fourth Edition

Orlando J. Miller
Wayne State University School of Medicine

Eeva Therman
Professor Emeritus, University of Wisconsin–Madison

With 108 Illustrations, 9 in Full Color

Springer

Orlando J. Miller
Center for Molecular Medicine and Genetics
Wayne State University School of Medicine
3216 Scott Hall
540 East Canfield Avenue
Detroit, MI 48201, USA
ojmiller@cmb.biosci.wayne.edu

Eeva Therman
Laboratory of Genetics
University of Wisconsin
Genetics Building
Madison, WI 53706, USA

Cover illustration: A metaphase spread (front cover) and a karyotype (back cover) after in situ hybridization and detection of multiplex probes. Red and green bars represent regions that hybridized to fragments in only one probe pool, and yellow bars represent regions that hybridized equally to fragments in both probe pools. Mixed colors are due to overrepresentation of fragments from one pool. Regions that fail to hybridize to fragments in either pool show background DAPI blue fluorescence. (Reproduced from Müller et al., Toward a multi-color chromosome bar code for the entire human karyotype by fluorescence in situ hybridization, Hum Genet 100, fig. 3, page 273, copyright Springer-Verlag 1997.)

Library of Congress Cataloging-in-Publication Data
Miller, Orlando J.
    Human chromosomes/Orlando J. Miller, Eeva Therman.—4th ed.
        p. cm.
    Includes bibliographical references and index.
    ISBN 0-387-95031-1 (alk. paper)—ISBN 0-387-95046-X (softcover: alk. paper)
        1. Human chromosomes.   2. Human cytogenetics.   3. Human chromosome abnormalities.
    I. Therman, Eeva.   II. Title.
    QH431 .T436 2000
    611'.01816—dc21   00-044007

Printed on acid-free paper.

Production coordinated by Chernow Editorial Services, Inc., and managed by Francine McNeill; manufacturing supervised by Jeffrey Taub.
Typeset by Best-Set Typesetter Ltd., Hong Kong.

9 8 7 6 5 4 3 2 1

ISBN 0-387-95031-1   SPIN 10764931   (hardcover)
ISBN 0-387-95046-X   SPIN 10768650   (softcover)

Springer-Verlag   New York Berlin Heidelberg
A member of BertelsmannSpringer Science+Business Media GmbH

Preface

This book presents a comprehensive introduction to the principles of human cytogenetics and provides examples of their applications, especially those that are important in diagnostic and preventive medicine. The authors have each worked in human cytogenetics for more than 40 years and have witnessed first-hand the enormous strides made in the field during this time. The many advances made since the third edition of this book reflect the rapidly growing application of molecular biological techniques and concepts by human cytogeneticists. Insertion of transposable elements, genomic imprinting, and expansion of tri-nucleotide repeats are only a few of the important cytogenetic mechanisms that have been discovered and shown to play a role in producing disease phenotypes. Molecular cytogenetic methods have taken center stage in cancer studies with the demonstration that cancers arise by chromosomal mechanisms such as gene amplification, oncogene activation by chromosome rearrangement, ectopic recombination leading to loss of heterozygosity, and multiple mechanisms leading to genome destabilization.

We present a comprehensive and relatively brief overview of the principles of cytogenetics, including the important new disease mechanisms previously mentioned. Examples are chosen that illustrate these principles and their application, thus preparing the reader to understand the new developments that constantly appear in the laboratory, the clinic, and the current literature of this very active field. At many points in the book, important unsolved problems in cytogenetics are mentioned with suggestions of how a solution might be sought, which acts as a stimulus to the reader to come up with his or her own suggestions. The book should be particularly useful for physicians who want to keep up with new developments in this field and students interested in a career in medical genetics or genetic counseling. It could easily serve as a text for a one-semester college or graduate-level course in human cytogenetics, or, with some supplementation, a course in general cytogenetics.

In this edition, the text has been extensively reorganized and almost completely rewritten to incorporate essential insights from cell and molecular genetics, along with other advances in cytogenetics, and to present them in a systematic way. Examples have been chosen that not only emphasize the underlying principles but also illustrate the growing clinical importance of molecular cytogenetics. Most of the tables and the majority of the figures are new, and virtually all are based on studies of human chromosomes. We are grateful to the colleagues and copyright holders who have generously permitted the use of their published and unpublished figures and tables.

Detroit, Michigan   Orlando J. Miller
Madison, Wisconsin   Eeva Therman

Contents

Contents

Contents

13 Chromosome Structural Aberrations   187

14 The Causes of Structural Aberrations   207

15 Syndromes Due to Autosomal Deletions and Duplications   223

Contents

1

Origins and Directions of Human Cytogenetics

Cytogenetics is the study of the structure, function, and evolution of chromosomes, the vehicles of inheritance that reside in the cell nucleus. Cytogenetics deals with chromosome behavior during the divisions of the somatic and early germline cells that produce identical daughter cells with two sets of chromosomes (*mitosis*) or the final two germline cell divisions that produce germ cells with a single set of chromosomes (*meiosis*). Human cytogenetics is particularly concerned with how these processes may go wrong and how structural changes arise, because changes in the number or structure of chromosomes are major causes of mental retardation, multiple malformations, cancer, infertility, and spontaneous abortions.

Origins: Cytology, Genetics, and DNA Chemistry

Human cytogenetics had its beginning in the nineteenth century, aided by the development of the compound microscope, fixatives for preserving cell structure, and chemical dyes that preferentially stain nuclei and chromosomes. These early studies in cytology (now called cell biology) led to the elaboration of the cell theory, that all living cells come from pre-existing cells. The first study of human chromosomes, by Fleming in 1889, provided limited information. However, along with the work of van Beneden, Strasburger, Waldeyer, and others, it led to a clear understanding that the behavior of chromosomes in mitosis and meiosis is consistent and that their key features are virtually identical in animals and plants. The evidence was so impressive that Weissmann, in 1892, claimed that chromosomes were the physical basis of heredity. A method of genetic analysis was needed to test this hypothesis. This was provided by the rediscovery of the principles of Mendelian inheritance in 1901. Within a year, Sutton and Boveri independently reported that the segregation of each pair of homologous chromosomes and the independent assortment of nonhomologous chromosomes in meiosis could account for the corresponding behavior of Mendelian unit factors (genes), if these were carried by the chromosomes. This was the first major theoretical contribution of cytogenetics to genetics, and it led to the realization that each chromosome must carry many different genes. Intensive efforts over the next 90 years showed that in each organism the genes were arranged in linear arrays along the length of each chromosome, with each gene having a unique location on a particular chromosome. But what was the chemical nature of the gene? The pioneering work of Avery, McLeod and MacCarty, published in 1944, established that genes are composed of deoxyribonucleic acid, or DNA, not protein.

The presence of DNA as a major component of cell nuclei had been known since the work of Miescher, late in the nineteenth century. In 1924, Feulgen developed a method for staining nuclei and chromosomes that was based on a specific chemical modification of DNA. This Feulgen reaction, or stain, allowed quantitative measurements of the DNA content of tissues and even of individual nuclei after cytophotometric methods were developed. The cells of each organism had a characteristic amount of DNA, with the amount of DNA in non-dividing diploid somatic cells (2C) twice that in haploid germ cells (C). The 3.65 picograms of DNA in haploid sperm provided an estimate of the size of the haploid *genome*, or complete set of human chromosomes: approximately 3.4 billion nucleotide base pairs of DNA.

The existence of nucleotide base pairing in DNA was unknown until 1953, when Watson and Crick, aided by the X-ray crystallographic data of Wilkins and Franklin, proposed that DNA is composed of two long, spiraling strands of nucleotides that are held together by hydrogen bonds between pairs of bases. Two hydrogen bonds linked each adenine (A) with a thymine (T), and three hydrogen bonds linked each guanine (G) with a cytosine (C). These were the only types of base pairs permitted by the proposed structure of DNA. As a consequence, the sequence of bases in one strand determines the sequence of bases in the complementary strand. Furthermore, the genetic information carried by each chromosome could be specified by the sequence of bases in its long DNA molecule. Crick showed that the information is, indeed, encoded in successive triplets of bases in DNA and in the messenger RNA (*mRNA*) that is transcribed from one strand of the DNA. Nirenberg, Mathai, and Ochoa worked out the complete genetic code—the particular nucleotide triplets (codons) that specify each amino acid in a growing polypeptide chain or protein and the stop codons that terminate chain growth. This completed the logic of genetics and explained the earlier fundamental discovery of Garrod, Beadle, and Tatum that each enzyme protein is generally the product of one gene: the "one gene-one enzyme" hypothesis.

The Midwives of Human Cytogenetics

The early cytological history of human cytogenetics has been reviewed comprehensively by Makino (1975). Few studies on human chromosomes were published before 1952. That of Painter in 1923 was responsible for the notion that the human chromosome number is 48, a mistake that went uncorrected for the next 33 years. His report was worded quite cautiously; after all, it was based almost entirely on the analysis of a few cells from one institutionalized individual! That number may have been correct for that individual, as rare institutionalized individuals with 48 chromosomes are now well known. However, it may simply have been the result of the inadequate methods available to the early investigators, who had to examine serial sections of testes because the badly overlapping chromosomes were not even in one focal plane. Accurate studies of human chromosomes became possible only after several technical developments. Improved cell culture methods provided a ready source of individual dividing

3

cells that could be squashed on a slide. Blakeslee and Eigsti showed in 1936 that colchicine destroys the mitotic spindle and blocks cells in metaphase, facilitating their accumulation and study. Hsu discovered in 1952 that treatment of cells with hypotonic salt solution before fixation gave a marked improvement in chromosome spreading, especially when combined with the use of single-cell suspensions and the addition of colchicine to the cells before hypotonic treatment.

Taking advantage of these new methods, Tjio and Levan (1956) established that the correct human diploid chromosome number is 46, based on their study of cultured embryonic lung cells from several individuals. The same year, Ford and Hamerton confirmed this in spermatogonia and showed that cells in meiosis have 23 paired chromosomes, or *bivalents*. Methods continued to improve. Air-drying cell suspensions directly on microscope slides gave better spreading and flattened the entire metaphase spread into a thin focal plane. An important innovation in cell culture technique came with the discovery in 1960 by Moorhead and his associates that peripheral blood lymphocytes can be induced to divide after a few days in culture in the presence of phytohemagglutinin, a bean extract. Because blood samples are so readily available, chromosome studies could be carried out quickly and easily on virtually anyone. Such cultures are still one of the most widely used sources of human chromosomes. An important additional source is amniotic fluid. In 1966, Steele and Breg reported that cells cultured from amniotic fluid could be used to determine the chromosome constitution of the fetus. This is the technique that is still most widely used for prenatal chromosome studies, although rapidly growing numbers of studies are now carried out on cells cultured from biopsies of chorionic villi taken from the placenta during the first trimester of pregnancy.

The Birth of Clinical Cytogenetics

The new techniques were soon applied to individuals who were mentally retarded or had multiple malformations. Miller (1995) gives a short review of this early phase, with extensive references. Lejeune et al. (1959) found that Down syndrome in several subjects was caused by the presence of three copies (*trisomy*), instead of the normal two, of number 21, one of the smallest human nonsex chromosomes, or *autosomes*. The same year, Jacobs and Strong found a male with Klinefelter syndrome who had an XXY complement, while Ford and his collaborators reported females with Turner syndrome who had a single X (XO or *monosomy* X) or were XO/XX *mosaics*, with both XO and XX cells. They also reported the first case of *double aneuploidy*: an extra sex chromosome and an extra chromosome 21 (XXY-trisomy 21) in a man with 48 chromosomes who had both Klinefelter and Down syndromes. These observations indicated that *sex determination* in humans depends upon the presence or absence of a Y chromosome and not on the ratio of X chromosomes to autosome sets, as it does in *Drosophila*. The Y chromosome is male determining even in the presence of as many as three or four X chromosomes, since XXXY and XXXXY individuals are male. Individuals who have a Y chromosome in only a fraction of their cells, such as XO/XY mosaics or XX/XY *chimeras*, which arise from fertilization by two sperm, often show mixed, or intersexual, development (Chapters 19 and 21).

The presence of multiple malformations in 21 trisomic patients led to the search for trisomy of other autosomes among patients with multiple malformations. Trisomy 13 and trisomy 18 were discovered in 1960 by groups headed by Patau and Edwards, respectively (Chapter 12). No additional trisomies were found in liveborns, so attention turned to a search for chromosome abnormalities in spontaneously aborted embryos or fetuses, based on the assumption that trisomy for these autosomes might act as embryonic lethals. Carr, and later Boue's group, carried out extensive studies of spontaneous abortuses and found that autosomal trisomies represented about 3% of all recognized pregnancies. In addition, XO and triploid (3n) abortuses were also extremely common: each made up about 1% of recognized pregnancies. Clearly, these chromosome constitutions were quite lethal, and human meiosis quite error-prone (Chapter 11).

Only structural aberrations that produced large changes in the length or arm ratio of a chromosome could be detected with the methods available before 1970. These included Robertsonian translocations, which involve the long arms of two acrocentric chromosomes, such as numbers 14 and 21. Their discovery by Penrose, Fraccaro, and others was the result of studies of exceptional cases of Down syndrome in which the mother was young or there was an affected relative (Chapter 13). A number of deletions were also detected. Nowell and Hungerford noted in 1960 the consistent presence of a deleted G group (Philadelphia) chromosome in chronic myelogenous leukemia cells. Later, using banded chromosomes, Rowley showed that the aberration was really a specific translocation (Chapter 27). The first example of a deletion of a D group chromosome was found in 1963 in a patient with retinoblastoma by Penrose's group, who pointed out that if the deletion was responsible for the disease, a gene for retinoblastoma must be in the deleted segment. Deletions were shown to cause some characteristic and previously unrecognized clinical syndromes: Lejeune's cri du chat (cat cry) syndrome by a deletion of the short arm of chromosome 5 and the Wolf-Hirschhorn syndrome by a deletion of chromosome 4 (Chapter 15). Characteristic phenotypes were also noted by de Grouchy and others in patients with deletions of either the long arm or the short arm of chromosome 18. In some families, the presence of a chromosome rearrangement in one parent was responsible for the deletion or other chromosome imbalance in one or more children (Chapter 16).

The Lyon Hypothesis

In 1949, Barr and Bertram described what is called the sex chromatin, or *Barr body*, which is visible in some cell nuclei in all mammalian females. Its frequently bipartite form led to the hypothesis that it arose from pairing of highly condensed (*heterochromatic*) segments of the two X chromosomes. The finding of three X chromosomes in individuals with two Barr bodies in some of their cells (for example, Jacobs et al., 1959) provided a key piece of information and made an alternative hypothesis more attractive: that a Barr body is a single heterochromatic X chromosome. This interpretation, first formulated by Ohno, played a role in the development of the single-active-X, or Lyon, hypothesis (Lyon, 1961). The hypothesis states that in mammalian females one X chromosome is inactivated in all the cells at an early embryonic stage. The original choice of which X is inactivated is random, but the same X remains inactive in all the descendants of that cell. If a cell has more than two X chromosomes, all but one are inactivated, and each inactive X chromosome may form a Barr body (Chapter 18). The Lyon hypothesis remains one of the most important theoretical concepts in human, and all mammalian, cytogenetics. It has directed research and led to many key insights in developmental and cancer biology and to even more in other areas of genetics, as described throughout later chapters.

Adolescence: The Chromosome Banding Era

Until 1970, chromosome identification, and particularly the identification of structural changes, was severely limited. Normal chromosomes could be sorted into seven groups on the basis of length and arm ratio, but only a few chromosomes can be individually recognized. Autoradiographic DNA replication patterns helped identify a few more, but this was time-consuming and still of very limited value. The introduction of chromosome banding techniques revolutionized human cytogenetics. In 1970, Caspersson et al. discovered that quinacrine mustard produces consistent fluorescent banding patterns along each human chromosome that are so distinctive that every chromosome can be individually identified. This discovery was followed by a flood of additional banding techniques, whose use greatly simplified chromosome studies and made possible the identification of an enormous range of chromosome abnormalities, especially structural aberrations, such as translocations, inversions, deletions, and duplications, that were previously undetectable. A fundamental discovery was that the DNA in each chromosome band replicated during a specific part of the DNA synthetic (S) phase of the cell cycle (Latt et al., 1973; see also Chapter 3). A standard system of chromosome nomenclature was developed through a series of conferences and publication of their recommendations. A standing committee now publishes comprehensive booklets incorporating the accepted nomenclature and modifications necessitated by new developments in the field. The most recent is ISCN (1995), an international system for human cytogenetic nomenclature.

Somatic Cell Genetics and Chromosome Mapping

Harris (1995) has published an excellent historical review of somatic cell genetics. The initial goals of this field were the development of methods for analyzing the segregation of mutant alleles and the recombination of linked genes using cultured somatic cells. Segregation of homologous chromosomes and the alleles they carry occurred in human-rodent somatic cell hybrids that had lost some of their human chromosomes. Thousands of genes were mapped to chromosomes using panels of these hybrids (Chapter 29). The linear order and physical distance separating these gene locations (*loci*) on a chromosome could be determined using hybrids carrying different segments of a particular chromosome. The most precise mapping was achieved using a panel of *radiation reduction hybrids* that contained different mixtures of the human chromosome complement (genome) or of any particular chromosome, because the human chromosomes were fragmented by massive irradiation of the cells before being hybridized with rodent cells (Chapter 23). These approaches were so successful that they continue to dominate gene mapping.

Maturity: The Molecular Era

The molecular era was ushered in by the development of methods for manipulating DNA. Marmur, Doty, Spiegelman, and Gillespie showed that the two strands of DNA fragments could be easily separated (denatured, or dissociated) and that complementary strands could be reannealed (renatured, or reassociated), even in the presence of large amounts of noncomplementary DNA. Such *molecular hybridization* became of fundamental importance in molecular biology, and *in situ hybridization* of labeled DNA probes to the DNA in cytological preparations of chromosomes and nuclei became a powerful tool in human cytogenetics. Several thousand loci have been mapped by *fluorescent in situ hybridization (FISH)*, a faster, more reliable, and more precise method for mapping genes than the autoradiographic detection of radioactively labeled DNA fragments (Chapter 8). Autoradiographic in situ hybridization led to the fundamental discovery that cytoplasmic mRNA molecules are much shorter than the genes that encode them, because of the presence of intervening sequences (*introns*) that break up the protein-coding portion of most genes into short segments called *exons*. The invention of methods to determine the sequence of nucleotide bases in DNA (*DNA sequencing*) by Gilbert and Sanger made it possible to characterize precisely the genes and other parts of the genome. Another powerful tool was the *polymerase chain reaction* (PCR), invented by Mullis. It permitted rapid amplification of any short fragment of DNA, yielding up to a million-fold increase in the number of copies, and this revolutionized many aspects of cytogenetics.

An important advance was the development of methods for cloning DNA fragments, including genes. This was based on the discovery that bacterial viruses are still infectious when they contain an inserted fragment of human DNA. Plasmids would accept inserts up to about 5 kilobase pairs (kb) in length, bacteriophages up to 15 kb, and cosmids up to 50 kb. A bacterial cell infected with a single recombinant virus could be grown into a clone of millions of cells each containing the same unique fragment of human DNA. Alternative techniques were developed to clone larger DNA fragments. Yeast artificial chromosomes (YACs), with a yeast centromere and telomeres, could accept human fragments of several hundred kb or rarely 1–2 megabase pairs (Mb) and functioned as fairly stable chromosomes in yeast. Bacterial artificial chromosomes (BACs), containing 160–235 kb of human DNA, were even more useful, being stable in bacteria and easier to purify. Any of these cloned fragments could be labeled to produce radioactive or fluorescent DNA probes. These could be used in molecular hybridizations to detect complementary DNA that has been size-separated by gel electrophoresis and transferred to nitrocellulose filters by a simple method called *Southern blotting*. The hybridized probe could also be detected in fixed cells or chromosome spreads by in situ hybridization (Chapter 8). For further information about molecular genetics, see Strachan and Read (1996) and Lewin (1997).

Molecular hybridization was widely used to construct genetic linkage maps (Chapter 29) and to determine the parent of origin of a deleted chromosome or of the extra chromosome in a trisomic individual (Chapters 11 and 15). It was also used to show that some individuals with a normal chromosome number received both their copies of a particular chromosome from the same parent (*uniparental disomy*). This led to the discovery of a previously unrecognized cause of disease and a novel mechanism of gene regulation, called *genomic imprinting*: the normal inactivation of either the maternal or the paternal copy of a gene (Chapter 21). Molecular methods were instrumental in the identification of many of the genes that regulate the cell cycle (Chapter 2) and clarification of the mechanisms by which chromosome breakage can lead to cancer (Chapters 24–28). Molecular methods led to the discovery of abundant interspersed sequence elements repeated many times in the genome. Many of these are still capable of acting as *transposable elements*, that is, moving to new locations in the genome, sometimes disrupting a gene or breaking a chromosome. Barbara McClintock described this behavior nearly 50 years ago in maize, and it appears to be universal in eukaryotes. For an interesting account of some of the fundamental discoveries made by this great cytogeneticist, see Federoff and Botstein (1992).

The genetic basis of sex determination was advanced by the molecular characterization of a gene on the Y chromosome, *SRY*, that is required for male sexual development. A number of genes on the autosomes are also involved in the complex process of male sex differentiation (Chapter 17). The molecular basis of X inactivation was advanced by the discovery of the *XIST* gene and the demonstration that it is expressed strongly only from the copy on the X chromosome that is inactivated (Chapter 18). Surprisingly, the *XIST* gene does not code for a protein; instead, its RNA product, or *transcript*, coats the X chromosome carrying it and mediates inactivation of this chromosome, in an unknown manner (Brown et al., 1992).

Ohno (1967) suggested that the gene content of the X chromosome has remained virtually unchanged throughout mammalian evolution—some 125 million years—because the dosage compensation mechanism associated with X inactivation would interfere with transfer of genes between the X chromosome and an autosome. However, the X chromosome of marsupials and monotremes, which split from placental mammals roughly 140–165 million years ago, is only about 60% as large as that of placental mammals. Molecular cytogenetic studies suggested that most of the short arm of the human X chromosome consists of genes that are autosomal in marsupials and monotremes but have managed to make the transition to the X chromosome and a dosage-compensated system (Graves et al., 1998).

Molecular cytogenetic methods led to spectacular successes in understanding cancer (Chapters 24–28). Major causes of genome destabilization were found, and chromosome instability was shown to be a major factor in carcinogenesis and tumor progression (Chapters 24 and 26). Molecular cytogenetic studies of chromosome aberrations specifically associated with particular types of cancer led to the discovery of many tumor suppressor genes (Chapter 28) and proto-oncogenes (Chapter 27), providing key insights into the origin of cancer.

Despite the impressive growth of knowledge about human chromosomes in the last 40 years, important questions remain unanswered. The enormous data-bases generated by the Human Genome Project, and powerful new technologies to generate and analyze data, can be used to address these (Chapter 31). Human cytogenetics promises to remain an exciting field, both for its scientific challenges and for its rapidly growing applications in the diagnosis, treatment, and prevention of human disease.

References

Brown CJ, Hendrick BD, Rupert JL, et al. (1992) The human *XIST* gene: analysis of a 17 kb inactive X-specific RNA that contains conserved repeats and is highly localized within the nucleus. Cell 71:527–542

Caspersson T, Zech L, Johansson C (1970) Differential banding of alkylating fluorochromes in human chromosomes. Exp Cell Res 60:315–319

Federoff N, Botstein D (1992) The dynamic genome. Barbara McClintock's ideas in the century of genetics. Cold Spring Harbor Laboratory, Plainview

Graves JAM, Disteche CM, Toder R (1998) Gene dosage in the evolution and function of mammalian sex chromosomes. Cytogenet Cell Genet 80:94–103

Harris, H (1995) The cells of the body. A history of somatic cell genetics. Cold Spring Harbor Laboratory, Plainview

ISCN (1995) An international system for human cytogenetic nomenclature (1995). Mitelman F (ed) Karger, Basel

Jacobs PA, Baikie AG, Court Brown WM (1959) Evidence for the existence of a human "superfemale." Lancet ii:423–425

Latt SA (1973) Microfluorometric detection of deoxyribonucleic acid replication in human metaphase chromosomes. Proc Natl Acad Sci USA 70:3395–3399

Lejeune J, Gautier M, Turpin R (1959) Etude des chromosomes somatiques de neuf enfants mongoliens. Compt Rend 248:1721–1722

Lewin B (1997) Genes VI. Oxford, New York

Lyon MF (1961) Gene action in the X-chromosome of the mouse (*Mus musculus* L.). Nature 190:372–373

Makino S (1975) Human chromosomes. Igaku Shoin, Tokyo

Miller OJ (1995) The fifties and the renaissance in human and mammalian cytogenetics. Genetics 139:489–494

Ohno S (1967) Sex chromosomes and sex-linked genes. Springer, Heidelberg

Strachan T, Read AP (1996) Human molecular genetics. Wiley-Liss, New York

Tjio JH, Levan A (1956) The chromosome number in man. Hereditas 42:1–6

2

The Mitotic Cell Cycle

Proliferating cells go through a regular cycle of events, the *mitotic cell cycle*, in which the genetic material is duplicated and divided equally between two *daughter cells*. This is brought about by the duplication of each chromosome to form two closely adjacent *sister chromatids*, which separate from each other to become two *daughter chromosomes*. These, along with the other chromosomes of each set, are then packaged into two genetically identical *daughter nuclei*. The molecular mechanisms underlying the cell cycle are highly conserved in all organisms with a nucleus (*eukaryotes*). Many of the genes and proteins involved in the human cell cycle have been identified because of their high degree of nucleotide and amino acid sequence similarity to homologous genes and proteins in the more easily studied budding yeast, *Saccharomyces cerevisiae*, in which the cell cycle is more fully understood.

The Cell Cycle: Interphase, Mitosis, and Cytokinesis

The interphase of the cell cycle (Fig. 2.1) is generally divided into three phases: G1 (Gap1), S (DNA synthetic), and G2 (Gap2). The mitotic part of the cycle (M) is divided into five phases: prophase, prometaphase, metaphase, anaphase, and telophase. Mitosis is followed by cytokinesis, the division of cytoplasm and cell membrane required to complete the formation of two daughter cells. All phases of the cell cycle are marked by an orderly progression of metabolic processes. Cell differentiation is generally associated with a loss of proliferative capacity (for example, in neurons or in muscle, kidney, or liver cells); these cells are in a resting state, called G0. Some G0 cells, such as lymphocytes or liver cells, can be induced by a mitogen or growth factor to enter the cell cycle and divide. Usually, extracellular mitogens bind to a membrane-bound receptor and trigger the "immediate early response" genes, *JUN*, *FOS*, and *MYC* (Fig. 2.2). These in turn trigger the sequential synthesis of cyclins and activation of specific cyclin-dependent kinases. These proteins (Fig. 2.1) and their genes (Table 2.1) regulate the cell cycle. They are also important in carcinogenesis, because cell cycle regulatory genes can function as oncogenes (or cease functioning as

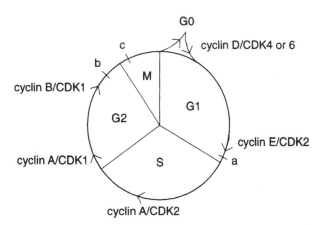

Figure 2.1. The cell cycle involves sequential activation of cyclin-dependent kinases (CDKs) by binding to cyclins. Cyclin D activates CDKs 4 and 6. The cycle can be blocked at G1 and G2 DNA damage checkpoints (a and b) or at a spindle assembly checkpoint (c).

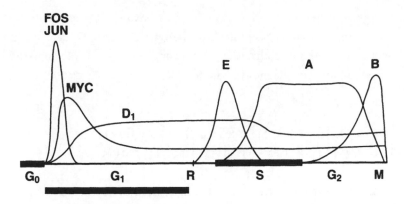

Figure 2.2. Changes in gene expression levels during the cell cycle. *FOS, JUN,* and *MYC* are immediate early response genes whose expression is triggered by extracellular stimuli. The ensuing sequential expression of cyclins D1, E, A, and B drives the cell cycle (Roussel, 1998, with permission, Academic Press).

Table 2.1. Major Cell Cycle Regulatory Proteins and the Genes That Encode Them

Activators		Inhibitors		
Protein/gene	Location*	Protein	Gene	Location
Cyclin A	4q26–q27	RB1	*RB1*	13q13
Cyclin B1	5q13–qter	HDAC1	*HDAC1*	
Cyclin D1	11q13	p53	*TP53*	17p13
Cyclin D2	12p13			
Cyclin D3	6p21	**General Cyclin/CDK inhibitors**		
Cyclin E	19q12–q13	p21	*CIP1/WAF1*	6p21.2
CDK1/CDC2	10q21	p27	*KIP1*	12p12–p13
CDK2	12q13	p57	*KIP2*	11p15.5[†]
CDK4	12q13			
CDK6	7p13–cen	**Specific Cyclin/CDK4 and 6 inhibitors**		
CDC25A	3p21	p15	*INK4B*	9p21–p22
		p16, p19	*INK4A*	9p21–p22
		p18	*INK4*	1p32

*Nomenclature of ISCN (1995)

[†]An imprinted gene (Chapter 21)

Source: Adapted from Hall and Peters (1997), with permission, Academic Press

tumor suppressor genes) when activated (or inactivated) by specific translocations, deletions, or other types of genetic change (Chapters 27–28). Their aberrant function can also destabilize the genome and lead to multiple numerical and
structural chromosome changes (Chapter 26).

Cell Cycle Progression: Cyclins and Cyclin-Dependent Kinases

The cell cycle is driven by the sequential activation of key proteins by *phospho-rylation*, the addition of phosphate groups to specific amino acid residues in the
proteins. Proteins that are called *cyclins*, because their synthesis is restricted to
particular parts of the cell cycle, play an important role (Figs. 2.1 and 2.2). They
activate a series of *cyclin-dependent kinases* (CDKs), the enzymes that carry out the
phosphorylations. Cyclins D1, D2, and D3 activate CDK4 and CDK6 and drive
G0 and G1 cells through G1. The cyclin E/CDK2 complex then drives the cells
through early S, cyclin A/CDK2 through mid-S, cyclin A/CDK1 (also called
CDC2, cell division cycle protein 2) through late S-early G2, and cyclin
B/CDK1 to the G2/M transition (Roussel, 1998). During cell differentiation, cell
lineages arise that are responsive to specific mitogenic signals, such as lym-
phokines in certain classes of lymphocytes and hormones in specific endocrine
tissues. Specific cell cycle proteins are involved in this process. Thus, ovarian
follicle cells proliferate in response to follicle stimulating hormone, but this is
dependent upon the presence of a specific cyclin, D2, which activates the necessary CDK4 or CDK6 (Sicinski et al., 1996). Mitogens of many types lead to
cell proliferation by binding to specific receptors and activating a cell cycle
signaling pathway (Fig. 2.3).

How does this cascade of phosphorylations lead to cell proliferation? Very
simply, the activated cyclin-dependent kinases add a few phosphate groups to
the retinoblastoma tumor suppressor protein, pRB (or RB). This reduces its
binding affinity for the E2F transcription factor, which is released in an active
form (Fig. 2.3). The underphosphorylated form of RB binds E2F, blocking its
function by a mechanism that was only recently discovered. Underphosphorylated RB in the RB/E2F complex binds a histone deacetylase (HDAC1), which
removes acetyl groups from the protruding tails of the adjacent nucleosomal histones H3 and H4 (Chapter 5), creating a localized region of heterochromatin
(Fig. 2.4). This reduces the access of transcription factors to the promoters of
the genes in this region and actively represses their transcription (Magnagni-

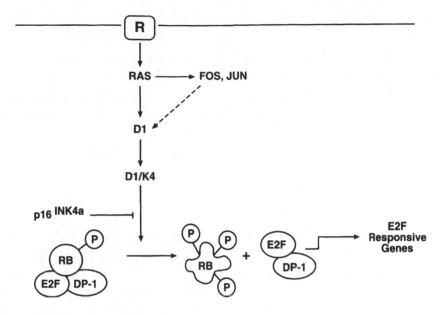

Figure 2.3. Signal transduction pathway from extracellular mitogen to the release of active E2F transcription factor. This step can be blocked by the inhibitory protein p16, a product of the *INK4A* gene (Roussel, 1998, with permission, Academic Press).

Jaulin et al., 1998). Other protein complexes with HDAC1 are involved in the tissue-specific repression of genes (Chapter 5), in X inactivation (Chapter 18), and in imprinting (Chapter 21).

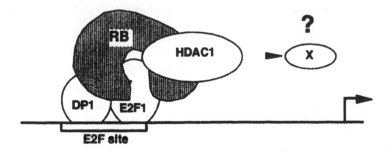

Figure 2.4. Model of repression of the transcription factor E2F1 in its E2F1/DP1 complex. RB binds to E2F and recruits HDAC1 to E2F sites in promoters. This allows HDAC1 to deacetylate a target protein (X) locally, blocking transcription (Magnagni-Jaulin et al., 1998, reprinted with permission from Nature, copyright 1998, Macmillan Magazines Limited).

When activated CDK4 and CDK6 phosphorylate RB, it releases E2F, which then activates the synthesis of genes required for cell proliferation. One of these, cyclin E, activates CDK2 kinase, leading to phosphorylation of still more proteins and the initiation of the DNA synthetic (S) phase. DNA replication is so important and has so many applications in cytogenetics that it is discussed more fully in Chapter 3. Progression from S to G2 is marked by phosphorylation of cyclin E, marking it for destruction by ubiquitin-mediated proteolysis. The same fate is met by other short-lived regulatory proteins, including E2A, p53, topoisomerase II, and the other cyclins in activated cyclin/CDK complexes, at the appropriate times (Murray, 1995). This ensures that DNA is replicated only once during each cell cycle.

Entry into mitosis (the G2/M transition) is mediated by the activation of a cyclin-dependent kinase, the cyclin B/CDK1 complex, which is also known as maturation factor or *mitosis promoting factor* (MPF). MPF was discovered when fusion of cells from different phases of the cell cycle revealed that cells in mitosis contain a factor (MPF) that can drive interphase cells into mitosis, producing *premature chromosome condensation*, or PCC (Fig. 23.3). G1 cells display long single chromosome fibers, G2 cells (in which the DNA strands have replicated) have long paired chromatids, and S-phase cells have fragmented chromosomes. MPF is activated by removal of phosphate groups by a phosphatase, CDC25, which appears in late S and peaks in G2. Microinjection of antibodies to CDC25 blocks the cell's entry into mitosis (Lammer et al., 1998).

The activated kinase, cyclin B/CDK1, as MPF is now called, is a component of mitotic chromatin that phosphorylates many proteins and thus triggers several independent mitotic processes: breakdown of the nuclear envelope, chromosome condensation, and spindle assembly. Phosphorylation of nuclear lamins leads to disassembly of the nuclear envelope (Gerace and Blobel, 1980); phosphorylation of histones H1 and H3 leads to chromosome condensation; and phosphorylation of stathmin, also called oncoprotein 18, leads to spindle assembly (Marklund et al., 1996). At the metaphase-anaphase transition, an evolutionarily highly conserved multi-protein complex called ubiquitin-protein ligase, or simply the anaphase-promoting complex (APC), catalyzes the conjugation of cyclin B and related regulatory proteins to ubiquitin, targeting them for proteolysis (King et al., 1995). In the absence of activated cyclin-dependent kinases to keep it phosphorylated, the RB protein loses its extra phosphate groups and again binds E2F, inactivating it and completing the cycle. Table 2.1 lists key cell cycle regulatory proteins and their gene loci. Many of these have been mapped by fluorescence in situ hybridization (Chapter 8; Demetrick, 1995). Mutation, disruption, and deletion of these genes play a major role in carcinogenesis (Chapters 26–28).

Cell Cycle Checkpoints

Genome integrity and the fidelity of mitosis are enhanced by a series of check-points that monitor the successful completion of the successive phases of the cell cycle and block further progression until each phase is completed (Fig. 2.1). Cells arrest in G1 and G2 if their DNA has been damaged (by ionizing radiation, say), and progression is delayed until DNA repair has occurred or cell death ensues. DNA damage triggers the sequential activation of proteins (Fig. 2.5), first the mutated in ataxia-telangiectasia (ATM) protein (Kastan et al., 1992), then the p53 tumor suppressor protein, and then the GADD45 protein, which binds to the proliferating cell nuclear antigen (PCNA) and blocks DNA replication (Chapter 3; Levine, 1997). There is also a spindle assembly checkpoint (Fig. 2.1). Cells arrest in G2 and delay their exit from mitosis in response to spindle microtubule disruption by spindle poisons such as colchicine, nocodazole, or benzimidazole.

A series of CDK inhibitors are involved in normal and genotoxic damage–induced checkpoint regulation of the cell cycle (Table 2.1). The *INK4A* gene product, p16, and two other proteins, p15 and p18, specifically inhibit the cyclin D–dependent kinases CDK4 and CDK6 and can lead to G1 arrest. The *KIP1* gene product, p27, as well as p21 and the *KIP2* gene product, p57, inhibit all cyclin/CDK kinases and can arrest the cycle at various points. The amount of the p27 protein is high in G0 cells but is reduced by mitogens. The tumor suppressor protein, p53, stimulates production of p21 and thus blocks cell proliferation (Levine, 1997). Mutation of the p53 gene is extremely

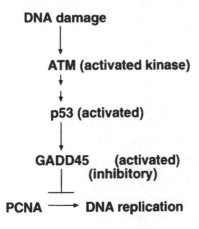

Figure 2.5. A signaling pathway by which DNA damage activates a cell cycle checkpoint and blocks DNA replication.

important in carcinogenesis because of its key role both in DNA damage checkpoints and at spindle assembly checkpoints. These checkpoints are discussed more fully in Chapter 26 because of the critical role they play in genome stability.

Prophase

In interphase nuclei, chromosomes are invisible except for occasional clumps of chromatin, called chromocenters, which vary in size and shape. One or more nucleoli are usually visible. At prophase, the chromosomes become visible as long thin threads that gradually shorten and thicken (condense) as mitosis progresses (Fig. 2.6). Chromosome condensation at the onset of prophase requires topoisomerase II (TOPO II), as shown by immunodepletion and complementation studies. TOPO II (scaffold protein 1, or SCI) is the most abundant component of the chromosome scaffold (Chapter 5). It reaches its peak level and activity at the G2/M transition. It binds to DNA at AT-rich sites called *scaffold attachment regions* (SARs) and appears to aggregate these, aiding in chromosome condensation (Adachi et al., 1991). As its name implies, TOPO II also introduces transient double-strand nicks that permit separation of the intertwined newly replicated DNA of sister chromatids. The second most abundant chromosome scaffold protein, SCII, is also involved in chromosome condensation. It belongs to the highly conserved SMC (stability of minichromosomes) family of proteins found in all eukaryotes. Their structure resembles that of mechanochemical (motor) proteins like myosin and kinesin, suggesting that chromosome condensation involves a motor function to bring chromosome regions more tightly together (Koshland and Strunnikov, 1996).

During mitosis, RNA synthesis is completely shut down, as first shown autoradiographically in the 1960s by the absence of incorporation of radioactively labeled [³H]uridine, which is a specific precursor of RNA. The mechanism of shutdown involves displacement of all the sequence-specific transcription factors from the promoter sequences of the genes, perhaps due in part to the action of the SMC protein complex, which can bring about renaturation of any single stranded regions in DNA (Sutani and Yanaguta, 1997). The SMC complex is also associated with the deacetylation and phosphorylation of the core histones that accompany chromosome condensation (Chapter 5). Nucleoli diminish in size as a result of the near cessation of ribosomal RNA synthesis, which is absent from prometaphase to late anaphase (Chapter 4).

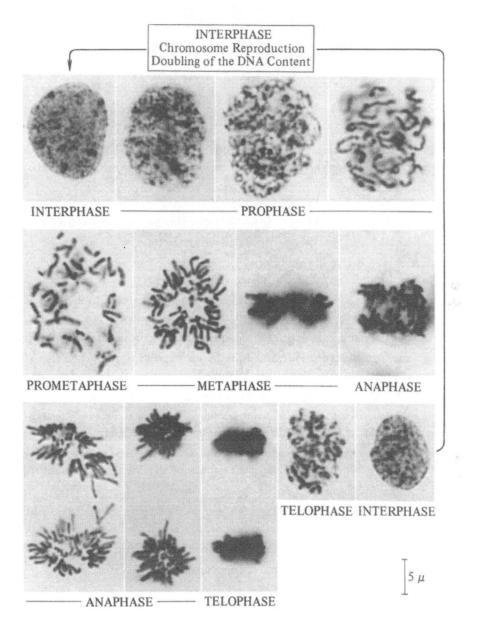

Figure 2.6. Mitotic cycle in cultured lymphocytes (Feulgen-stained squash preparations).

Prometaphase, Centrosomes, and the Mitotic Spindle

Prometaphase is associated with the breakdown of the nuclear membrane and the disappearance of nucleoli. The chromosomes continue their condensation, and a new structure, the mitotic spindle, is organized between the two centrosomes. The *centrosome* is the major microtubule organizing center (MTOC) in both interphase and mitotic cells. It consists of the highly structured *centriole* and its surrounding proteinaceous matrix, and contains specific proteins such as γ-tubulin and centrosomins A and B (Petzelt et al., 1997). The γ-tubulin anchors the minus (slow-growing) end of microtubules to the pericentriolar material of the centrosome. The plus (fast-growing) end of each microtubule projects into the cytoplasm during interphase and into the spindle during mitosis (Zheng et al., 1995).

The centrosome duplicates itself once per cell cycle, starting at the G1/S transition. Cyclin E/CDK2 is essential for centrosome duplication in S, ensuring proper coordination of spindle assembly with other events of the cell cycle (Fig. 2.1; Hinchcliffe et al., 1999). At the G2/M transition (prophase), the duplicated centrosomes separate and go to opposite poles of the nucleus. A kinesin-related motor protein, Eg5, is essential for this process, as microinjection of antibodies to Eg5 arrests cells in mitosis with a monopolar spindle. To act as a motor protein, Eg5 must first be phosphorylated by the cyclin-dependent kinase CDK1; this coordinates its activity with other events at the G2/M transition (Blangy et al., 1995). The separated centrosomes reorganize the microtubules of the cell by the coordinated dissolution of all the cytoplasmic microtubular apparatus and assembly of the mitotic spindle, with one centrosome at each pole. A complex of the nuclear mitotic apparatus protein (NuMA) and the cytoplasmic motor protein, dynein, is essential for assembly of the spindle. NuMA tethers bundles of microtubules and anchors them to the spindle poles. At telophase, NuMA is released from each pole as part of the dissolution of the spindle (Merdes et al., 1996).

Metaphase, Anaphase, and Telophase

The chromosomes reach their maximal state of compaction during metaphase. The spindle apparatus mediates the orchestrated movement of chromosomes during mitosis. Microtubules emanating from the spindle pole are captured by a special structure, the *kinetochore* (Chapter 4), which forms at the centromere of each chromatid and attaches it to the mitotic spindle. The kinetochore protein CENP-E belongs to the kinesin family of motor proteins. After phosphorylation by activated CDK1, it participates in the chromosome movements at metaphase and early anaphase (Liao et al, 1994). CENP-E and other motor proteins of the spindle fibers pull the kinetochores to a plane, called the *metaphase plate*, that is midway between the two poles of the spindle. The chromosomes reach their maximum condensation, 10,000-fold shorter than the DNA molecule itself, as anaphase begins.

The onset of anaphase is controlled by a specific mitotic *spindle assembly checkpoint* (Chapter 26) that monitors kinetochore attachment to the spindle and is mediated by an inhibitory signal produced by kinetochores that have not yet captured the necessary spindle microtubules (Wells, 1996). Apparently, it is the tension produced at the kinetochore by the bipolar attachment of spindle fibers that is monitored (Nicklas, 1997). The proteins required for proper checkpoint function and the genes that encode them are highly conserved during evolution. Mutations that enable cells to proceed through mitosis despite depolymerization of the spindle by nocodazole or benzimidazole have recently defined three MAD (mitosis arrest deficient) and three BUB (budding uninhibited by benzimidazole) spindle assembly checkpoint genes in budding yeast, and several of their human homologs have already been identified. One of them, MAD2, forms a complex with APC and the mitotic regulator CDC20 to block the activation of APC (Fang et al., 1998). Another, BUB1, may be the tension-sensitive kinase that provides the signal for anaphase to begin.

Once the spindle is assembled and attachment of each chromosome to spindle fibers from both poles is completed, producing tension, the transition from metaphase to anaphase occurs. The process is still very poorly understood. Two cell division cycle (CDC) proteins that co-localize to the spindle, CDC16 and CDC27, are required. When antibodies to human CDC27 are injected into HeLa (human cervical cancer) cells, the transition from metaphase to anaphase is blocked (Tugendreich et al., 1995).

During anaphase, the kinetochores of sister chromatids separate and the movement of motor proteins along the shortening spindle microtubules drags the daughter chromosomes towards opposite poles of the elongating spindle (Fig. 2.6). One motor protein, MKLP-1 (mitotic kinesin-like protein-1), associates specifically with the mitotic spindle. It bundles antiparallel microtubules and moves them apart at about the same speed as spindle elongation, suggesting that it provides the force for anaphase B, spindle elongation (Nislow et al., 1992).

The separation of sister chromatids requires ubiquitin-mediated breakdown of proteins by APC. It also requires topoisomerase II (TOPO II), to untangle the tangled, or catenated, strands of newly replicated DNA, and the TOPO II-interacting protein, BRRN1, the human homologue of the *barren* protein in *Drosophila* (Bhat et al., 1996). Inhibitors of TOPO II block sister chromatid separation; partial inhibition can lead to nondisjunction (Downes et al., 1991).

The regulation of anaphase is still rather poorly understood, but study of further mutations should be helpful. Two families have been described in which sister chromatids frequently separated prematurely in two unrelated infants, leading to nondisjunction, multiple mosaic trisomies and monosomies, and severe malformations. All four parents had a milder form of premature chromatid separation and were presumably heterozygous for the mutation that was homozygous in their severely affected offspring (Kajii et al., 1998).

During telophase, the mitosis-to-interphase transition, the chromosomes decondense, the mitotic spindle dissipates, the microtubules reorganize into a cytoplasmic microtubule network, the nuclear envelope reassembles, and nucleoli reappear. Nucleoli tend to fuse, but the largest number observed at telophase indicates the number and chromosomal location of active ribosomal RNA gene clusters, the *nucleolus organizers* (Chapter 4). Telophase is usually followed by cytoplasmic division (cytokinesis), after which the cells are once more in interphase. A novel human protein, PRC1, associates with the mitotic spindle and localizes to the cell midbody during cytokinesis. Microinjection of anti-PRC1 antibodies into human HeLa cells does not affect nuclear division (mitosis) but blocks cytokinesis and cell cleavage (Jiang et al., 1998).

Nondisjunction, Loss of Chromosomes, and Mosaicism

The orderly segregation of daughter chromosomes in anaphase may fail for various reasons, such as mutation of one of the many genes required for mitosis or advancing maternal age, which has a profound effect of unknown cause. The inclusion of both daughter chromosomes in the same nucleus, by whatever mechanism, is called *nondisjunction*. In this process, one daughter cell receives an extra copy of a chromosome, becoming *trisomic* for this chromosome, and the other daughter cell loses a copy, becoming *monosomic* for this chromosome. One or both daughter chromosomes may lag behind at anaphase and reach neither pole. Such laggards form *micronuclei* in telophase. Micronuclei usually do not enter mitosis, and consequently any chromosomes included in a micronucleus are lost from the complement.

If the trisomic and monosomic cells that arise in somatic tissues through mitotic nondisjunction or chromosome loss are viable, the result is *chromosomal mosaicism*, the presence in one individual of clones of cells with differing chromosome constitutions. Mosaicism may be found in a single tissue, multiple tissues, or the entire organism, depending on the time in development at which nondisjunction occurred. Given the frequencies of nondisjunction and chromosome loss throughout life, we are all mosaics to some degree.

References

Adachi Y, Luke M, Laemmli UK (1991) Chromosome assembly in vitro: topoisomerase II is required for condensation. Cell 64:137–148

Bhat MA, Philip AV, Glover DM, et al. (1996) Chromatid segregation at anaphase requires the barren product, a novel chromosome-associated protein that interacts with topoisomerase II. Cell 89:1159–1163

Blangy A, Lane HA, d'Hérin P, et al. (1995) Phosphorylation by p34(CDC2) regulates spindle association of human Eg5, a kinesin-related motor essential for bipolar spindle formation in vivo. Cell 83:1159–1163

Demetrick DJ (1995) Fluorescence in situ hybridization and human cell cycle genes. In: Pagano M (ed) Cell cycle—materials and methods. Springer-Verlag, Berlin, pp 29–45

Downes CS, Mullinger AM, Johnson RT, et al. (1991) Inhibitors of topoisomerase II prevent chromatid separation in mammalian cells but do not prevent exit from mitosis. Proc Natl Acad Sci USA 88:8895–8899

Fang G, Yu H, Kirschner MW (1998) The checkpoint protein MAD2 and the mitotic regulator CDC20 form a ternary complex with the anaphase-promoting complex to control anaphase initiation. Genes Dev 12:1871–1883

Gerace L, Blobel G (1980) The nuclear envelope lamina is reversibly depolymerized during mitosis. Cell 19:277–287

Hall M, Peters G (1997) Genetic alterations of cyclins, cyclin-dependent kinases, and Cdk inhibitors in human cancer. Adv Cancer Res 68:67–108

Hinchcliffe EH, Li C, Thompson EA, et al. (1999) Requirement of Cdk2-cyclin E activity for repeated centrosome reproduction in *Xenopus* egg extracts. Science 283:851–854

Jiang W, Jimenez G, Wells NJ, et al. (1998) PRC1: a human mitotic spindle-associated CDK substrate protein required for cytokinesis. Mol Cell 2: 877–885

Kajii T, Kawai T, Takumi T, et al. (1998) Mosaic variegated aneuploidy with multiple congenital abnormalities: homozygosity for total premature chromatid separation trait. Am J Med Genet 78:245–249

Kastan MB, Zhan Q, El-Deiry WS, et al. (1992) A mammalian cell cycle checkpoint pathway utilizing p53 and GADD45 is defective in ataxia-telangiectasia. Cell 71:587–597

King RW, Peters J-M, Tugendreich S, et al. (1995) A 20S complex containing CDC27 and CDC16 catalyzes the mitosis-specific conjugation of ubiquitin to cyclin B. Cell 81:279–288

Koshland D, Strunnikov A (1996) Mitotic chromosome condensation. Annu Rev Cell Dev Biol 12:305–333

Lammer C, Wagerer S, Safrich R, et al. (1998) The cdc28B phosphatase is essential for the G2/M transition in human cells. J Cell Sci 111:2445–2453

Levine A (1997) p53, the cellular gatekeeper for growth and division. Cell 88:323–331

Liao H, Li G, Yen TJ (1994) Mitotic regulation of microtubule cross-linking activity of CENP-E kinetochore protein. Science 265:394–395

Magnagni-Jaulin L, Groisman R, Naguibneva I, et al. (1998) Retinoblastoma protein represses transcription by recruiting a histone deacetylase. Nature 391:601–605

Marklund U, Larson N, Gradin H, et al. (1996) Oncoprotein 18 is a phosphoprotein-responsive regulator of microtubule dynamics. EMBO J 15:5290–5298

Merdes A, Remyar K, Vechio JD, et al. (1996) A complex of NuMA and cytoplasmic dynein is essential for mitotic spindle assembly. Cell 87:447–458

Murray AW (1995) Cyclin ubiquitination: the destructive end of mitosis. Cell 81:149–152

Nicklas RB (1997) A tension-sensitive kinase at the kinetochore regulates the onset of mitosis. Science 275:632–637

Nislow C, Lombillo VA, Kuriyama R, et al. (1992) A plus-end-directed motor enzyme that moves antiparallel microtubules in vitro localizes to the interzone of mitotic spindles. Nature 359:543–547

Petzelt C, Joswig G, Mincheva A, et al. (1997) The centrosomal protein centrosomin A and the nuclear protein centrosomin B derive from one gene by post-transcriptional processes involving RNA editing. J Cell Sci 110:2573–2578

Roussel MF (1998) Key effectors of signal transduction and G1 progression. Adv Cancer Res 74:1–24

Sicinski P, Donaher JL, Geng Y, et al. (1996) Cyclin D2 is an FSH-responsive gene involved in gonadal proliferation and oncogenesis. Nature 384: 470–474

Sutani T, Yanaguta M (1997) DNA renaturation activity of the SMC complex implicated in chromosome condensation. Nature 388:798–801

Tugendreich S, Tomkiel J, Earnshaw W, et al. (1995) CDC27Hs colocalizes with CDC16Hs to the centrosome and mitotic spindle and is essential for the metaphase to anaphase transition. Cell 81:261–268

Wells AD (1996) The spindle assembly checkpoint: aiming for a perfect mitosis, every time. Trends Cell Biol 6:228–234

Zheng Y, Wong ML, Alberts B, et al. (1995) Nucleation of microtubule assembly by a γ-tubulin-containing ring complex. Nature 378:578–583

3

DNA Replication and Chromosome Reproduction

Replication Is Semiconservative

In each DNA synthetic (S) phase, the two strands of the DNA double helix separate by unwinding. Each strand serves as a template for synthesis of a completely new complementary strand from the deoxyribonucleotides dA, dG, dC, and dT, hereafter called A, G, C, and T. Since the new DNA double helix consists of one conserved strand and one newly synthesized strand, replication is called semi-conservative. This was first demonstrated at the chromosomal level by autoradiography, growing cells in the presence of [³H]thymidine during one cell cycle and in the absence of this radioactive DNA precursor during the next cycle. If replication is semi-conservative and each chromatid contains a single DNA molecule, label will be incorporated into the newly replicated strand of DNA in each chromatid at the first cycle but into neither new strand of DNA at the second cycle. The result will be a radiolabeled strand in only one of the two sister chromatids. This was confirmed autoradiographically by placing a photographic emulsion on metaphase chromosome preparations. Electrons produced by radioactive decay of the tritium (^{3}H) produced silver grains in the emulsion, mostly within 1 micrometer (μm) of the source, and these were concentrated over one of the two chromatids (Fig. 3.1). Occasionally, however,

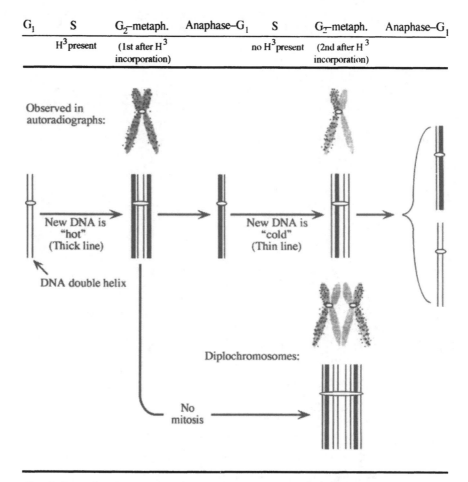

G_1	S	G_2–metaph.	Anaphase–G_1	S	G_2–metaph.	Anaphase–G_1
	H^3 present	(1st after H^3 incorporation)		no H^3 present	(2nd after H^3 incorporation)	

Observed in autoradiographs:

New DNA is "hot" (Thick line)

New DNA is "cold" (Thin line)

DNA double helix

Diplochromosomes:

No mitosis

Conclusions: 1. at least one double helix per chromatid; replication is semiconservative
2. strand continuity is not interrupted at the centromere
3. in diplochromosomes, the "oldest" strands are in the inner pair of chromatids

Figure 3.1. Semiconservative replication of chromosomes demonstrated with tritiated thymidine and autoradiography.

a region of the second chromatid would be labeled while the corresponding region of the other chromatid lacked label. This reciprocal labeling pattern is the result of breakage and rejoining of sister chromatids, resulting in an exchange between them (Taylor, 1963).

Sister chromatid exchanges (SCEs) can be induced by radioactivity, but they also occur spontaneously, as shown by the formation of double-sized dicentric ring chromosomes from simple ring chromosomes (Wolff, 1977; see also Chapter 15). In *diplochromosomes*, which have undergone two rounds of DNA replication without an intervening mitosis, there are four chromatids (see, for example, Fig. 15.2). The outer two chromatids are radioactive (Fig. 3.1). This shows that the diplochromosomes are ordered in some as yet unknown way, with the older, template DNA strands in the two inner chromatids and the newly replicated strands in the two outer chromatids. While semi-conservative replication and SCEs were first visualized autoradiographically, they can be visualized more quickly and the exchange points determined with higher resolution using bromodeoxyuridine (BrdU) incorporation into newly synthesized DNA strands and a non-radioactive detection system (Fig. 24.1). For this approach, one can use enzyme- or fluorescein-labeled antibodies to BrdU. Alternatively, on can use ultraviolet (UV)-enhanced photolysis (degradation) of the BrdU-containing DNA strands, which reduces the amount of DNA and thus the intensity of staining with DNA-binding fluorochromes or Giemsa stain. SCE frequencies are widely used as a test for exogenous or endogenous genotoxic, or chromosome-breaking, agents (*clastogens*).

The Chemistry of Replication

Replication of each new strand of DNA proceeds only in a 5′ to 3′ direction (Fig. 3.2). That is, single nucleotide monophosphates, attached to the 5′ OH of the deoxyribose sugar, are added sequentially to the 3′ end by forming a covalent chemical bond with a 3′ OH. When the 3′ to 5′ strand serves as template, its complementary strand is synthesized from a single short RNA primer in one continuous 5′ to 3′ process. However, when the 5′ to 3′ strand serves as template, a series of thousands of short sequences (*Okazaki fragments*) are synthesized 5′ to 3′, each from a short RNA primer (dotted line in Figs. 3.2 and 3.3), which is then removed. The adjacent fragments are ligated together by an enzyme called *DNA ligase*. The stepwise addition of nucleotides to a growing (nascent)

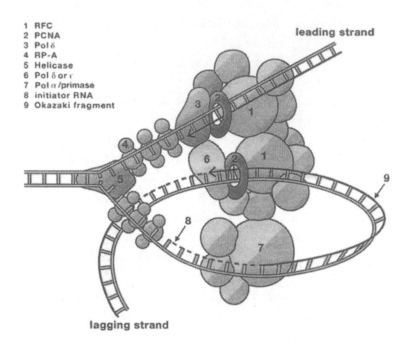

Figure 3.2. A DNA replication fork and components of a replication factory. Note that replication along the leading strand moves continuously towards the fork, whereas replication along the lagging strand proceeds from the fork along a short Okazaki fragment. See text for further explanation (Jónsson and Hübscher, 1997, BioEssays, Vol. 19, p 968; copyright 1997, John Wiley & Sons; reprinted by permission of Wiley-Liss, Inc., a subsidiary of John Wiley & Sons, Inc.).

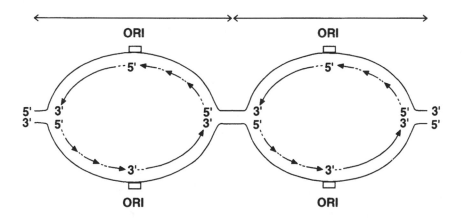

Figure 3.3. Diagram showing synchronous replication in two adjacent units of replication (replicons). DNA synthesis begins at an origin (ORI) and proceeds in both directions, always 5′ to 3′. Synthesis is continuous along the leading strand and discontinuous (occurring in Okazaki fragments) along the trailing strand. Adjacent replication bubbles coalesce to yield two identical double helices of DNA.

DNA strand is an error-prone process. Several error-correcting mechanisms have evolved and been incorporated into the replication machinery, resulting in remarkably, though not completely, accurate duplication of the DNA molecule in each chromosome of the genome.

DNA replication takes place in nuclear foci called replication factories (Fig. 3.2). These are anchored to the nuclear matrix (Chapter 4) and are enormous multiprotein complexes. They contain DNA polymerases α and δ, DNA ligase, several nucleotide mismatch repair proteins, a DNA methylase, and a fascinating protein called proliferating cell nuclear antigen (PCNA), which plays a key role in DNA replication and its control (Jónsson and Hübscher, 1997). The initiation of DNA replication requires the activation of cyclin-dependent kinases (CDKs) by binding to cyclins. When CDK2 is bound to cyclin E or A, it can phosphorylate CDC6 (cell division cycle protein 6), which is essential for the initiation of replication (Chapter 2). RCC1, a DNA-binding protein that regulates chromosome condensation, is also essential. It forms a complex with the Ran/TC4 protein and targets it to the nucleus, where the complex somehow monitors the progress of replication and couples its completion to the onset of mitosis (Ren et al., 1993).

Figure 3.2 presents a diagrammatic model of the key workers in a replication factory at a replication fork. Two molecules of PCNA join to form a doughnut-shaped homodimer through which DNA is threaded during replication. This requires the interaction of PCNA with a protein complex called replication factor C (RF-C). The two sides of the PCNA doughnut have different features, which may provide the basis for distinguishing the newly replicated strand from the conserved strand. This is essential in order that nucleotide mismatches produced by replication errors in the new strand can be correctly repaired by the mismatch repair proteins, MSH2 and MLH1. PCNA is also important for cell cycle regulation. The WAF1 gene product, p21, is a cell cycle inhibitor that acts by binding to PCNA and blocking DNA replication, thus arresting the cell cycle (Chuang et al., 1997).

Initiation at Many Sites: Origins of Replication

When cells are incubated with [³H]thymidine for very short intervals during the S phase, many sites of DNA synthesis are seen by autoradiography, both in interphase nuclei and along the length of metaphase chromosomes. That is, the single giant DNA molecule in each chromosome is not replicated continuously from one end to the other, which would take months, but from a large number of initiation sites, called origins of replication (Fig. 3.3). Fiber autoradiography, the analysis of DNA fibers isolated on microscope slides from cells grown for varying periods in [^{3}H]thymidine, has shown that replication proceeds in both directions from almost every origin. The replication fork progresses at a rate of about 0.6 µm, or 2 *kilobases* (kb), per minute. Fiber autoradiography has shown that units of replication (*replicons*) range in size from about 15 to 100 µm, or 50 to 330 kb, with an average of about 100 kb (Edenberg and Huberman, 1975). More recent technological developments enabled Tomilin et al. (1995) to visualize the elementary units of DNA replication and show that they correspond in size to the DNA loop domains described in Chapter 5. Since replication of the roughly 3.4 billion base pairs of DNA in the haploid genome proceeds at about 2 kb per minute, or 120 kb per hour, and takes approximately eight hours (S phase), there must be about 34,000 replicons. Replicons that initiate synthesis at the same time are clustered in linear tandem arrays of four or more, as shown by the apparent continuity of adjacent silver grain tracks in fiber autoradiographs. In fact, these clusters can contain 10–25 or more replicons, or 1–3 megabases (Mb) of DNA, each perhaps representing an entire chromosome band (Chapter 6).

The nature of human origins of replication is still unclear. Some origins appear to be very short DNA sequences, while in other regions of the genome replication may begin anywhere within a long stretch of DNA. Methods to define origins are still limited. Vassilev and Johnson (1990) used PCR (polymerase chain reaction) amplification of very short, newly replicated (*nascent*) strands of DNA to localize the origin of replication of the cellular oncogene c-*MYC*. This fell within a 2-kb region centered about 1.5 kb upstream of the first coding region, exon 1, and replication from this origin was bidirectional. Using the same technique, Kumar et al. (1996) localized an origin on chromosome 19 within a 500-bp segment at the 5' end of the gene for the nuclear envelope protein, lamin B2. They showed that the same origin was used in six different cell types, of myeloid, neural, epithelial, and connective tissue origin. Aladjem et al. (1995) similarly localized the β-globin origin. An alternative approach using DNA fibers from yeast artificial chromosomes (YACs) and fluorescence in situ hybridization (fiber-FISH; see Chapter 8) allowed identification of two origins in a 400-kb region of the huge 2.4-Mb dystrophin (Duchenne muscular dystrophy, or *DMD*) gene on the X chromosome (Rosenberg et al., 1995).

Replication Is Precisely Ordered: Replication Banding

DNA replication follows a precise order, with a corresponding progression and arrangement of replication foci and a consistent pattern of replication along each chromosome. Some origins consistently initiate replication early in S and others late in S; the mechanism underlying this is not clear. The initial autoradiographic studies of human DNA replication by Morishima, Grumbach, and Taylor showed that one X chromosome in XX cells and two in XXX cells terminate replication much later than do the autosomes and the other X chromosome. This indicates that the heteropycnotic, Barr body–forming X chromosomes are late replicating. With the development of a simple method (C-banding) to recognize constitutive heterochromatin, it became clear that constitutive as well as facultative heterochromatin is consistently late replicating. Even with the limited resolution of tritium autoradiography, the distinctive patterns of replication of a few autosomes permitted a distinction to be made between morphologically similar chromosomes, such as numbers 4 and 5, as demonstrated in patients with 5p- and 4p- deletion syndromes by German and Wolf, respectively.

BrdU incorporation makes DNA sensitive to photolysis in the presence of DNA-binding fluorochromes, enabling sites of replication to be visualized at high resolution. Latt et al. (1976) grew cells for 40 to 44 hours in a medium containing BrdU and then substituted thymidine for BrdU during the last 6 to 7 hours before fixation. After this treatment, the late-replicating X was more intensely stained with the fluorochromes Hoechst 33258 or coriphosphine O than was the other X (Fig. 3.4). More important, this procedure produced highly consistent and detailed replication banding patterns along metaphase and prometaphase chromosomes (Dutrillaux et al., 1976). These *replication banding* patterns closely resemble Q-, G-, and R-banding patterns (Figs. 3.4 and 6.4). Limiting the incorporation of BrdU to narrow windows (fractions) of the S phase has allowed detailed analyses of the replication timing of individual bands to be carried out. These indicate that the DNA in each band is replicated independently of the DNA in the adjacent bands and that chromosomes are replicated in 8–10 or more successive waves of replication, with each band always replicated in the same wave (Drouin et al., 1991). Dutrillaux et al. (1976) recognized as many as 18 successive replication times.

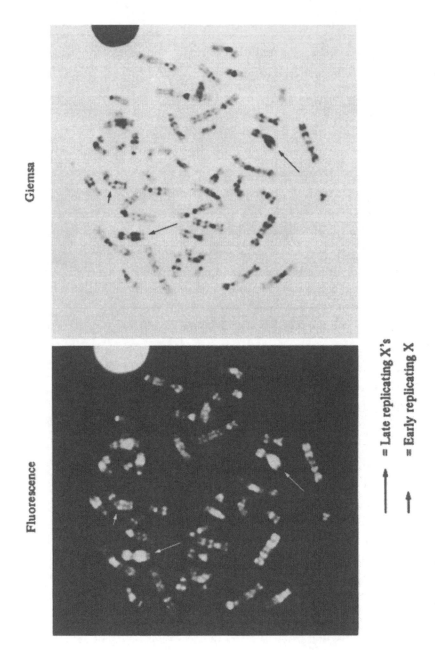

Figure 3.4. Two late-replicating X chromosomes (long arrows) and one early-replicating X (short arrow) in a 47,XXX cell, demonstrated using BrdU (left, stained with Hoechst 33258; right, with Giemsa) (Latt et al., 1976).

These results fit well with the earlier demonstration, using fiber autoradiography, of the synchronous replication of long tandem arrays of replicons (Edenberg and Huberman, 1975). They also fit with more recent findings using confocal laser-scanning microscopy to examine sites of replication in interphase nuclei throughout the S phase. These sites, called replication factories, are anchored to the nuclear matrix, whose structure and functions are reviewed by Berezny et al. (1995). The sites can be identified by the incorporation of BrdU during a limited fraction of S and by staining of the metaphase spreads with fluorescein isothiocyanate (FITC)–labeled antibodies to BrdU. At the beginning of S, about 150 sites, or factories, are visible. Replication proceeds for 45–60 minutes in these sites, and then secondary replication sites appear, to be replaced in another 45–60 minutes by tertiary sites, and so on throughout the roughly 8-hour S phase (Jackson and Pombo, 1998). Fluorescence laser-scanning confo-

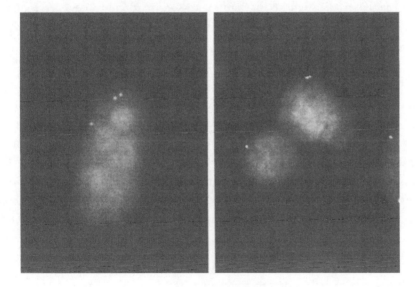

Figure 3.5. Fluorescence in situ hybridization (FISH) with a plasmid *XIST* probe (left) or a cosmid *FMR1* probe (right). Each nucleus shows a doublet signal (fully replicated) and a singlet signal (not fully replicated). This indicates the asynchronous replication of the two *XIST* alleles (left) and the two *FMR1* alleles (right) in XX cells (reproduced from the Am J Hum Genet, Torchia et al., copyright 1994, the American Society of Human Genetics, with permission of the University of Chicago Press).

cal microscopy has revealed that individual sites of replication appear to be grouped into some 22 higher-order domains, each containing many of the individual sites (Wei et al., 1998).

Methods have been developed to determine the time of replication of specific genes. The standard method for this is filter hybridization. DNA is isolated from cells that had incorporated BrdU at different intervals of S using flow cytophotometry to separate cells that completed a different fraction of S in the presence of BrdU. Antibodies to BrdU can be used to isolate the BrdU-containing (newly replicated) strand of DNA and determine its genic content. Alternatively, photolysis can destroy the BrdU-containing strand and single-strand-specific nuclease used to digest the exposed regions of the other strand. The remaining DNA, which was not replicating during the BrdU exposure interval, is digested with a restriction endonuclease; the fragments are separated by size by gel electrophoresis and transferred to a nitrocellulose filter. Gene probes are hybridized to these Southern blots. These studies have shown that most *housekeeping genes* (those that are expressed in all cell types) are replicated early, while genes whose expression is tissue specific are replicated early in the tissues in which they can be expressed and replicated late in other tissues (Chapter 7).

It is still unclear just how large a segment of DNA is involved in such developmental switches in replication timing, although it may be considerably less than a standardsized band. In the case of the β-globin gene cluster, more than 200 kb of DNA, encompassing the entire gene cluster, is early replicating in erythroid cells and late replicating in other cells. A more recent technique takes advantage of the fact that after a gene has been replicated, a copy will be present on each chromatid. In later S or G2 nuclei, these may appear by FISH (Chapter 8) as paired dots instead of the single dot seen in earlier S or G1 nuclei (Fig. 3.5; Selig et al., 1992).

The Control of DNA Replication

There are various levels of control of DNA replication, most of them not well understood. For example, what mediates the synchronous initiation of replication of all the replicons in a band? How are all the bands organized into 8–10 or more cohorts that initiate sequentially? What determines which cohort will initiate first? How does the completion of replication by each synchronous group of replicons trigger the initiation of replication in the next group? How is cell cycle regulation mediated, so that every replicon is replicated once, and only once, in each cycle? What happens when these controls break down?

Cell fusion studies by Johnson and Rao in 1970 provided the first major step in understanding how DNA replication is controlled. Fusion of a G1-phase cell with an S-phase cell leads to rapid initiation of replication in the G1 nucleus. Fusion of a G2-phase cell with an S-phase cell has no such effect. Thus, cells in G1, but not cells in G2, are competent to initiate replication. This competence is gained late in mitosis, when a replication licensing complex containing seven MCM (minichromosome maintenance) proteins and the CDC6 protein binds to the six-protein origin of replication complex (ORC) that is permanently associated with each origin to form an even larger pre-replication complex, or pre-RC. CDC6 expression is turned on by the transcription factor E2F when this is released from its inactive complex with the RB protein (Chapter 2). Phosphorylation of the CDC6 protein by cyclin E/CDK2 or cyclin A/CDK2 and of the MCM proteins by the CDC7 protein kinase leads to activation of an origin. After an origin is activated (fires), the CDC6 and MCM proteins dissociate from the origin, leaving it unable to fire again until a new pre-RC is assembled late in mitosis. This may be the mechanism ensuring that each origin fires once, and only once, in each cycle (Stillman, 1996).

Povirk (1977) provided one of the earliest clues to the mechanism of synchronous replication of a group of adjacent replicons. Taking advantage of the fact that ultraviolet (UV) irradiation damages DNA that contains BrdU and suppresses the initiation of replication, Povirk demonstrated that one UV-induced lesion per 100–500 μm of DNA could suppress initiation in a region 500–750 μm (1.6–2.4 Mb) long, or about 10–15 replicons. This suggests that a single critical site may control the initiation of replication of all the replicons in a synchronized cluster, or *band*. Using more precise methods, Aladjem et al. (1995) showed that, while replication in the β-globin gene cluster in erythroid cells is initiated in the δ–β region 50 kb downstream of the locus control region (LCR), deletion of the LCR abolished initiation within this cluster. This indicates that the LCR controls replication in this region, which is replicated early in erythroid cells in which the genes are transcriptionally competent and late in cells in which the globin genes cannot be expressed. The LCR also controls the transcription and chromatin structure of the entire gene cluster (Chapter 5).

Previous studies support the view that sites of early initiation of replication tend to be close to transcribed sequences and that the earliest sites of replication in interphase nuclei are adjacent to sites of transcription (Jackson and Pombo, 1998). It is possible that the more open chromatin conformation in these regions facilitates access by the massive replication factory. If so, sequential alterations in the chromatin structure of the various replication cohorts might play a role in the temporal regulation of replication. It is not clear how such alterations might be triggered.

Replication of Chromosome Ends: Telomerase and Cell Aging

DNA replication proceeds 5′ to 3′ from a short RNA primer. When this primer is removed after replication is completed, it leaves an unreplicated stretch about 50–200 nucleotides long at the 5′ end of each new strand. Therefore, one might expect that DNA synthesis at the very end of the chromosome would be incomplete, and it is. In fact, chromosome duplication in normal somatic cells leads to progressive shortening of the chromosome ends (*telomeres;* Chapter 4), until the chromosomes begin to fuse after a number of cell divisions and the cell dies. However, in early embryonic cells and in the stem cells that provide a constant supply of bone marrow, gut and skin epithelium, spermatogonia, and so on, telomeres compensate for this limitation of DNA end-replication.

Telomeres consist of a tandemly repetitive six-base-pair unit, TTAGGG in the G-rich strand and CCCTAA in the complementary C-rich strand. The G-rich strand is oriented 5′ to 3′ toward the end of the chromosome. The C-rich strand ends somewhat short of the G-rich strand, so that the latter forms a single-stranded tail at the end of the chromosome. Telomeres are synthesized not by the usual replication mechanism but by a special ribonucleoprotein complex called *telomerase*, which can add new repeat units to the 3′-end of the G-rich strand. The enzyme is a *reverse transcriptase*, synthesizing DNA from an RNA rather than a DNA template, in this case the RNA oligonucleotide, CUAACCCUAAC, which is an intrinsic part of this multisubunit enzyme (Harrington et al., 1997). The enzyme uses the end of the G-rich strand as its primer, thus replacing the DNA repeat units that are lost during chromosome replication.

Telomerase is abundant in embryonic and cancer cells but is absent from non-embryonic cells other than stem cells. Consequently, telomeric shortening occurs in diploid fibroblasts, which lose 50–200 bp of telomeric DNA per cell doubling (Levy et al., 1992). This may account for the well-known phenomenon of cell senescence, with diploid cells losing their ability to divide and ultimately dying after a certain number of cell doublings, as first noted in 1961 by Hayflick and Moorhead. This number, about 50, is sufficient for the cell to lose all its telomeric repeats from a few chromosomes and for telomeric fusions to lead to chromosome loss or cell death. Evidence that telomere shortening does trigger cell senescence has come from studies in which a telomerase gene was introduced into telomerase-negative normal human cells and the transfected cells were able to grow in culture for many more cell doublings than usual (Bodnar et al., 1998).

Certain viruses or chemical mutagens can transform diploid, telomerase-deficient cells with a limited life span into aneuploid cells that have reactivated their telomerase, have stable telomeres, and are immortal, being able to grow indefinitely. Cancer cells are also immortal, and more than 90% of them show a high level of telomerase activity compared to adjacent normal cells (Kim et al., 1994). The product of the MYC oncogene induces expression of the catalytic subunit of telomerase in cultured human epithelial cells and fibroblasts and extends their proliferative life span (Wang et al., 1998; see also Chapter 27).

Postreplication Steps: DNA Methylation and Chromatin Assembly

DNA methylation occurs almost immediately after replication of each DNA segment. DNA (cytosine 5) methyltransferase (MCMT, or DNA methylase) is part of the PCNA replication complex referred to earlier in this chapter. MCMT methylates newly replicated DNA only at CpG sites where the CpG in the complementary (template) strand is methylated. In this way, the pattern of DNA methylation can be maintained without change through successive mitotic cell divisions, and even throughout life. When the cell cycle is arrested by p21, the product of the WAF1 gene, the MCMT-PCNA complex is also disrupted and DNA methylation is inhibited (Chuang et al., 1997).

Chromosomal proteins are added to the newly replicated DNA so as to reproduce the chromatin conformation present before mitosis began. Chromatin assembly factor 1 (CAF-1) assembles nucleosomes in a replication-dependent manner. CAF-1 has three protein subunits: p150, p60, and p48; the last is a histone acetylase. This protein complex acts on histones H3 and H4, which are added first (Verreault et al., 1996). Histones H2A and H2B are then added, with histone H1 and the nonhistone proteins following. Histones can be acetylated or phosphorylated. This maintains the tissue-specific chromatin structure (and thus the state of differentiation) unchanged throughout successive mitotic divisions, except during the differentiation process itself. DNA methylation and chromosomal protein acetylation and phosphorylation are called *epigenetic* processes because they can be maintained through a series of mitotic divisions but do not involve an alteration of the basic nucleotide sequence.

References

Aladjem MI, Groudine M, Brody LL, et al. (1995) Participation of the human β-globin locus control region in initiation of DNA replication. Science 270:815–819

Berezny R, Mortillaro MJ, Ma H, et al. (1995) The nuclear matrix: a structural milieu for genomic functions. Int Rev Cytol 162:1–65

Bodnar AG, Oullette M, Frolkis M, et al. (1998) Extension of the life span by introduction of telomerase into normal human cells. Science 279:349–352

Chuang LS-H, Ian H-I, Koh T-W, et al. (1997) DNA (cytosine 5) methyltransferase-PCNA complex as a target for p21 (WAF1). Science 277:1996–1999

Drouin R, Lemieux N, Richer C-L (1991) Chromosome condensation from prophase to late metaphase: relationship to chromosome bands and their replication time. Cytogenet Cell Genet 57:91–99

Dutrillaux B, Couturier J, Richer C-L, et al. (1976) Sequence of DNA replication in 277 R- and Q-bands of human chromosomes using a BrdU treatment. Chromosoma 58:51–61

Edenberg HJ, Huberman JA (1975) Eukaryotic chromosome replication. Annu Rev Genet 9:245–284

Harrington L, Zhou W, McPhail T, et al. (1997) Human telomerase contains evolutionarily conserved catalytic and structural subunits. Genes Dev 11:3109–3115

Jackson DA, Pombo A (1998) Replicon clusters are stable units of chromosome structure: evidence that nuclear organization contributes to the efficient activation and propagation of S phase in human cells. J Cell Biol 140: 1285–1295

Jónsson ZO, Hübscher U (1997) Proliferating cell nuclear antigen: more than a clamp for DNA polymerases. BioEssays 19:967–975

Kim NW, Piatyszek MA, Prowse KR, et al. (1994) Specific association of human telomerase activity with immortal cells and cancer. Science 266: 2011–2015

Kumar S, Giacca M, Norio P, et al. (1996) Utilization of the same DNA replication origin by human cells of different derivation. Nucleic Acids Res 24:3289–3294

Latt SA, Willard HF, Gerald PS (1976) BrdU-33258 Hoechst analysis of DNA replication in human lymphocytes with supernumerary or structurally abnormal X chromosomes. Chromosoma 57:135–153

Levy MZ, Allsop RC, Futcher AB, et al. (1992) Telomere end-replication problem and cell aging. J Mol Biol 225:951–960

Povirk LF (1977) Localization of inhibition of replicon initiation to damaged regions of DNA. J Mol Biol 114:141–151

Ren M, Drivas G, D'Eustachio P, et al. (1993) Ran/TC4: a small nuclear GTP-binding protein that regulates DNA synthesis. J Cell Biol 120:313–323

Rosenberg C, Florijn RJ, Van De Rijke FM, et al. (1995) High resolution DNA fiber-FISH in yeast artificial chromosomes: direct visualization of DNA replication. Nat Genet 10:477–479

Selig S, Okumura K, Ward DC, et al. (1992) Delineation of DNA replication time zones by fluorescence *in situ* hybridization. EMBO J 11:1217–1225

Stillman B (1996) Cell cycle control of DNA replication. Science 274: 1659–1664

Taylor JH (1963) The replication and organization of DNA in chromosomes. In: Taylor JH (ed) Molecular genetics I. Academic, New York, pp 65–111

Tomilin N, Solovjeva L, Krutilina R, et al. (1995) Visualization of elementary DNA replication units in human nuclei corresponding in size to DNA loop domains. Chrom Res 3:32–40

Torchia BS, Call LM, Migeon BR (1994) DNA replication analysis of FMR1, XIST, and factor 8C loci by FISH shows nontranscribed X-linked genes replicate late. Am J Hum Genet 55:96–104

Vassilev L, Johnson EM (1990) An initiator zone of chromosomal DNA replication located upstream of the c-*myc* gene in proliferating HeLa cells. Mol Cell Biol 10:4899–4904

Verreault A, Kaufman PD, Kobayashi R, et al. (1996) Nucleosome assembly by a complex of CAF-1 and acetylated histones H3/H4. Cell 87:95–104

Wang J, Xie LY, Allan S, et al. (1998) Myc activates telomerase. Genes Dev 12:1769–1774

Wei X, Samarabandu J, Devdhar RS, et al. (1998) Segregation of transcription and replication sites into higher order domains. Science 281:1502–1505

Wolff S (1976) Sister-chromatid exchange. Annu Rev Genet 11:183–201

4

General Features of Mitotic Chromosomes

Metaphase Chromosomes

Chromosomes are most often studied at mitotic metaphase, when the chromosomes are shortest and thickest, or most condensed. To obtain suitable metaphases, cells are grown in culture, treated with colcemid to destroy the highly viscous spindle (preventing anaphase), and treated with a hypotonic saline solution to swell the cells. The cell suspension is fixed with methanol-acetic acid, air-dried on a glass microscope slide to achieve optimal spreading and flattening of the chromosomes, and stained with a dye that binds to DNA. At metaphase the duplicated chromosomes each consist of two *sister chromatids*, which become daughter chromosomes after their separation at anaphase (Chapter 2). Each chromatid has two arms separated by a *primary constriction*, or unstained gap (Fig. 4.1). This marks the location of the *centromere*, the site of *spindle*

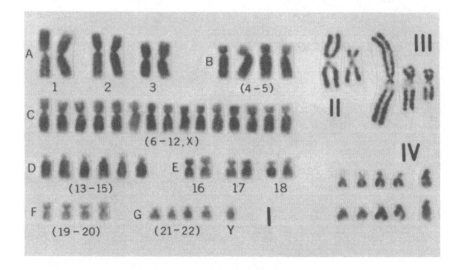

Figure 4.1. (I) Unbanded normal male karyotype from a lymphocyte. (II) Chromosomes 1 and 9 have fuzzy regions of heterochromatin (III) Chromosomes 1, 9, and 16, with fuzzy regions. (IV) G group and Y chromosomes from a father (above) and son (orcein staining) (courtesy of E. Therman).

microtubule attachment, which is essential for the normal movements of the chromosomes during mitotic and meiotic cell divisions. A chromosome without a centromere is *acentric* and either is lost or drifts passively towards a pole of the spindle. Some chromosomes have a *secondary constriction*, called a *nucleolus organizer*. A segment of variable size, a cytological *satellite*, is sometimes visible distal to a secondary constriction.

Each chromosome has a characteristic length (1–10 μm or so) and position of the centromere, which divides the chromosome into a short arm, designated p (petite) at the Paris Conference: 1971 (1972), and a long arm, designated q (since all geneticists know that p + q = 1!). That is, if a gene exists in only two forms (alleles), the fraction of each type, specified by "p" and "q," must together equal 1 (100%). A *metacentric* chromosome has its centromere near the middle. A *telocentric* chromosome has its centromere at the very end and is found only as a result of a structural change. *Submetacentric* chromosomes are intermediate between these two types. *Acrocentric* chromosomes have markedly unequal arms. The *arm ratio* (q/p) is the length of the long arm divided by that of the short arm. The *centromere index* is the length of the short arm divided by the total chromosome length. The length and arm ratio or centromeric index are rarely sufficient to permit unambiguous identification of a chromosome, whether it is normal or abnormal. Chromosome identification is therefore almost always based on the use of chromosome banding or in situ hybridization with chromosome-specific DNA probes (Chapters 6 and 8).

The Chromosome Complement and Karyotype

Somatic cells have two complete sets of chromosomes, so the chromosome number, 46, is referred to as *diploid*, or 2n. The gametes have one complete set of 23 chromosomes, with the *haploid* number, n. The chromosome complement consists of 22 pairs of *autosomes* (non-sex chromosomes) and one pair of sex chromosomes (XX in females, XY in males). Their analysis is aided by constructing a *karyotype*, a display in which the chromosomes of a single metaphase spread are aligned in pairs, generally from longest to shortest, with the short arm of each chromosome at the top. The chromosome constitution of an individual is referred to as his or her karyotype. A diagrammatic karyotype, or *ideogram*, is usually based on the analysis of multiple cells. Figure 4.1 shows a karyotype prepared from an unbanded metaphase spread from a normal male. Chromosomes that can sometimes be individually distinguished by their length and arm ratio are so listed in the figure. Chromosome 9 can sometimes be identified by the presence of a less-stained region on the long arm near the centromere. The rest of the chromosomes can be classified only as belonging to one or another group. The X chromosome belongs to the large C group of eight chromosome pairs. Chromosomes 1 and 3 are metacentric; chromosome 2 and the chromosomes in the B, C, and E groups are submetacentric. The D and G group chromosomes are acrocentric and all have a nucleolus organizer (NOR) on the short arm.

Human chromosome banding was discovered by Caspersson et al. (1970). Banding techniques (Chapter 6) enable each chromosome to be distinguished by its pattern of darker and lighter bands. Consequently, an enormous number of structural changes undetectable by the earlier methods became identifiable. A G-banded karyotype is presented in Fig. 4.2. The chromosomes are arranged numerically according to length, with one exception; chromosome 22 is actually longer than 21. Since the chromosome that in the trisomic state causes Down syndrome had long been called 21, this number has been retained. The overwhelming importance of chromosome banding in human cytogenetics is amply demonstrated in many chapters of this book.

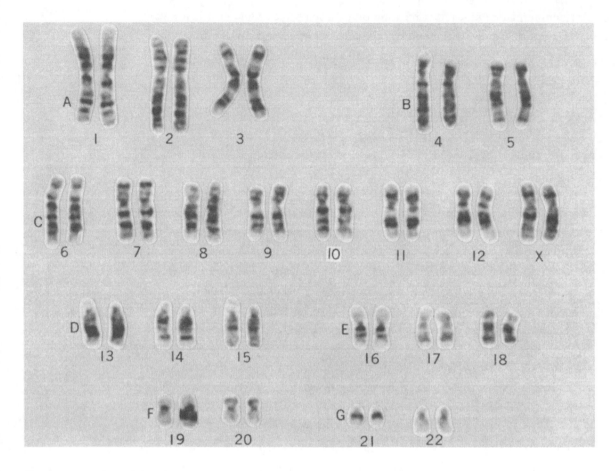

Figure 4.2. Normal female karyotype by G-banding.

DNA Content and DNA-Based Flow Cytometric Karyotypes

The DNA content of Feulgen-stained nuclei or fluorochrome-stained chromosomes (Mayall et al., 1984) and chromosome bands (Caspersson et al., 1970) can be determined by cytophotometry; automated scanning methods facilitate such measurements. However, for many purposes flow cytometry has replaced cytophotometry. The flow cytometer is an instrument that can be used to measure the DNA content of single chromosomes at a rate of several thousand per minute. The flow cytometer converts a suspension of chromosomes that have been doubly stained with two fluorochromes into microdrops and shoots these through a detector that measures the intensity of fluorescence of each microdrop at two wavelengths and prepares a statistical profile of all the measurements. The suspension is made so dilute that virtually no droplet contains more than one chromosome. The thousands of individual measurements of the DNA content of each droplet then fall into rather sharply defined clusters, forming what is called a flow karyotype (Fig. 4.3). A fluorescence-activated cell sorter is able to isolate one class of microdrops, based on DNA content, and this technique has been used to prepare chromosome fractions so enriched for a single chromosome that quite workable PCR-amplified libraries can be constructed for use as chromosome-specific painting probes (Chapter 8). These techniques have exciting phylogenetic and clinical diagnostic applications (Ferguson-Smith, 1997).

Centromeres and Kinetochores

Centromeres, or kinetochores, are essential to attach chromosomes to the spindle so they can be distributed correctly to the daughter cells in mitosis and meiosis. Centromere malfunction leads to nondisjunction or other maldistribution of chromosomes. The terms centromere and kinetochore are sometimes used interchangeably. To avoid ambiguity, we shall use *centromere* to refer to the chromatin core (DNA plus histones) at the primary constriction and *kinetochore* to refer to the complex proteinaceous structure at the centromere that mediates attachment of spindle microtubules and chromosome movement in metaphase and anaphase. Choo (1997) and Lee et al. (1997) have reviewed the kinds of DNA sequences found at human centromeres, which contain large amounts of simple-sequence DNA, just like other constitutive heterochromatin. The most abundant, and the only one found on every chromosome, is called α-satellite DNA. It is made up of long tandem repeats of a basic monomeric sequence that is approximately 170 bp long. Its total length varies from chromosome to chromosome, but the amounts at all centromeres are very large, ranging from about 300 to 5000 kb. Thus, the 170-bp sequence is repeated over and over some 1700 to 29,000 times at a centromere. This satellite can serve as a molecular probe for centromeric sequences in intact chromosomes. Most cloned α-satellite fragments hybridize preferentially to the centromeric region of a specific chromosome. In fact, chromosome-specific α-satellite probes for every chromosome except 13 and 21 are now in common use for rapid analysis, by fluorescence in

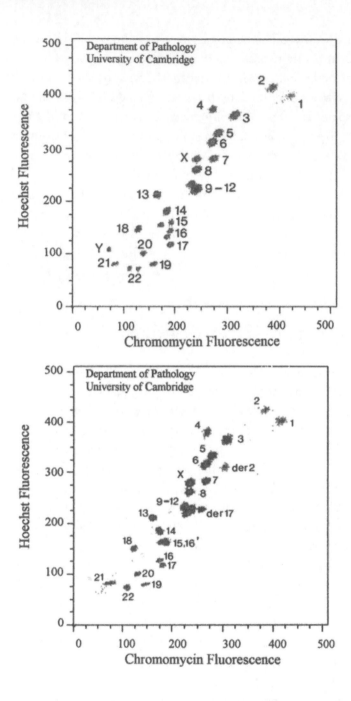

Figure 4.3. Two-fluorochrome bivariate flow cytometric karyotypes. (top) In this individual, all the chromosomes except numbers 9–12 differ in their DNA content or GC richness, and so do the homologues of chromosomes 15 and 16. (bottom) In this individual, a reciprocal 2; 17 translocation has produced derivative der 2 and der 17 chromosomes of novel sizes (Ferguson-Smith, 1997, reproduced with permission of S. Karger AG, Basel).

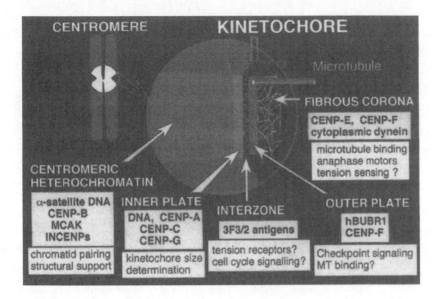

Figure 4.4. Model of an active centromere, or kinetochore; see text for details (courtesy of William Earnshaw).

situ hybridization (FISH), of the chromosome content of cells (Choo, 1997; see also Chapter 8). For examples, see Figures 8.1 and 9.6.

The kinetochore can be visualized by electron microscopy as a trilaminar structure resting on the surface of the heterochromatin at the primary constriction. Heterochromatin protein 1 (HP1) binds firmly to centromeric heterochromatin and serves to bind other specific proteins to the site. For example, the inner centromeric protein, INCENP, has a 13-amino-acid sequence that targets it to HP1 (Ainsztein et al., 1998). The molecular structure of the centromere/kinetochore is beginning to be understood in considerable detail (reviewed by Pluta et al., 1995). Figure 4.4 is a model showing the location of the major functional components. Most of these are described in the following paragraphs. The kinetochore has dense outer and inner plates separated by a less dense plate (*interzone*). Spindle microtubules insert into the outer plate or the fibrous corona surrounding it, which contains the motor protein dynein. Almost half the microtubules insert into the outer plate, the rest into the surrounding region or the inner plate. The inner plate contains chromatin, as shown by its disruption by DNAse and its abundant phosphate, detectable by electron spectroscopic imaging (Rattner and Bazell-Jones, 1989). Kinetochores form attachments to the spindle fibers and contain the cytoplasmic motor protein dynein, which is involved in propelling kinetochores and their attached chromosomes to the spindle poles.

Kinetochores can be divided into subunits that retain the essential functions of the intact structure (Zinkowski et al., 1991). When DNA replication is blocked with hydroxyurea, cells blocked at the transition from the Gl to the S phase accumulate. When caffeine, an inhibitor of the DNA damage checkpoint discussed in Chapters 2 and 26, is added to these cells, they undergo mitosis even though the chromosomes have not replicated, and the chromosomes fall to pieces. The kinetochores break up into fragments, but these are still able to function as kinetochores. They line up on the metaphase plate, form attachments to the spindle fibers, and migrate to the spindle poles despite the absence of attached chromosomes. Kinetochores vary in size, with large sizes more common. Chromosomes with small kinetochores are more likely to be involved in nondisjunction than chromosomes with large kinetochores; that is, there appears to be a lower size limit for effective kinetochore function (Cherry and Johnston, 1987).

Antibodies in the serum of patients with the CREST form of an autoimmune disease called scleroderma bind specifically to the centromere and kinetochore. Consequently, fluorescent CREST antibodies will "light up" the centromeric regions of both metaphase and interphase chromosomes. These antibodies identify at least seven centromeric proteins (CENPs). In addition, there are inner centromeric proteins (INCENPs) that may play a role in holding sister chromatids together until the beginning of anaphase, and other proteins that may just be passengers. The known CENPs include the following:

1. CENP-A, a homologue of histone H3, has been immunolocalized to the inner kinetochore plate. CENP-A is present in centromeric nucleosomes in place of histone H3 and alters the chromatin structure. How is this achieved despite the high affinity of histone H3 for DNA? The answer appears to be that histone H3 synthesis is limited to the S phase and none is available when centromeric heterochromatin is replicated in very late S, whereas CENP-A is available even into early G2 (Shelby et al., 1997).
2. CENP-B binds to a conserved 17-bp DNA sequence, the "CENP-B box," that is present in many α-satellite repeats (Masumoto et al., 1989).
3. CENP-C binds to DNA and localizes to the inner kinetochore plate (Yang et al., 1996). In a dicentric chromosome in which one centromere has been

inactivated, immunostaining shows that CENP-B is present at both the active and the inactive centromeres, whereas CENP-C is present only at the active centromere and is thus diagnostic of active centromeres (Page et al., 1995). Microinjection of antibodies to CENP-C during interphase causes the tri-laminar kinetochore at the next metaphase to be smaller and unable to bind microtubules. CENP-C is thus necessary for kinetochore function (Tomkiel et al., 1994). Cells from some scleroderma patients who produce antibodies to CENP-C appear to have a somewhat increased incidence of aneuploid cells (Jabs et al., 1993).

4. CENP-D may be the RCC1 protein, a regulator of chromosome condensation (Bischoff et al., 1990).
5. CENP-E is also found at active but not inactive centromeres (Page et al., 1995). Like dynein, it is close to the surface of the outer kinetochore plate (Wordeman and Mitchison, 1995). Its role is to cross-link microtubules (Liao et al., 1994).
6. CENP-F (mitosin) has a poorly understood role in microtubule binding to the outer kinetochore plate.
7. CENP-G binds to the same α-1 satellite family as CENP-B (He et al., 1998).

Additional proteins that are present in kinetochores include tubulin (the monomer from which microtubules are assembled) and cytoplasmic dynein (an ATPase that provides the motive force for moving materials along microtubules in cells). Another motor protein, mitotic centromere-associated kinesin (MCAK), is found throughout the centromeric region and between the kineto-chore plates from prophase through telophase (Wordeman and Mitchison, 1995). The 3F3/2 antigen is in the interzone between the two kinetochore plates. It may function in sensing tension and in the cell cycle signaling that initiates anaphase (Chapter 2).

The assembly of the kinetochore proceeds by unknown mechanisms. It usually occurs at sites of large arrays of tandemly repetitive α-satellite DNA. The formation of a kinetochore at only one of the two α-satellite arrays on dicentric chromosomes with one inactive centromere (Chapter 22) indicates that this repetitive centromeric DNA is not sufficient for kinetochore assembly. Is it even necessary? The α-satellite sequences vary from chromosome to chromosome, although the short 17-bp CENP-B box, which binds the CENP-B protein, does seem to be conserved in at least some of the 170-bp repeats of this satellite on each chromosome. The strongest evidence that α-satellite is not essential for centromeric function is the occurrence of neocentromeres that contain *no* α-

satellite (Tyler-Smith et al., 1999). CENP-C and CENP-E, which are present at active but not inactive centromeres, are also present at these neocentromeres (Depinet et al., 1997). What is unclear is whether potential neocentromeres are restricted to sequences that share some higher-order structure with the normal centromeric DNA that enables it to bind HP1 protein so that CENP-C and INCENP can bind and thus anchor the inner kinetochore plate. Much more work is needed to understand kinetochore assembly and function.

Another function of centromeres is to carry passenger proteins into the daughter nuclei. Some proteins that are centromeric during mitotic prophase are later found attached to spindle fibers. Other proteins move from the chromosome and become concentrated under the cell membrane in the region where the cleavage furrow (for cytokinesis) will appear. Some of the passengers are nucleolar proteins, which are involved in ribosomal RNA synthesis and processing in the nucleolus. The storage of nucleolar proteins on the chromosomes may be necessary for the very rapid reconstitution of nucleoli and reinitiation of ribosomal RNA synthesis in late anaphase. Ribosomal RNA synthesis is still active in prophase and is absent only from prometaphase to late anaphase. Some of the ribosomal RNA genes remain decondensed throughout mitosis and are still associated with the nucleolar transcription factor, upstream binding factor (UBF) (Gebrane-Younes et al., 1997). This accounts for the presence of secondary constrictions and the silver staining capability of nucleolus organizer regions, described below.

Telomeres

Both ends of every chromosome are capped by special structures called telomeres. The tandemly repetitive telomeric unit, TTAGGG in one strand and CCCTAA in the other, is repeated several thousand times to give a stretch of DNA about 10 kb long at each end of each chromosome (Moyzis et al., 1988). Telomeres are critical for the complete replication of chromosome ends, as described in Chapter 3. They also play an essential role in the pairing of homologous chromosomes in prophase of meiosis (Chapter 10). Telomeres are protected from exonucleases that attack free ends of DNA by a protein that binds specifically to the single-stranded tail. Other telomere-specific proteins may play a role during meiosis in the association of telomeres with the nuclear envelope, and with each other, and may have additional functions (Broccoli et al., 1997).

Nucleolus Organizers and Ribosomal RNA Genes

The nucleolus is a nuclear organelle that is not bounded by a membrane. It is the site of transcription and processing of the 45S ribosomal RNA (rRNA) precursor into 28S, 18S, and 5.8S rRNAs and the site of the assembly of these rRNAs and over 80 different proteins into the two ribosomal subunits. These join to form the cytoplasmic ribosomes that play a key role in translating mRNAs into proteins. Nucleoli, which disappear during mitosis, are formed at telophase at specific chromosome sites called nucleolus organizer regions (NORs). Telophase is the time that rRNA synthesis starts up again after having been shut down from prometaphase to late anaphase (Gebrane-Younes et al., 1997). NORs can be strongly stained with a silver nitrate solution (Goodpasture and Bloom, 1975), because a specific nucleolar protein quickly reduces the silver ions to native silver. NORs are located on the short arms of the acrocentric chromosomes of the D and G groups, numbers 13, 14, 15, 21, and 22 (Fig. 20.3).

Immunofluorescence studies have shown that the proteins of the rRNA transcriptional machinery remain associated with active NORs, but not inactive NORs, throughout mitosis (Roussel et al., 1996). As interphase progresses, the number of nucleoli decreases, because nucleoli tend to fuse together into larger nucleoli, by an unknown mechanism and for unknown reasons. Even at metaphase there is a residual trace of this fusion, called *satellite association*, in which the short arms of 2–10 acrocentric chromosomes are touching or very close to one another. The tandemly repetitive ribosomal RNA gene clusters (rDNA) are heteromorphic, showing variations in the length of the secondary constrictions, the amount of silver staining of the NORs, and the amount of rDNA. The frequency with which a chromosome is involved in satellite association is strongly correlated with the size of its silver-stained NOR and less strongly with its number of rRNA genes, reflecting the importance of rRNA gene activity for nucleolar fusion (Miller et al., 1977).

NORs are usually visible either as secondary constrictions or as faintly stained regions, depending upon the amount of chromatin distal to the NOR. However, any rRNA gene clusters that have been inactivated, either as a result of the absence of a species-specific transcription factor or by intense methylation of the rRNA genes, do not form secondary constrictions or stain with silver nitrate (Miller et al., 1976; Tantravahi et al., 1981). The number of rRNA genes in the cluster on each acrocentric chromosome is variable, usually ranging from perhaps 10 to 100 copies, although accurate estimates are few. Srivastava et al. (1993) isolated DNA from a human-rodent somatic cell hybrid (Chapter 23) that contained a single chromosome 22 as its only human acrocentric chromosome. They digested the DNA with the restriction enzyme *EcoRV*, which does not cut rDNA, and size-separated the resulting DNA fragments by pulsed-field gel electrophoresis. The human rDNA fragment was 1.6 Mb in size. Since each rRNA gene is 43 kb long, the cluster contained, at most, 39 contiguous rRNA genes. The chromatin distal to an NOR may be visible as a cytological satellite. This appears to consist solely of short, tandemly repetitive sequences called satellite DNA, or constitutive heterochromatin, which contains no genes and is highly variable in amount. Much of it appears to consist of long runs of (GACA)n (Fig. 20.3; Guttenbach et al., 1998).

Constitutive and Facultative Heterochromatin

In any type of cell, only a fraction of the genes are transcribed. The rest of the genome is maintained in an inactive configuration, called *heterochromatin*, in which transcription cannot take place (Hennig, 1999). Heterochromatin tends to be clumped in interphase nuclei and is replicated late in the S phase (Chapter 3). Its generally out-of-phase behavior throughout the cell cycle, or *allocycly*, has long been known. One type, *constitutive heterochromatin*, can be visualized by C-banding techniques (Chapter 6) and consists of satellite DNA. The other type of transcriptionally silent chromatin is called *facultative heterochromatin*. It contains potentially transcribable sequences that are specifically inactivated in certain cell types or at certain phases of development. The best-known example of facultative heterochromatin is the inactive X chromosome (Chapter 18).

References

Ainsztein AM, Kandels-Lewis SE, Mackay AM, et al. (1998) INCENP centromere and spindle targeting: identification of essential conserved motifs and involvement of heterochromatin protein 1. J Cell Biol 143: 1763–1774

Bischoff FR, Maier G, Tilz G, et al. (1990) A 47-kDa human nuclear protein recognized by antikinetochore autoimmune sera is homologous with the protein encoded by *RCC1*, a gene implicated in onset of chromosome condensation. Proc Natl Acad Sci USA 87:8617–8621

Broccoli D, Smogorzewska A, Chong L, et al. (1997) Human telomeres contain two distinct Myb-related proteins, TRF1 and TRF2. Nat Genet 17: 231–235

Caspersson T, Zech L, Johansson C (1970) Differential banding of alkylating fluorochromes in human chromosomes. Exp Cell Res 60:315–319

Cherry LM, Johnston DA (1987) Size variation in kinetochores in human chromosomes. Hum Genet 75:155–158

Choo KHA (1997) The centromere. Oxford, Oxford

Depinet TW, Zackowski JL, Earnshaw WC, et al. (1997) Characterization of neo-centromeres in marker chromosomes lacking detectable alpha-satellite DNA. Hum Mol Genet 6:1195–1204

Ferguson-Smith MA (1997) Genetic analysis by chromosome sorting and painting: phylogenetic and diagnostic applications. Eur J Hum Genet 5: 253–265

Gébrane-Younès J, Fomproix N, Hernandez-Verdun D (1997) When rDNA transcription is arrested during mitosis, UBF is still associated with non-condensed rDNA. J Cell Sci 110:2429–2440

Goodpasture C, Bloom SE (1975) Visualization of nucleolar organizer regions in mammalian chromosomes using silver staining. Chromosoma 53:37–50

He D, Zeng C, Woods K, et al. (1998) CENP-G: a new centromeric protein that is associated with the α-1 satellite DNA family. Chromosoma 107: 189–197

Hennig W (1999) Heterochromatin. Chromosoma 108:1–9

Jabs EW, Tuck-Muller CM, Anhalt GJ, et al. (1993) Cytogenetic survey in systemic sclerosis: correlation of aneuploidy with the presence of anticentromere antibodies. Cytogenet Cell Genet 63:169–175

Lee C, Wevrick R, Fisher RB, et al. (1997) Human centromeric DNAs. Hum Genet 100:291–304

Liao L, Li G, Yen TJ (1994) Mitotic regulation of microtubule cross-linking activity of CENP-E kinetochore protein. Science 265:394–398

Masumoto H, Masukata H, Muro Y, et al. (1989) A human centromere antigen (CENP-B) interacts with a short specific sequence in alphoid DNA, a human centromere satellite. J Cell Biol 109:1963–1973

Mayall BH, Carano AV, Moore DH, et al. (1984) The DNA-based human karyotype. Cytometry 5:376–385

Miller DA, Tantravahi U, Dev VG, et al. (1977) Frequency of satellite association of human chromosomes is correlated with the amount of Ag-staining of the nucleolus organizer region. Am J Hum Genet 29:490–502

Miller OJ, Miller DA, Dev VG, et al. (1976) Expression of human and suppression of mouse nucleolus organizer activity in mouse-human somatic cell hybrids. Proc Natl Acad Sci USA 73:4531–4535

Moyzis RK, Buckingham JM, Cram LS, et al. (1988) A highly conserved repetitive DNA sequence, TTAGGGn, present at the telomeres of human chromosomes. Proc Natl Acad Sci USA 85:6622–6626

Page SL, Earnshaw WC, Choo KHA, et al. (1995) Further evidence that CENP-C is a necessary component of active centromeres: studies of a dic(X;15) with simultaneous immunofluorescence and FISH. Hum Mol Genet 4:289–294

Paris Conference: 1971 (1972) Standardization in human cytogenetics. Birth defects: original article series, VII:7. New York: The National Foundation

Pluta AF, Mackay AN, Ainsztein AM, et al. (1995) The centromere: hub of chromosomal activities. Science 270:1591–1594

Rattner JB, Bazell-Jones DP (1989) Kinetochore structure: electron spectroscopic imaging of the kinetochores. J Cell Biol 108:1209–1219

Roussel P, André C, Comai L, et al. (1996) The rDNA transcription machinery is assembled during mitosis in active NORs and absent in inactive NORs. J Cell Biol 133:235–246

Shelby RD, Vafa O, Sullivan KF (1997) Assembly of CENP-A into centromere chromatin requires a cooperative array of nucleosomal contact sites. J Cell Biol 136:501–513

Srivastava AK, Hagino Y, Schlessinger D (1993) Ribosomal DNA clusters in pulsed-field gel electrophoretic analysis of human acrocentric chromosomes. Mammal Genome 4:445–450

Tantravahi U, Breg WR, Wertelecki W, et al. (1981) Evidence for methylation of inactive human rRNA genes in amplified regions. Hum Genet 56: 315–320

Tomkiel J, Cooke CA, Saitoh H, et al. (1994) CENP-C is required for maintaining proper kinetochore size and for a timely transition to anaphase. J Cell Biol 125:531–545

Tyler-Smith C, Gimelli G, Giglia S, et al. (1999) Transmission of a fully functional human neocentromere through three generations. Am J Hum Genet 64:1440–1444

Wordeman L, Mitchison TJ (1995) Identification and partial characterization of mitotic centromere-associated kinesin, a kinesin-related protein that associates with centromeres during mitosis. J Cell Biol 128:95–105

Yang CH, Tomkiel J, Saitoh H, et al. (1996) Identification of overlapping DNA-binding and centromere-targeting domains in the human kinetochore protein CENP-C. Mol Cell Biol 16:3576–3586

Zinkowski RP, Meyne J, Brinkley BR (1991) The centromere-kinetochore complex: a repeat subunit model. J Cell Biol 113:1091–1110

5

The Chemistry and Packaging of Chromosomes

Microscopists of the late nineteenth century observed that material in the nucleus and chromosomes stained avidly with certain dyes, and they called this material *chromatin*. Chromatin is a highly structured complex of DNA and various histone and nonhistone proteins, including the rather insoluble scaffold proteins. The analytical methods of molecular biology and the discovery of model systems in which to study chromosome function have greatly expanded our understanding of the structural organization of chromatin and chromosomes. (For reviews, see Alberts et al., 1994; Lewin, 1997).

DNA Content and the Estimated Number of Base Pairs in the Genome

Quantitative measurements of the amount of DNA in sperm heads provide the best estimates of the size of the haploid nuclear genome, because somatic cells contain several thousand cytoplasmic mitochondria, each containing one or more circular DNA molecules over 18 kb in size, whereas sperm heads have none. The nuclear genome of a haploid human cell consists of approximately 3.65 picograms of DNA, which is equivalent to about 3.4 billion base pairs (3.4 million kb, or 3400 Mb). Since 3 kb of DNA is about 1 μm in length, a diploid cell contains about 2 meters of DNA and an average chromosome about 5 cm of DNA. An average chromatid in a human cell is approximately 5 μm long and contains a single continuous DNA double helix. Two central questions concerning chromosome structure are how a 5-cm-long strand of DNA is packaged into a 10,000-fold shorter metaphase chromosome, and how this is selectively unfolded to allow hundreds and sometimes thousands of genes to be expressed in a particular type of cell. This appears to be mediated by a variety of chromosomal proteins. The main features of this remarkable feat of packaging are presented in this chapter, but many important details remain to be discovered.

Histone Proteins and the Nucleosome

Histones are low molecular weight, *basic* proteins that have a high binding affinity for deoxyribonucleic *acid*, or DNA. There are five major types of histones, called H1, H2A, H2B, H3, and H4. They are of fundamental importance in packaging the enormous length of DNA in the cell nucleus into forms in which its genes can be turned on or off and the genome can be distributed accurately to daughter cells. These proteins, especially H3 and H4, display a remarkable similarity of structure throughout all eukaryotes. This evolutionary conservation of structure reflects the constraints imposed on amino acid substitutions in any one of these proteins by the three-dimensional structural interactions of each protein with the other proteins and DNA in chromatin. The best-characterized component of chromatin structure is the nucleosome, which represents the first level of packaging. The DNA is wound around a protein disk consisting of histones H2A, H2B, H3, and H4. The disk itself is a sandwich, the center consisting of two molecules of H3 and two of H4 and each of the faces consisting of an H2A-H2B complex. The DNA is wound twice around this histone octamer to form a *nucleosome core*. The two coils of DNA, containing 146 bp are thus compacted into a length of 5.7 nm, which is the thickness of the histone disk. This produces about a nine-fold compaction of the chromosomal DNA. Running between the successive nucleosome cores are DNA *linkers* of 90–100 nucleotide pairs. The linker length varies depending on the tissue from which the chromatin is extracted. A nucleosome thus consists of the core and the linker and contains about 200 bp of DNA. When chromatin is dispersed, the chromosome strands appear in electron micrographs as 11-nm disk-like beads on a DNA string (Fig. 5.1).

At the next level of compaction of chromatin, single molecules of histone 1 (H1) attach to the nucleosomes at the position of the linkers. This brings together successive nucleosome cores and twists them into a 30-nm fiber, producing another six-fold DNA compaction. The formation of the 30-nm strand thus shortens the DNA double helix by a factor of nearly 50. The compactness of a metaphase chromosome can be achieved by further shortening the 30-nm fiber by a factor of 200 to 250. The exact structure of the 30-nm strand remains to be determined. A strong possibility is that the strand can be a solenoid, a simple helical coil of successive nucleosome cores arranged so that each left-handed turn contains about six nucleosome disks. The next level of packing leads to the formation of a thicker fiber, 130–300 nm in diameter. The structure of this thick, contorted fiber is unclear, but it is made up of loops, called *domains*, that are anchored to a scaffold of chromosomal proteins. Figure 5.1 includes an early representation of the widely accepted scaffold-loop model of chromosome structure.

The Chromosome Scaffold and Chromatin Loops

DNA and histones make up about two-thirds of the chromosome mass. Nonhistone proteins make up most of the rest. They include various HMG (high-mobility-group) proteins, named for their rapid electrophoretic migration, and a few rather insoluble scaffold proteins, which make up about 5% of chromosome mass. The scaffold proteins form a unique structure called the *chromosome core*, or *scaffold*, which plays a critical role in the higher-order structure of chromatin. Above the level of the 30-nm fiber, chromatin is constrained by scaffold proteins

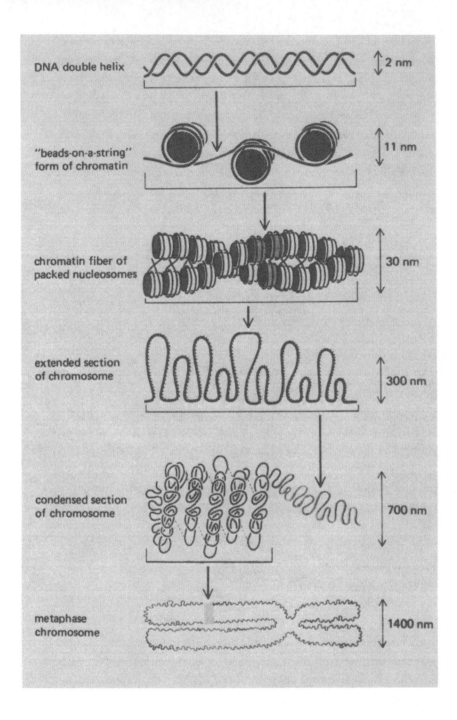

Figure 5.1. Diagram of the different orders of chromatin packing assumed to give rise to a metaphase chromosome (from Figure 8–24 of Alberts et al., Molecular Biology of the Cell, 1983).

into loop domains about 50–150 kb long, which can be visualized by removing the soluble chromosomal proteins (Paulson and Laemmli, 1977). The result is remarkable. The DNA spreads out from the chromosome to form a halo of naked DNA fibers (Fig. 5.2). At the center is a denser structure that is a ghostly echo of the original metaphase chromosome: the chromosome scaffold. It is composed mainly of three proteins: topoisomerase II (TOPO II), SCII (an ATPase), and HMG1/Y (high mobility group 1/Y). The most abundant is the enzyme TOPO II. Treatment of histone-depleted chromosomes with fluorescent antibody against TOPO II results in bright staining of both mitotic and meiotic chromosome cores, or scaffolds (Moens and Earnshaw, 1989). TOPO II can catalyze the passage of one double-stranded DNA molecule through another. It can thus link and unlink circles of DNA. More important, it can decrease or increase the amount of super-coiling within a single twisted DNA molecule and thus relieve torsional stress within the loops that is generated during changes in chromatin condensation. TOPO II is involved in the final stages of chromosome condensation and in the decondensation that follows telophase. A surprising, and still unexplained, function of TOPO II is the suppression of mitotic crossing over (Wang, 1996).

ScII is the second most abundant protein in the chromosome scaffold, where it forms complexes with TOPO II, but it is absent from the interphase nuclear matrix. ScII belongs to the highly conserved SMC family of genes, named for their involvement in the stability of minichromosomes in yeast. ScII appears to be essential for mitotic chromosome condensation and for sister chromatid separation (Saitoh et al., 1995). The least abundant scaffold protein is HMG1/Y; its function is not understood. Other proteins bind to the scaffold, for example, metaphase chromosome protein 1 (MCP1). Its name was given because during mitosis MCP1 binds exclusively to condensed chromosomes, suggesting it may have a role in chromosome condensation. It is also tightly bound to the nuclear matrix during interphase, raising more questions about its role or roles (Bronze-da-Rocha et al., 1998). Less abundant proteins found at the scaffold include CENP-C and CENP-E, which are required for kinetochore function, and INCENPs (inner centromeric proteins), thought to play a role in maintaining chromatid adhesion until the beginning of anaphase and later in cytokinesis (Chapter 4; Saitoh et al., 1995).

The chromatin loops are tethered to the scaffold at special highly AT-rich scaffold attachment regions (SARs). SARs have been called the cis-acting determinants of chromatin structural loops and functional domains. The SARs at the two ends of each loop are anchored very close to each other in the scaffold, perhaps to the same protein complex. Somehow, the scaffold proteins at the

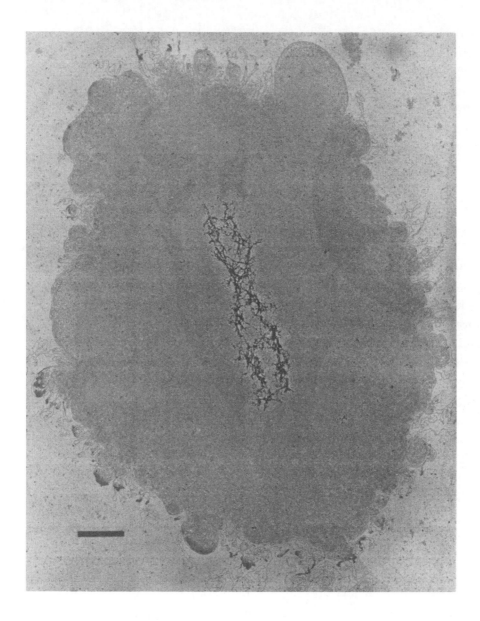

Figure 5.2. Histone-depleted metaphase chromosome showing a portion of the fibrillar halo of DNA released from the still chromosome-shaped scaffold; bar = 2 μm (from Figure 1 of Paulson and Laemmli, 1977, Cell 12: p 819, copyright 1977, M.I.T.).

bases of adjacent loops of chromatin are organized into a linear array and brought into a larger helical arrangement. The key to understanding how each long DNA molecule and its associated proteins are packaged into a recognizable chromosome lies in the solution of these problems. What is already known is that at metaphase the chromosome scaffold exhibits helical folding and the sister chromatids generally have opposite helical handedness (Boy de la Tour and Laemmli, 1988). Baumgartner et al. (1991) confirmed this using fluorescence in situ hybridization (FISH). They found that each gene occupies symmetrical lateral positions in relation to the longitudinal axis of the sister chromatids. Moreover, this position remains invariant from very early prometaphase through metaphase, indicating that the final stages of chromosome condensation occur by tighter packing, not by further coiling.

Most models of chromosome banding attribute the banding to variations in the base composition of adjacent bands, although it has long been a source of concern that the variations in base composition are so very slight in comparison to the marked variations in staining (Chapters 6 and 7). Saitoh and Laemmli (1994) have proposed a different model, suggesting that the structural basis for chromosome bands may be a differential organization, or folding path, of the very AT-rich sequences they call the AT queue. Using three-dimensional reconstructions, they developed a model in which the AT queue is tightly coiled (with short loops of DNA) in AT-rich, Q-bright (quinacrine-bright or daunomycin-bright) Q- or G-bands and more loosely coiled (with longer loops of DNA) in the GC-rich, Q-dull R-bands. This model is supported by immunolocalization of the scaffold proteins TOPO II and HMG1/Y, which shows that the AT queues correspond to the locations of the chromosome scaffold. The "random walk/giant loop model" is an attempt to develop a corresponding model for interphase chromosomes (Sacks et al., 1995).

Chromosome Domains in the Interphase Nucleus

Virtually all the genetically controlled events in the life of an individual, other than cell division, take place in the interphase nucleus. It is therefore critical to understand the functional organization of interphase chromosomes, or chromatin. It has long been known that constitutive (C-band) heterochromatin is late replicating and tends to form condensed clumps adjacent to the nuclear envelope or the nucleolus. The late-replicating inactive X chromosome (facultative heterochromatin) is also preferentially located at the nuclear envelope, or adjacent to the nucleolus in some neurons, and has a smaller surface area than the less condensed active X chromosome. In contrast to constitutive heterochromatin, the multiple inactive X chromosomes in individuals with more than two X chromosomes do not tend to clump together. The studies on heterochromatin that revealed these characteristics provided little insight into the functional aspects of chromatin organization in the nucleus. However, powerful methods for locating specific DNA sequences, DNA-binding proteins, and gene transcripts within the three-dimensional nucleus are beginning to resolve its functional compartments with considerable accuracy and to determine the mechanisms by which compartmentalization is accomplished.

The interphase nucleus has about four times the volume of the corresponding fully condensed metaphase chromosomes (Belmont et al., 1989), but it is almost filled by the somewhat decondensed interphase chromosomes. Individual chromosomes occupy discrete territories within the interphase nucleus, as shown by FISH with whole chromosome paints (Cremer et al., 1993; see also Chapter 8). The territories are divided into discrete compartments, such as arm domains, bandlike domains, centromere domains, and telomere domains, as shown using the appropriate FISH probes (Dietzl et al., 1998; Zhao et al., 1995). Using two-color FISH with 42 pairs of unique sequence probes for loci that were 0.1–1.5 Mb apart, Yakota et al. (1997) found that the distances between the sites detected by each probe pair were distributed as would be expected if the chromatin of G0/G1 nuclei were organized in megabase-sized loops and arranged according to the "random walk/giant loop" model mentioned earlier. They also showed that the fluorescent signals were closer together when the loci were from a band that stained intensely by G-banding than when the loci were from a band that stained weakly by G-banding. That is, the degree of compaction of chromatin is greater in G-dark bands than in G-light bands, which have a more open conformation. Genes are localized preferentially in the periphery of chromosome territories (Kurz et al., 1996).

The Nuclear Matrix: Replication and Transcription Complexes

Interphase nuclei contain numerous proteins that are insoluble in 2M salt solution. This fraction is called the *nuclear matrix*. It contains lamins (nuclear envelope proteins), nucleolar proteins, and ribonucleoproteins not found in the chromosome scaffold. The nuclear matrix plays a major role in the organization of functions within the nucleus (Berezny et al., 1995). An interesting review of early evidence for nonrandom organization within the nucleus is that of Haaf and Schmid (1991). Recent evidence of the striking compartmentalization of nuclear functions is well covered by Lamond and Earnshaw (1998). The first evidence of this was the discovery that the transcription and processing of ribosomal RNA (rRNA) and the assembly of ribosomes take place only in nucleoli (Chapter 4). Other nuclear functions are also compartmentalized. This is seen most clearly for DNA replication.

Nascent DNA is tightly associated with the nuclear matrix, the structural milieu for both replication and transcription. Immunocytochemical studies with antibodies to bromodeoxyuridine (BrdU) have shown the presence of nearly 150 foci of incorporation of BrdU at any time point during the S phase (Nakamura et al., 1986). The foci can be seen by electron microscopy using immunogold labeling of protein components of the replication machinery (Hozak et al., 1993). The foci appear as dots at the beginning of the S phase and become more diffuse as it progresses, reflecting the movement of newly replicated DNA away from each replication factory. At the beginning of the S phase, early replication is initiated at sites that are adjacent to sites of transcription, suggesting that the open conformation of transcribing genes is required for early replication. Replication proceeds in a very orderly manner, with most replicating units in a particular site duplicated within an hour and their completion triggering, in some way, the next wave at adjacent sets of replicating units (Chapter 3; Nakamura et al., 1986).

RNA transcription shows several distinctive types of functional compartmentalization. Transcription of ribosomal RNA is carried out by RNA polymerase I and is restricted to the nucleolus (Chapter 4). Transcription of messenger RNAs is mediated by RNA polymerase II and takes place in a relatively small number of transcription factories that are attached to the nuclear matrix. Transcription of the intronless histone genes and most of the *small nuclear RNA* (snRNA) genes by RNA polymerase II takes place in a small subset of transcription factories called *coiled bodies* (CBs). These are nuclear organelles that are highly enriched for transcription factors and the *small nuclear ribonucleoproteins* (snRNPs) that bind snRNAs. Transcription of 5S ribosomal RNA and the U6 and U7 snRNAs is mediated by RNA polymerase III and is not associated with CBs (Jacobs et al., 1999).

Active genes tend to lie at or near the surface of chromosome territories, and so do transcription factories. The final processing and transport of the RNA takes place in the interchromosomal domains, or channels, that separate chromosome territories. The nuclear matrix is particularly important in regulating gene expression. Special DNA segments called *matrix attachment regions* appear to bind expressed genes to the matrix. The matrix attachment regions function as domain boundaries that partition chromosomes into independently regulated units, enabling them to impart position-independent regulation of a tissue-specific gene (McKnight et al., 1992).

Locus Control Regions and Functional Domains

The expression of the genes in the *β-globin* cluster is under the control of a 20-kb segment of DNA called the *locus control region* (LCR), which is located about 60 kb upstream (5′) of the *β-globin* gene. This LCR is necessary, and sufficient, to keep the entire cluster in an open chromatin configuration in erythroid cells, the only cells that express globin genes. A deletion of this LCR causes a major alteration in chromatin structure and time of replication across the entire *β-globin* gene cluster (Forrester et al., 1990). A general method to study this phenomenon anywhere in the genome uses *transfection*, the transfer of a cloned gene into a host genome. The cloned gene can be constructed so as to include a *reporter gene* that induces green fluorescence or another signal when it is expressed. Ordinarily, when such a gene construct is transfected into a host cell, it integrates at random into a chromosomal site in either euchromatin or heterochromatin. Its level of expression is thus quite variable and depends upon the site of integration.

The LCR protects globin genes from position effect inactivation when a *β-globin* transgene integrates into heterochromatin, where it would otherwise be inactivated (Ellis et al., 1996). LCRs direct high-level and tissue-specific expression of closely linked genes; that is, they are *cis*-acting insulators, preventing spreading of inactivation from adjacent heterochromatin into the gene loci they control. Such upstream regulators can be quite far away. For example, the *SOX9* gene, which is essential for male sex differentiation (Chapter 17) is controlled by a still undefined region more than 1000 kb upstream. Translocations with breakpoints anywhere in this 1-Mb region impair the expression of the downstream *SOX9* allele (Pfeifer et al., 1999).

Euchromatin and Heterochromatin: Regulation of Gene Function

Regions of the chromosome in which genes can be transcribed differ structurally from inactivated regions. For many years, it has been known that transcription takes place from euchromatin, not heterochromatin. Moreover, the DNA in active regions is more sensitive to digestion by DNase I than is the DNA in inactive regions (Eissenberg et al., 1985). This accessibility to nuclease attack is believed to result from the more open conformation of chromatin containing active (or inducible, i.e., potentially active) genes. Transcription factors cannot bind to target DNA sequences in compact nucleosomes, so no transcription can take place. The DNA loops in active or inducible domains appear to have an open conformation. This makes them available to transcription factors, RNA polymerases, and even DNase I. The activation of genes may require the interaction of regulatory proteins with DNA immediately after DNA synthesis, that is, in the short interval between replication and histone binding. Highly condensed chromatin containing inactive but potentially active genes is called facultative heterochromatin. It is compared to constitutive heterochromatin in Table 5.1.

For many years, the only known effects of constitutive heterochromatin have been prolongation of the S phase (since it is so late replicating) and a slight, poorly understood enhancement of genetic recombination frequencies in

Table 5.1. Comparison of Constitutive and Facultative Heterochromatin

Constitutive heterochromatin	Facultative* heterochromatin
Mostly satellite DNAs	All kinds of DNA
C-band positive	Mostly C-band negative
Always late replicating	Late replicating
Never transcribed	Not transcribed
No meiotic crossing over	Meiotic crossing over
Stickiness and ectopic pairing	No corresponding behavior
May aid autosomal gene silencing	Silences genes on X and autosomes

*Limited to some types of cells, and with the chromatin condensation (hetero chromatinization) potentially reversible

euchromatin (there is no meiotic crossing over in heterochromatin). Recently, a number of reports have suggested that constitutive heterochromatin plays an active role in gene regulation. During B lymphocyte differentiation, several genes are inactivated. Using confocal microscopy and immunofluorescence-FISH (Chapter 8), Brown et al. (1997) demonstrated that this transcriptional silencing is mediated by the Ikaros family of proteins, which selectively target the genes to regions of centromeric heterochromatin. This explains why the Ikaros proteins are required for B cell development. Immunofluorescence studies have shown that three different human Polycomb proteins form a complex that associates with pericentromeric heterochromatin (Saurin et al., 1998). The genes encoding these proteins were cloned on the basis of their homology to the *Drosophila* Polycomb genes, whose products are the main transcriptional silencers of the homeotic genes so important in early embryonic development. It is not yet known whether the human Polycomb proteins have a role in silencing the abundant, and highly homologous, homeobox genes (Chapter 31). If they do, it will strengthen the notion that constitutive heterochromatin plays an active role in the silencing process.

Histone Modifications, DNA Methylation, and Chromosome Condensation

During the cell cycle, marked changes in chromatin or chromosome condensation occur. The addition of phosphate groups (phosphorylation) to histone H1 was long considered to drive chromosome condensation and their removal (dephosphorylation) to bring about decondensation (Bradbury, 1992). However, topoisomerase II is also required for changes in chromosome condensation because of the need to relieve torsional stress in the DNA molecule. Furthermore, it has become increasingly clear that the addition or removal of acetyl groups from histones (acetylation or deacetylation) plays a major role in regulating chromosome condensation and gene expression. Histones are generally hyperacetylated in transcriptionally active chromatin and hypoacetylated in transcriptionally silent chromatin. The amino terminus of each histone H3 and H4 molecule extends out from the nucleosome core and is modified by acetyltransferase and deacetylase during the cell cycle. The deacetylated tails displace transcription factors and thus block transcription (Grunstein, 1997). A protein

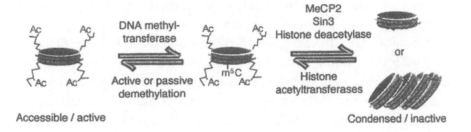

Figure 5.3. Reversible repression of gene expression by chromatin condensation induced by DNA methylation at the fifth carbon of cytosine residues (m^5C) and histone deacetylation. MeCP2 is methylated DNA binding protein 2; Sin3 is a transcription repressor; see text for details (reproduced, with permission from Nature 393:311–312, Bestor, copyright 1998, Macmillan Magazines Limited).

complex containing histone deacetylase and several other proteins mediates repression of transcription (Nagy et al., 1997).

A critical question is, what targets acetyltransferase or histone deacetylase to the specific DNA sequences or genes that are to be activated or silenced? Little is known in this regard for acetyltransferase, which belongs to the ATM gene superfamily (Grant et al., 1998, and Vassilev et al., 1998). However, new findings have provided an unexpected and very exciting answer regarding histone deacetylase. Nan et al. (1998) have shown that one multiprotein complex containing a histone deacetylase also contains methyl-cytosine binding protein 2 (MeCP2), a protein that selectively binds to methylated DNA sequences. DNA methylation involves the addition of methyl groups to cytosine residues 5′ to guanine residues, that is, at CpG sites. DNA methylation has long been known to be associated with transcriptional silencing, both at the level of single genes and in an entire X chromosome (Chapter 18). Two proteins, MeCP1 and MeCP2, specifically bind to methylated DNA. Boyes and Bird (1991) showed that DNA methylation inhibits transcription only if MeCP1 is present. Now a role for MeCP2 in gene inactivation is also clear. Histone deacetylase is targeted to chromatin domains in which the DNA is methylated, and by removing acetyl groups from the histones, it alters the structure of the chromatin at that site, resulting in transcriptional silencing (Fig. 5.3). Thus, the well-established methylation-dependent inactivation of genes is mediated by histone deacetylation. X-ray crystallographic evidence suggests that the removal of acetyl groups from histones alters chromatin structure, leading to compaction of adjacent nucleosomes (Luger et al., 1997). Inhibitors of histone deacetylase, such as the fungal metabolite depudecin, modify cellular phenotypes in ways just beginning to be explored (Kwon et al., 1998).

Mutations of the *MECP2* gene at Xq28 produce Rett syndrome (Amir et al., 1999). This is a severe, progressive neurological disorder marked by loss of speech, seizures, ataxia, stereotyped hand movements, and secondary microcephaly. It is one of the most common causes of marked mental retardation in females and is inherited as an X-linked dominant disorder that is lethal in males. MeCP2 is expressed throughout the body, so it is unclear why the phenotypic abnormalities in Rett syndrome are mainly in the brain.

References

Alberts B, Bray D, Lewis J, et al. (1994) Molecular biology of the cell, 3rd edn. Garland, New York

Amir RE, Van der Veyver IB, Wan M, et al. (1999) Rett syndrome is caused by mutations in X-linked *MECP2*, encoding methyl-CpG binding protein 2. Nat Genet 23:185–188

Baumgartner M, Dutrillaux B, Lemieux N, et al. (1991) Genes occupy a fixed and symmetrical position in sister chromatids. Cell 64:761–766

Belmont AS, Braunfeld MB, Sedat JW, et al. (1989) Large-scale chromatin structural domains within mitotic and interphase chromosomes in vivo and in vitro. Chromosoma 98:129–143

Berezny R, Mortillaro MJ, Ma H, et al. (1995) The nuclear matrix: a structural milieu for genomic function. Int Rev Cytol 162A:1–65

Bestor TH, Tycko B (1996) Creation of genomic methylation patterns. Nat Genet 12:363–367

Boy de la Tour E, Laemmli UK (1988) The metaphase scaffold is helically folded: sister chromatids have predominantly opposite helical handedness. Cell 55:937–944

Boyes J, Bird AP (1991) DNA methylation inhibits transcription indirectly via a methyl-CpG binding protein. Cell 64:1123–1134

Bradbury EM (1992) Reversible histone modifications and the chromosome cell cycle. BioEssays 14:9–16

Bronze-da-Rocha E, Catita JA, Sunkel CE (1998) Molecular cloning of metaphase chromosome protein 1 (MCP1), a novel human autoantigen that associates with condensed chromosomes during mitosis. Chrom Res 6:85–95

Brown KE, Guest SS, Smale ST, et al. (1997) Association of transcriptionally silent genes with Ikaros complexes at centromeric heterochromatin. Cell 91:845–854

Cremer T, Kurx A, Zirbel R, et al. (1993) Role of chromosome territories in the functional compartmentalization of the cell nucleus. Cold Spring Harbor Symp Quant Biol 58:777–792

Dietzl S, Jauch A, Kienle D, et al. (1998) Separate and variably shaped chromosome arm domains are disclosed by chromosome arm painting in human cell nuclei. Chrom Res 6:25–33

Eissenberg JC, Cartwright IL, Thomas GH, et al. (1985) Selected topics in chromatin structure. Annu Rev Genet 19:485–536

Ellis J, Tan-Un KC, Harper A, et al. (1996) A dominant chromatin-opening activity in 5′ hypersensitive site 3 of the human β-globin locus control region. EMBO J 15:562–568

Forrester WC, Epner E, Driscoll MC, et al. (1990) A deletion of the human β-globin locus activation region causes a major alteration in chromatin structure and replication across the entire β-globin locus. Genes Dev 4:1637–1649

Grant PA, Schieltz D, Pray-Grant MG, et al. (1998) The ATM-related cofactor Tra1 is a component of the purified SAGA complex. Mol Cell 2:863–867

Grunstein M (1997) Histone acetylation in chromatin structure and transcription. Nature 389:349–352

Haaf T, Schmid M (1991) Chromosome topology in mammalian interphase nuclei. Exp Cell Res 192:325–332

Hozak P, Hassan AB, Jackson DA, et al. (1993) Visualization of replication factories attached to a nucleoskeleton. Cell 73:361–373

Jacobs EY, Frey MR, Wu W, et al. (1999) Coiled bodies preferentially associate with U4, U11, and U12 small nuclear RNA genes in interphase HeLa cells but not with U6 or U7 genes. Mol Biol Cell 10:1653–1663

Kurz A, Lampel S, Nickolenko JE (1996) Active and inactive genes localize preferentially in the periphery of chromosome territories. J Cell Biol 135:1195–1202

Kwon HJ, Owa T, Hassig CA, et al. (1998) Depudecin induces morphological reversion of transformed fibroblasts via the inhibition of histone deacetylase. Proc Natl Acad Sci USA 95:3356–3361

Lamond AI, Earnshaw WC (1998) Structure and function in the nucleus. Science 280:547–553

Lewin B (1997) Genes VI. Oxford, New York

Luger K, Mader AW, Richmond RK, et al. (1997) Crystal structure of the nucleosome core particle at 2.8 Å resolution. Nature 389:251–260

McKnight RA, Shamay A, Sankaran L, et al. (1992) Matrix-attachment regions can impart position-independent regulation of a tissue-specific gene in transgenic mice. Proc Natl Acad Sci USA 89:6943–6947

Moens PB, Earnshaw WC (1989) Anti-topoisomerase II recognizes meiotic chromosome cores. Chromosoma 98:317–322

Nagy L, Kao H-Y, Chakravarti D, et al. (1997) Nuclear receptor repression mediated by a complex containing SMRT, mSin3A, and histone deacetylase. Cell 89:373–380

Nakamura H, Morita T, Sato C (1986) Structural organization of replicon domains during DNA synthesis phase in the mammalian nucleus. Exp Cell Res 165:291–297

Nan X, Ng H-H, Johnson CA, et al. (1998) Transcriptional repression by the methyl-CpG-binding protein MeCP2 involves a histone deacetylase complex. Nature 393:386–389

Paulson JR, Laemmli UK (1977) The structure of histone-depleted metaphase chromosomes. Cell 12:817–828

Pfeifer D, Kist R, Dewar K, et al. (1999) Campomelic dysplasia translocation breakpoints are scattered over 1 Mb proximal to SOX9: evidence for an extended control region. Am J Hum Genet 65:111–124

Sacks RK, van den Engh G, Trask B, et al. (1995) A random walk/giant loop model for interphase chromosomes. Proc Natl Acad Sci USA 92:2710–2714

Saitoh N, Goldberg I, Earnshaw WC (1995) The SMC proteins and the coming of age of the chromosome scaffold hypothesis. BioEssays 17:759–766

Saitoh Y, Laemmli UK (1994) Metaphase chromosome structure: bands arise from a differential folding path of the highly AT-rich scaffold. Cell 76:609–622

Saurin AJ, Shiels C, Williamson J, et al. (1998) The human Polycomb group complex associates with pericentromeric heterochromatin to form a novel nuclear domain. J Cell Biol 142:887–898

Vassilev A, Yamauchi J, Kotani T, et al. (1998) The 400 kDa subunit of the PCAF histone acetylase complex belongs to the ATM superfamily. Mol Cell 2:869–875

Wang JC (1996) DNA topoisomerases. Annu Rev Biochem 65:636–692

Yakota H, Singer MJ, van den Engh GJ, et al. (1997) Regional differences in the composition of chromatin in human G0/G1 interphase nuclei. Chrom Res 5:157–166

Zhao K, Hart CM, Laemmli UK (1995) Visualization of chromosomal domains with boundary element-associated factor BEAF-32. Cell 81:879–889

6

Chromosome Bands

A decisive step forward in human cytogenetics was the discovery of banding techniques that reveal distinctive and reproducible patterns of transverse bands along the chromosomes. These permit accurate identification of all the chromosomes, recognition of a host of structural rearrangements, and identification of the breakpoints in most of these. Chromosome bands have great theoretical and practical significance. They are fundamental units of chromosome organization and play a key role in gene regulation, as described in Chapter 7. They have made possible the rapid identification of an enormous range of karyotypic abnormalities and the construction of increasingly comprehensive physical and genetic linkage maps of the chromosomes. New banding techniques continue to enrich our understanding of the complex human genome. Banding techniques and their applications have been reviewed extensively (for example, Verma and Babu, 1995; Bickmore and Craig, 1997).

Q-banding

Caspersson et al. (1970) set out to develop a DNA-binding fluorochrome that would show base specificity in intensity of fluorescence; they achieved their goal with the guanine-alkylating agent quinacrine mustard. When chromosomes are stained with quinacrine mustard (or quinacrine) and examined with a fluorescence microscope, they show bands of different fluorescent intensity, or brightness, called Q-bands (Fig. 6.1). Quinacrine, like most chromosome stains (including acridine orange and methylene blue, the DNA-staining component of the complex Giemsa stain), is a tricyclic hydrocarbon whose three rings lie in one plane. This thin, flat molecule is just the thickness to fit (intercalate) between adjacent base pairs in DNA.

The extremely detailed subdivision of human chromosomes into hundreds of differentially stained bands came as a huge surprise, because no one thought there were as large regional differences in base composition as these results seemed to imply. However, in 1972, Weisblum and de Haseth provided a partial explanation. They found that synthetic polynucleotides rich in AT base pairs fluoresce brightly with quinacrine, whereas those rich in GC base pairs quench fluorescence. The degree of GC interspersion is thus the critical factor: Only stretches with three or more AT base pairs in a row fluoresce brightly. That is, two stretches of DNA with the same ratio of GC to AT base pairs but different degrees of interspersion of GC with AT base pairs can have different intensities of quinacrine fluorescence. The DNA in Q-bright bands is thus rich in AT clusters, while the DNA of Q-dull bands could either be GC-rich or simply have closely interspersed AT and GC base pairs. Other methods have shown that adjacent bands do, in fact, differ to some extent in their base ratios (Chapter 7). Most striking are the brilliant Q-bands containing highly AT-rich satellite DNAs. The one on the distal long arm of the Y chromosome is particularly striking and is usually visible as a fluorescent Y-body in nuclei (Fig. 6.1) and sperm (Fig. 17.1).

With Q-banding, the chromosomes are stained without any pretreatment, so their morphology is retained. The fluorescence intensity of the bands can be measured (Caspersson et al., 1970). However, slides stained with quinacrine do not last, as the fluorescence fades rather quickly. Chromosomes are usually analyzed in detail only from photographs, although one could now use a sensitive charge-coupled device (CCD) digital camera and computer-assisted analysis, as is used for fluorescence in situ hybridization, or FISH (Chapter 8). However,

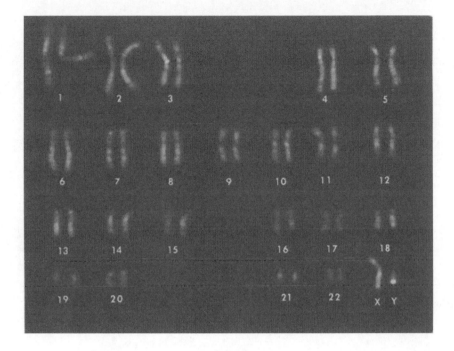

Figure 6.1. Q-banded karyotype of a cell from an XY male (Breg, Quinacrine fluorescence for identifying metaphase chromosomes, with special reference to photomicrography. Stain Technology 47:87–93, copyright 1972, Williams & Wilkins).

for routine work Q-banding has been largely replaced by nonfluorescent G- and R-banding techniques.

C-banding

C-banding was discovered by accident when chromosome spreads were heated to denature the DNA for in situ hybridization. The Giemsa staining of the chromosomes was greatly reduced except in the centromeric regions of most chromosomes and the distal part of the Y (Fig. 6.2). C-banding can be produced in a number of ways. The most common is to treat chromosomes briefly with acid and then with an alkali such as barium hydroxide prior to Giemsa staining. Prominent C-bands are found on chromosomes 1, 9, 16, and the distal Y. Differences in the types of satellite DNAs in the various C-bands are respon-

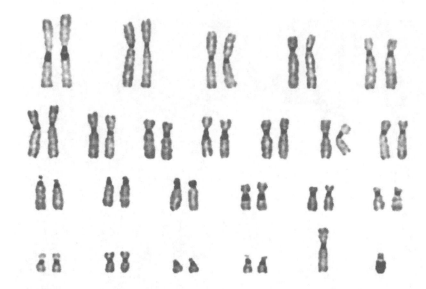

Figure 6.2. C-banded karyotype of an XY cell (courtesy of Arvind Babu).

sible for the differences observed using various stains. For example, the distal end of the Y chromosome is brightly fluorescent with Q-banding, whereas the centromeric heterochromatin of chromosomes 1, 9, and 16 is dark. Giemsa staining at pH 11 (the G11 technique) stains the C-band of chromosome 9 only.

C-bands vary considerably in size in the population. The greatest variability is demonstrated in the satellites and short arms of the acrocentric chromosomes. These variations are generally called *heteromorphisms*, because the term *polymorphism* is restricted by geneticists to a heritable variant that has a frequency of at least 1 per cent in the population. Inversions that involve the large blocks of heterochromatin of chromosomes 1 and 9 and the Q-bright centric band on chromosome 3 are fairly common heteromorphisms. Even extreme variations in the sizes of C-bands do not affect the phenotype. The reason for this is that C-banding stains constitutive heterochromatin, which contains no genes and is never transcribed. Changes in the size of a C-band do not take place very often, since the C-band variants are generally constant from one generation to the next and show normal Mendelian inheritance. Chromosomal heteromorphisms have been used as markers in gene mapping, paternity testing, and distinguishing monozygotic from dizygotic twins or the parent of origin of trisomy and triploidy. They have aided our understanding of hydatidiform moles and the origins of the different cell lines in chimeras, including persons with bone marrow transplants. However, most of these applications have been supplanted by more precise molecular cytogenetic methods.

G-banding, R-banding, and T-banding

G-banding was discovered by accident during attempts to improve C-banding. G-bands are obtained when the chromosomes are pretreated with a salt solution at 60°C or with a proteolytic enzyme, such as trypsin, before staining with Giemsa or a comparable chromatin stain. G-banding (Fig. 4.2) yields essentially the same information as Q-banding (Fig. 6.1). The bands that fluoresce brightly with quinacrine stain intensely with Giemsa stain (are G-dark), whereas the Q-dull regions are G-light. Each method has its advantages. G-banding is permanent and therefore more suitable than the evanescent Q-banding for routine work. By either technique, some 300 bands are readily distinguishable in the haploid genome at metaphase (Paris Conference: 1971 [1972]) and 850–1250 at prometaphase and prophase (ISCN, 1995). Interestingly, the Q-bright, G-dark bands correspond to the chromomeres seen at the pachytene stage of meiosis (Chapter 9). Chromosome bands have shown remarkable constancy during mammalian evolution (Chapter 30).

Reverse banding, or R-banding, was discovered by Dutrillaux and Lejeune (1971). Their technique involves pretreatment with hot (80–90°C) alkali and subsequent staining with Giemsa stain or the fluorochrome acridine orange. Acridine orange intercalates between base pairs in double-stranded DNA and fluoresces bright yellow. It can also bind to single-stranded (heat-denatured) DNA by base stacking, and then it fluoresces pale red. As the name indicates, the R-banding pattern (Fig. 6.3) is the reverse of the Q- or G-banding pattern; in other words, the bands that are intense by R-banding are faint by Q- or G-banding, and vice versa. R-banding stains chromosome ends more distinctively and is often useful for the study of structural changes involving chromosome ends that might go undetected with Q- or G-banding. A modification of R-banding, called T-banding, stains mainly regions near the tips of many chromosomes (Dutrillaux, 1973).

All Q-, G-, and R-bands, whether they fluoresce or stain strongly or weakly, contain abundant DNA and genes, and all bands are counted in determining the total number of bands. However, for ease of communication, one generally refers to Q-bright bands as Q-bands, G-dark (intensely stained) bands as G-bands, and

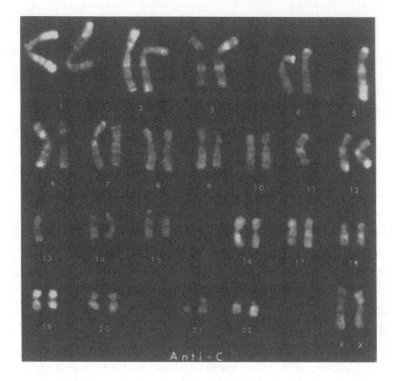

Figure 6.3. R-banded karyotype produced by immunofluorescence detection of anti-cytosine antibodies after denaturing GC-rich DNA in an XX metaphase spread by photooxidation (Schreck et al., 1973).

Q-dull, G-light, or R-intense bands as R-bands. That is, the terms Q-, G-, R-, T-, and C-band commonly refer to bands that are more intensely stained or fluorescent by the particular method used.

High-Resolution and Replication Banding

Prometaphase and prophase chromosomes are much longer than metaphase chromosomes, but they too can be banded. Such high-resolution banding can increase the number of visible bands to 850–1250 (Yunis, 1980; ISCN, 1995). For high-resolution G- or R-banding, dividing cells are blocked in the S phase with amethopterin (methotrexate). When the block is released with thymidine-rich medium, the cell cycle is synchronized in a large fraction of the cells, which can then be studied at the desired stage. Prophase chromosomes show some natural banding, but this can be enhanced with banding techniques. The longer the chromosomes are, the more bands they show. However, longer chromosomes overlap more, and the analysis becomes tedious. The most important uses of high-resolution banding are the recognition of subtle structural changes (Chapters 14–16) and the mapping of genes by in situ hybridization (Chapters 8 and 29).

Chromosome banding comparable to that achieved with other banding techniques can be induced using BrdU to study DNA replication timing (Chapter 3). The replication bands are detected either in the standard way (Latt, 1973) or by using anti-BrdU antibodies (Fetni et al., 1996). The R-bands replicate during the first half of the S period. The Q- or G-bands replicate during the second half of S. Completely reciprocal replication R- and G-bands are seen at high resolution (550–1250 bands per haploid genome) when BrdU is incorporated during either early or late S phase (Fig. 6.4). The individual bands show a constant order of replication (Drouin et al., 1994). Differential staining of the two chromatids in the C-band regions has been observed after BrdU incorporation during one cycle of replication. This *lateral asymmetry* of staining may reflect the presence of one T-rich strand and one A-rich strand in some satellite DNAs (Galloway and Evans, 1975), because BrdU is incorporated in place of T and subsequent photolysis destroys the BrdU-substituted DNA. Lateral asymmetry of C-bands may not be seen with high-resolution replication banding, for reasons that are unclear (Drouin et al., 1994).

Banding with Other Fluorochromes and Nonfluorochromes

Some fluorochromes bind specifically to AT base pairs, for example DAPI (4′,6-diamidino-2-phenylindole) and its close relatives, DIPI and Hoechst 33258. Some fluorochromes bind specifically to GC base pairs, for example, chromomycin. Nonfluorescent dyes may also show base-specific binding. Methyl green and distamycin A bind to AT base pairs, while actinomycin D binds preferentially to GC base pairs. The AT specific fluorochromes produce a fluorescent G-banding pattern (lacking the intense Q-brightness of AT-rich satellite DNA), but not a very precise one, because the intensity of fluorescence exactly parallels the AT-richness of each band. Similarly, GC-specific fluorochromes produce a mediocre R-band pattern. The fluorescent G-band patterns are

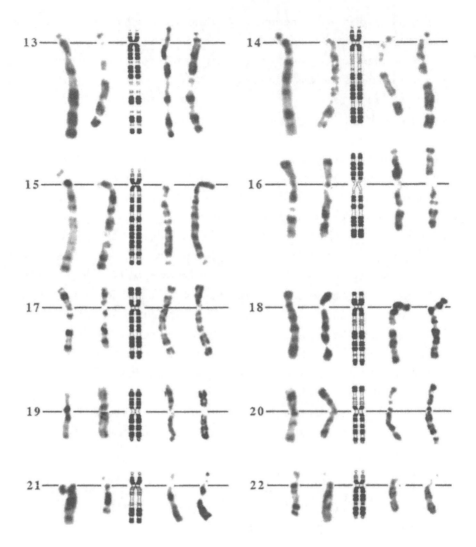

Figure 6.4. High-resolution (850–1250 bands) partial karyotypes. From left to right: GBG (G-bands by BrdU using Giemsa stain), GB-AP (G-bands by BrdU using anti-BrdU and peroxidase), RBG 1250 ideogram, RB-AP (R-bands by BrdU using anti-BrdU and peroxidase), and RBG (R-bands by BrdU suing Giemsa stain) (Fetni et al., 1996, reproduced, with permission of S. Karger AG, Basel).

enhanced when the chromosomes are stained jointly with an AT-specific fluorochrome (DAPI) and a GC-specific nonfluorescent dye (actinomycin D). In the same way, very precise fluorescent R-band patterns are produced when the GC-specific fluorochrome chromomycin is used in combination with the nonfluorescent dye methyl green. The reason such *counterstaining* is effective is that the nonfluorescent dye quenches fluorescence of the fluorochrome by a process of energy transfer that takes place only when the different types of dye molecules are extremely close together.

These counterstain-enhanced banding methods (Schweitzer, 1981) can yield beautifully sharp bands. They have also confirmed earlier evidence that G-bands contain abundant stretches of uninterrupted AT base pairs and that R-bands contain similar uninterrupted stretches of GC base pairs. Furthermore, when chromosomes are stained with two AT-specific chemicals, the fluorescent DAPI and the nonfluorescent distamycin A, a subset of fluorescent C-bands is observed, particularly those on chromosomomes 9, 15, and the Y, the same ones that are revealed using antibodies to methylated DNA (see below).

Antibody Banding

The first direct demonstration that chromosome bands reflect differences in base composition of the DNA in different bands came from immunocytochemical studies using antibodies to adenosine and cytosine (Miller et al., 1973). These antibodies bind to the bases in single-stranded (denatured) but not double-stranded DNA. When AT-rich DNA is denatured by heat or UV-irradiation, anti-A produces Q- or G-banding. When GC-rich DNA is dena-tured by photo-oxidization of G residues, anti-C produces sharp R-banding (Fig. 6.3). R-banding is also produced by antibodies to double-stranded GC polymers (Magaud et al., 1985) and antibodies to acetylated forms of histone proteins (Jeppesen and Turner, 1993). When chromosomal DNA is denatured by UV-irradiation, antibodies to 5-methylcytosine (anti-5MeC) produce C-banding (Miller et al., 1974); this was the first evidence that CpGs in the AT-rich satellite 2 and 3 DNAs are highly methylated. The same regions bind antibodies to the methylated CpG-binding protein, MeCP2 (Lewis et al., 1992) and antibodies to the heterochromatin-binding protein, HP1 (Wreggett et al., 1994).

Weak R-banding is produced by antibodies to Z-DNA, suggesting that a tiny amount of this special form of DNA is present in R-bands (Viegas-Pequignot et al., 1983). Left-handed DNA, or Z-DNA, is transiently present in cells. It tends to form just behind advancing RNA polymerase II molecules. Most genes have about three short regions near the 5′ promoter region of the gene that can take a Z conformation when strongly supercoiled. This may have functional significance, because specific proteins bind to Z-DNA. One such protein, double-stranded RNA adenosine deaminase, is an RNA editing enzyme that deaminates adenosine in the RNA to produce inosine, which is translated as if it were guanine. This can lead to the incorporation of a different amino acid in the resultant protein (Herbert and Rich, 1996).

Nuclease Banding

A large number of restriction endonucleases have been identified that cut DNA only at specific sequences that are 4–8 base pairs long and palindromic, that is, the sequence is the same, from 5' to 3', in each strand. Thus, HaeIII cuts at GGCC sites, EcoRI at GAATTC, MspI at CCGG or C^{5Me}CGG, and HpaII at CCGG but not C^{5Me}CGG. Miller et al. (1983) showed that the generally abundant sites of a number of restriction enzymes are not distributed at random along the chromosomes. Some restriction enzymes produce a standard G-band or C-band pattern on fixed chromosomes, but some produce unusual patterns. The most informative involve the C-bands, which contain several types of satellite DNA. If the very short repeating sequence of a particular satellite DNA contains the cutting site for a particular restriction enzyme, the DNA will be cut into tiny fragments and lost from the chromosomes. If the satellite DNA lacks the site, it will remain intact and stain strongly against a background of generally reduced staining. A series of restriction enzymes can be used to divide the C-bands into several classes (Miller et al., 1983), and analyze heteromorphisms (Babu, 1988). MseI cuts at TTAA sites and removes DNA mainly from G-bands and some R-bands, leaving the C-bands, T-bands, and some R-bands (Ludeña et al., 1991).

Nucleases have also been used to produce banding, or labeling, that is related to the functional state of the DNA rather than its DNA sequence. By comparing the effects of HpaII and MspI on fixed chromosome spreads, Miller et al. (1983) demonstrated that amplified but inactive ribosomal RNA genes are hypermethylated by showing they are easily cut by MspI but resist cutting by HpaII. Molecular studies have shown that transcriptionally active (or potentially active) genes are in DNAase I–sensitive chromatin in nuclei, reflecting its more open conformation (Chapter 5). Remarkably, variations in nuclease sensitivity are still present in metaphase chromosomes. Kerem et al. (1984) showed that the active X chromosome is much more DNAase I sensitive than the inactive X and that R-bands are, in general, more DNAase I sensitive than G-bands. To do this, they used nick-translation to incorporate BrdU at sites of nuclease nicking of DNA in chromosome spreads and labeled antibodies to detect the BrdU.

In Situ Hybridization Banding

In situ hybridization of the very abundant *Alu* repetitive sequences produces an R-banding pattern, while hybridization of LINE1 repeats produces a G-banding pattern. This indicates that each of these two classes of repetitive DNA is more abundant in a different fraction of the genome (Korenberg and Rykowski, 1988). By using the polymerase chain reaction with oligonucleotide primers complementary to parts of the *Alu* or LINE1 sequence, hybridization probes can be generated that enable a banding pattern to be produced simultaneously with localization of gene probes (Lichter et al., 1990). Hybridization of a highly GC-rich fraction of DNA produces T-bands (Fig. 7.2), and so does hybridization of the unmethylated CpG-rich tiny fraction of genomic DNA (Craig and Bickmore, 1994). There is lateral asymmetry of G and C residues in telomeric repeats, with the Gs in one strand and the Cs in the other (Chapter 4). This is the basis for a novel method to detect pericentric inversions, using a single-stranded probe that hybridizes only to one strand of telomeric DNA (Bailey et al., 1996; Chapter 16).

Nomenclature of Banded Chromosomes and Abnormal Karyotypes

At the Paris Conference: 1971 (1972) a system of nomenclature was proposed for banded human chromosomes and chromosome abnormalities. Figure 6.5 shows an *ideogram* of a banded karyotype according to this system. Telomeres, centromeres, and a number of prominent bands are used as landmarks. A section of a chromosome between two landmarks is called a *region*, and these regions are numbered 1, 2, 3, and so on, in both p and q directions, starting from the centromere. The bands within the regions are numbered according to the same rule. Thus, the first band in the second region of the short arm of chromosome 1 is lp21.

The increasing use of high-resolution banding has led to an extension of this system. For example, 14q32 (Fig. 6.6, left) indicates chromosome 14, long arm

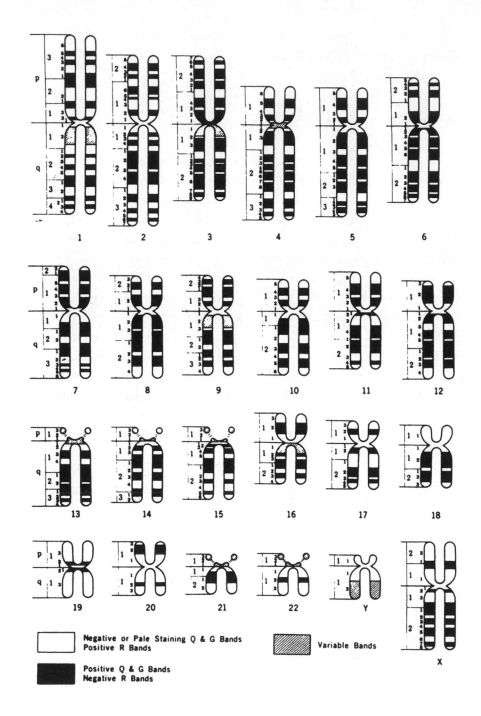

Figure 6.5. Paris conference ideogram of a banded karyotype.

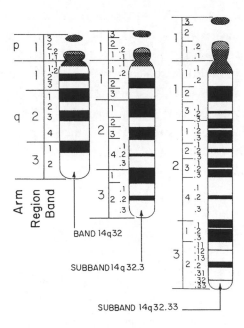

Figure 6.6. Ideograms of chromosome 14 at the 320- (left), 500- (center), and 900-band (right) stages, illustrating the Paris nomenclature. The sub-bands seen at the 500-band stage are designated with decimals, those seen at the 900-band stage with decimals and digits (Yunis, 1980).

region 3, band 2. High-resolution banding reveals three subbands in this band. To indicate a subband a dot is used, followed by the number of the subband (they are numbered sequentially from the centromere). The most distal subband in chromosome 14 (Fig. 6.6, middle) is thus 14q32.3. When the subband is further subdivided by still higher-resolution banding (Fig. 6.6, right), an additional digit is added, the last subband thus being 14q32.33. The most recent update of this international system of nomenclature is ISCN (1995). Francke (1994) has prepared a digitized and differentially shaded ideogram for genomic applications.

For the designation of chromosome abnormalities, two systems, one short and one detailed, are provided. For the actual use of both systems the reader is referred to the current standardization committee report (ISCN, 1995). In the following discussion only a few basic examples of the short system are given. An extra or a missing chromosome is denoted with a plus or a minus sign, respectively, before the number of the chromosome. The designation of a female with trisomy 13 is 47,XX,+13, and that of a male with monosomy 21 is 45,XY,−21. A female with a deletion of the short arm of chromosome 5 and the cri du chat syndrome is designated 46,XX,del(5p). The karyotype of a female carrier of a Robertsonian translocation (centric fusion) between chromosomes 14 and 21 is 45,XX,der(14;21)(q10;q10). For historical reasons, the designation *rob* is also used: 45,XX,rob(14;21)(q10;q10). The karyotype of a male translocation heterozygote (carrier) in whom chromosome arms 3p and 6q have exchanged segments, the breakpoints being 3p12 and 6q34, is 46,XY,t(3;6)(p12;q34).

References

Babu A (1988) Heterogeneity of heterochromatin of human chromosomes as demonstrated by restriction endonuclease treatment. In: Verma RS (ed) Heterochromatin: molecular and structural aspects. Cambridge University Press, Cambridge, pp 250–275

Bailey SM, Meyne J, Cornforth MN, et al. (1996) A new method for detecting pericentric inversions using COD-FISH. Cytogenet Cell Genet 75: 248–253

Bickmore WA, Craig J (1997) Chromosome bands: patterns in the genome. Chapman & Hall, New York

Caspersson T, Zech L, Johansson C (1970) Differential binding of alkylating fluorochromes in human chromosomes. Exp Cell Res 60:315–319

Craig JM, Bickmore WA (1994) The distribution of CpG islands in mammalian chromosomes. Nat Genet 7:376–382

Drouin R, Holmquist GP, Richer C-L (1994) High-resolution replication bands compared with morphologic G- and R-bands. In: Harris H, Hirschhorn K (eds) Advances in Human Genetics 22. Plenum Press, New York, pp 47–115

Dutrillaux B (1973) Nouveau système de marquage chromosomique les bandes T. Chromosoma 41:395–402

Dutrillaux B, Lejeune J (1971) Sur une nouvelle technique d'analyse du caryotype humaine. CR Acad Sci Paris D 272:2638–2640

Fetni R, Drouin R, Richer C-L, et al. (1996) Complementary replication R- and G-band patterns induced by cell blocking at the R-band/G-band transition, a possible regulatory checkpoint within the S phase of the cell cycle. Cytogenet Cell Genet 75:172–179

Francke U (1994) Digitized and differentially shaded human chromosome ideograms for genomic applications. Cytogenet Cell Genet 57:91–99

Galloway SM, Evans HJ (1975) Asymmetrical C-bands and satellite DNA in man. Exp Cell Res 94:454–459

Herbert A, Rich A (1996) The biology of left-handed Z-DNA. J Biol Chem 271:11595–11598

ISCN (1995) An international system for human cytogenetic nomenclature. Mitelman F (ed); S Karger, Basel

Jeppesen P, Turner BM (1993) The inactive X chromosome in female mammals is distinguished by a lack of histone H4 acetylation, a cytogenetic marker for gene expression. Cell 74:281–289

Kerem B-S, Goitein R, Diamond G, et al. (1984) Mapping of DNAase I sensitive regions on mitotic chromosomes. Cell 38:493–499

Korenberg JR, Rykowski MC (1988) Human genome organization. Alu, Lines, and the molecular structure of metaphase chromosome bands. Cell 53: 391–400

Latt SA (1973) Microfluorometric detection of deoxyribonucleic acid replication in human metaphase chromosomes. Proc Natl Acad Sci USA 70: 3395–3399

Lewis JD, Meehan RR, Henzel WJ, et al. (1992) Purification, sequence, and cellular localization of a novel chromosome protein that binds to methylated DNA. Cell 69:905–914

Lichter P, Ledbetter SA, Ledbetter DH, et al. (1990) Fluorescence *in situ* hybridization with *Alu* and L1 polymerase chain reaction probes for rapid characterization of human chromosomes in hybrid cell lines. Proc Natl Acad Sci USA 87:6634–6638

Ludeña P, Sentis C, De Cabo SF, et al. (1991) Visualization of R-bands in human metaphase chromosomes by the restriction endonuclease *Mse*I. Cytogenet Cell Genet 57:82–86

Magaud J-P, Rimokh R, Brochier J, et al. (1985) Chromosomal R-banding with a monoclonal antidouble-stranded DNA antibody. Hum Genet 69: 238–242

Miller DA, Choi Y-C, Miller OJ (1983) Chromosome localization of highly repetitive human DNAs and amplified rDNA by use of restriction enzymes. Science 219:395–397

Miller OJ, Schreck RR, Beiser SM, et al. (1973) Immunofluorescent studies of chromosome banding with anti-nucleoside antibodies. In: Caspersson T, Zech L (eds) Nobel Symposium XXIII. Chromosome identification— technique and applications in biology and medicine. Academic, New York, pp 43–48

Miller OJ, Schnedl W, Allen J, et al. (1974) 5-Methylcytosine localized in mammalian constitutive heterochromatin. Nature 251:636–637

Paris Conference: 1971 (1972) Standardization in human cytogenetics. Birth defects: original article series, VIII: 7. The National Foundation, New York

Schreck RR, Warburton D, Miller OJ, et al. (1973) Chromosome structure as revealed by a combined chemical and immunochemical procedure. Proc Natl Acad Sci USA 70:804–807

Schweizer D (1981) Counterstain-enhanced chromosome banding. Hum Genet 57:1–14

Verma RS, Babu A (1995) Human chromosomes: principles and techniques. McGraw-Hill, New York

Viegas-Pequignot E, Derbin C, Malfoy B, et al. (1983) Z-DNA immunoreactivity in fixed metaphase chromosomes of primates. Proc Natl Acad Sci USA 80:5890–5894

Wraggett KA, Hill F, James PS, et al. (1994) A mammalian homologue of Drosophila heterochromatin protein 1 (HP1) is a component of constitutive heterochromatin. Cytogenet Cell Genet 66:99–103

Yunis JJ (1980) Nomenclature for high resolution human chromosomes. Cancer Genet Cytogenet 2:221–229

7

Molecular Correlates of Chromosome Bands

GC- and AT-rich Isochores

Bernardi and his coworkers have demonstrated that the human genome is organized into alternating blocks, each over 300 kb long, of rather homogeneous DNA of quite different GC richness, called *isochores* (Saccone et al., 1996). These can be separated from one another on the basis of their buoyant density in a cesium chloride ultracentrifugal gradient (Fig. 7.1). This reflects the average nucleotide composition of the isochores, because GC base pairs and GC-rich DNA are denser than AT base pairs and AT-rich DNA. There are five classes of isochores, called L1, L2 (pooled in Fig. 7.1), H1, H2, and H3, in order of increasing buoyant density and thus GC-richness. In situ hybridization shows that the H3 isochores are gene-richest (especially in housekeeping genes) and

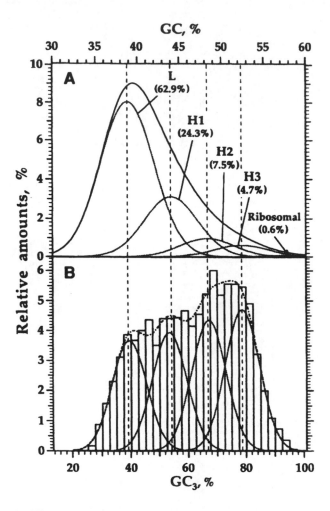

Figure 7.1. (A) The cesium chloride density gradient profile of human DNA can be resolved into four major components and a small ribosomal RNA gene component. (B) A histogram based on the GC richness of the third codon position of 4270 human genes similarly resolved into four components. The GC-riches fractions (H3 and H2) are the richest in genes (reprinted from Gene 174:95–102, 1996, Zoubak et al., The GC-richest (non-ribosomal) component of the human genome, with permission of Elsevier Science).

are present in highest concentration in 28 T-bands and at somewhat lower concentration in 31 additional R-bands (Fig. 7.2; see color insert).

The remaining R-bands at a 400-band resolution contain no H3 isochores but abundant H1 and H2 isochores. They contain more than half the tissue-

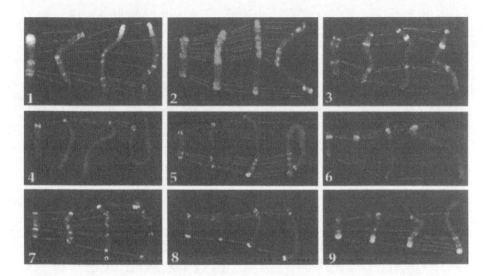

Figure 7.2. Hybridization of GC-rich H3 isochore DNA to chromosomes 1–9 at different stages of contraction. Biotinylated H3 isochore DNA was detected with avidin-FITC (yellow) on propidium iodide-stained chromosomes (red) (Saccone et al., Identification of the gene-richest bands in human prometaphase chromosomes, Chrom Res 7: figure 1, p 382, copyright 1999, with kind permission from Kluwer Academic Publishers) (See color insert).

restricted (tissue-specific) genes, as well as many housekeeping genes. The G-bands contain mainly the L1 and L2 (AT-rich) isochores and nearly half of the tissue-specific genes. The GC-richest component of the genome is at least 17 times as rich in genes as the GC-poor regions are. There is about one gene per 9 kb in H3 isochore DNA, one gene per 15 kb in H2 isochore DNA, one gene per 54 kb in H1 isochore DNA, and one gene per 150 kb, on average, in the combined L1 and L2 isochore DNAs (Zoubak et al., 1996). Adjacent isochores need not be functionally distinct. Thus, the SOX9 gene belongs to a GC-rich H2 isochore, but almost all of the 1063 kb of DNA upstream (5') of the gene belongs to the GC-poor L1 and L2 isochore families. However, when this region is disrupted at any point by a translocation, the expression of the SOX9 gene is impaired (Pfeifer et al., 1999).

How sharp are the boundaries between adjacent isochores, and what is their structure and functional significance? There is growing evidence that chromosome bands may have specific boundary sequences that separate a

GC-rich band from a less GC-rich band. The still-homologous pseudoautosomal region on Xp and Yp (Chapter 17) has a boundary sequence, and pseudoautosomal boundary-like sequences (PABLs) have been found throughout the genome. They have a highly conserved 650-bp consensus sequence. One example is provided by the major histocompatibility complex (*MHC*) gene locus at 6p21.3 (Fig. 29.3). This is a wide R-band that includes a very GC-rich T-band. There is also an intervening G-band only 200 kb in length at 6p21.32, which separates the *MHC* class II genes from class I and class III genes. This tiny region contains a PABL1 element and a cluster of LINE1 elements (Fukagawa et al., 1996).

Unmethylated CpG Clusters and Housekeeping and Tissue-Specific Genes

The restriction endonuclease *Msp*I cleaves double-stranded DNA at CCGG sites in DNA whether or not the internal C is methylated, whereas another restriction enzyme, *Hpa*II, cleaves CCGG sites in DNA only if the internal C is not methylated. Bird and his collaborators have shown that a tiny fraction of the genome, about 1%, is highly enriched in methylatable sites (CGs, usually referred to as CpGs) that are not methylated. These were first detected as clusters of *Hpa*II-generated tiny fragments, or HTF islands. Analysis of 375 sequenced genes indicated that all housekeeping genes and about 40% of tissue-restricted genes are associated with CpG clusters. The majority of these clusters, which are generally 200–1400 bp long, are at the 5′ ends of housekeeping genes, but there is no such bias towards the 5′ end of tissue-restricted genes (Larsen et al., 1992).

Antiqua and Bird (1995) showed that there are about 45,000 CpG islands in the haploid human genome. Combining this with the data of Larsen et al. (1992) and data indicating there is one CpG island per 36 kb in R-(including T-) bands, Antiqua and Bird estimated the total number of genes in the human genome to be about 80,000, made up of 22,000 housekeeping and 58,000 tissue-restricted genes. The housekeeping genes and about half of the tissue-specific genes are in the T- and R-bands, with the remainder of the tissue-specific genes situated in the Q- or G-bands (Holmquist, 1992; Saccone et al., 1996; Zoubak et al., 1996).

Interspersed Repetitive DNA Sequences and Transposable Elements

The human genome contains an enormous number of interspersed repetitive elements that have spread through the genome by some sort of transposition process. They make up almost 35% of the genome (Kazazian and Moran, 1998). Short interspersed elements (SINEs), such as the 340-bp *Alu* sequences, whose name reflects the presence of an *Alu*I restriction endonuclease cutting site in each one, are especially abundant; they make up more than 10% of the genome. They are located mainly in the R-bands (Korenberg and Rykowski, 1988). A small subset of the *Alu* repeats can be transcribed and a DNA copy made by a special DNA polymerase called *reverse transcriptase*, because it makes DNA from an RNA template. This DNA copy can be transposed into a new genomic site. If this disrupts a gene (insertion mutagenesis), the latter may be inactivated, as exemplified by a cholinesterase gene mutation (Muratani et al., 1991).

Retroelements tend to integrate into actively transcribed DNA, in part because its more open conformation makes its DNA more accessible. *Alu* sequences are thus most abundant in the gene-rich T- and R-bands. For example, the *RCC1* gene, a regulator of chromosome condensation (Chapter 2), contains 37 *Alu* sequences within its 35 kb of DNA (Furuno et al., 1991). The consensus sequence of one *Alu* subfamily contains a binding site for the retinoic acid receptor that is present as part of the 5' promoter region of many genes. This provides a mechanism by which retrotransposition of *Alu* sequences can alter the expression of many genes and contribute to evolutionary potential (Vansant and Reynolds, 1995).

Long interspersed elements (LINEs), with about 50,000–100,000 copies, make up nearly 15% of the genome. They are mainly located in the Q- (G-)bands and are particularly abundant on the X chromosome, especially in the pericentromeric region (Korenberg and Rykowski, 1988). Full length LINE sequences are over 6 kb long and, like similar transposable elements, contain several genes; however, most copies are truncated to a variable extent at the 5' end and are much shorter. A significant number, estimated at 30–60, of the 3,000–4,000 full-length LINE L1 elements are able to initiate transposition (Sassaman et al., 1997). These encode an endonuclease that can nick DNA and a reverse transcriptase that can make a usually truncated DNA copy of an RNA transcript.

This can integrate into a new chromosomal site. Insertion of a truncated LINE L1 element into the X-linked factor VIII gene has been seen in a number of patients with hemophilia A (Dombroski et al., 1991).

Sometimes DNA copies of unrelated mRNAs are generated using reverse transcriptase and are integrated into the genome. These *processed pseudogenes* have no introns, because these are removed during the processing of RNA transcripts into mature mRNA. Usually, there are only a few copies of these pseudogenes, which are thus a minor component of interspersed repeats. Transposable elements are much more abundant. One, called *mariner*, or MITE (*mariner* insert-like transposable element), is present in about 1000 copies, and so are several other such elements. The complete *mariner* transposon contains inverted terminal repeats that enable it to bind a transposase and thus to move. There are more than 100 of these complete *mariner* elements scattered throughout the genome, except on the Y, as shown by primed in situ labeling (PRINS) with primers matching the right and left inverted repeats. Many of them are at sites where a disease-producing deletion, duplication, or inversion has occurred, suggesting that they are involved in initiating homologous recombination events (Reiter et al., 1999).

An updated list of retroelements can be obtained from the repetitive element data base (Repbase.http://www.girinst.org), or those on chromosome 6 from the chromosome 6 database (http://www.sanger.ac.uk/chr6). Their distribution in relation to bands is not well established. Given the frequency and amount of transposition characteristic of the human and other genomes, how have such striking differences in GC-richness of isochores been achieved, and maintained? One exciting hypothesis is that integration of exogenous DNA tends to be isopycnic, that is, the GC level of the incoming DNA sequence matches that of the host sequences around the integration site. This has been demonstrated for multiple integration sites of hepatitis B (HBV), AIDS (HIV-1), and human T-cell leukemia, type 1 (HTLV-1) viruses (Glukhova et al., 1999).

Tandemly Repetitive Elements: Telomeres, Centromeres, and Satellite DNAs

At the molecular level, chromosomes consist of highly repeated sequences (most of 100,000 copies or more), middle repeated sequences (most of 100–10,000 copies) and unique sequences (one or several copies). The boundaries between these classes are somewhat arbitrary. In the highly repetitive sequences, the repeating units vary in length from 2 to 2000 base pairs. In simple-sequence DNA, such as that of constitutive heterochromatin, short sequences are repeated over and over again to make long, meaningless stretches of DNA. These simple DNAs show extensive sequence variation. Simple-sequence DNA may have a buoyant density distinctly different from that of the bulk DNA of the organism, forming satellite bands in density gradient centrifugation; hence the name *satellite* DNA. There are three types of satellite DNA. Megabase pair satellite DNAs are limited mainly to centromeric heterochromatin and distal Yq. Kilobase-pair *minisatellite* DNAs are preferentially (90%) located in subtelomeric T-bands. The roughly 10 to 40-base-pair *microsatellite* DNAs are widely distributed throughout the genome.

Multigene families, which encode ribosomal RNA, histones, and a few other classes of genes, represent middle repetitive DNA. These genes, which encode RNAs or proteins needed in large quantities in the cell, are repeated hundreds of times in the genome. Many of them are located in T-bands and some in R- or G-bands. The clusters of ribosomal RNA genes are located within satellite DNA and are virtually the only type of gene able to avoid being inactivated by such close contact with heterochromatin. Transposable elements, which do not have a fixed location on the chromosomes, make up a considerable proportion of middle repetitive DNA. Other multigene families, of which globin and immunoglobulin genes are typical examples, bridge the gap between unique sequences and middle repeated sequences. Characteristic of genes in such multigene families are multiplicity, close linkage, sequence homology, and similar or overlapping functions. Unique sequences can be divided into transcribed genes and noncoding sequences, including spacers and nonfunctional pseudogenes.

Chromosomal Proteins and Chromatin Conformation

The usual technique for labeling DNA for molecular hybridization is called *nick translation*. It involves nicking DNA with DNAase I in the presence of DNA polymerase and radioactive nucleotides. The polymerase fills in the gaps produced by the DNAase I with nucleotides, producing a labeled probe. Kerem et al. (1984) applied nick translation in situ to fixed chromosome spreads, using a limiting concentration of DNAase I, and found that DNAase I–sensitive regions were concentrated on the active X chromosome in preference to the inactive X and in R-bands in preference to G-bands. This confirms the more open conformation of the chromatin on the active X and in the R-bands.

This difference in chromatin structure is associated with striking differences in the proteins of the various types of bands. Histone H4 is hyperacetylated in both T- and R-bands but not in G- or C-bands, as shown by immunofluorescence studies (Jeppesen and Turner, 1993). Nucleosomal core histones are also hyperacetylated in interphase chromatin fractions that contain unmethylated CpG islands, which come from T- and R-bands. Histone H1, an inhibitor of gene activity, is depleted in these fractions (Tazi and Bird, 1990). In contrast, histone H1 is enriched in nucleosomes that contain methylated DNA, is more abundant in G-bands than in R-bands, and can be depleted from nucleosome arrays by hyperacetylation. The high-mobility-group protein 1 (HMG1) localizes to G- and C-bands (Disney et al., 1989). The chromosome scaffold proteins topoisomerase II and SCII are especially abundant in centromeric heterochromatin, and topoisomerase II is also found in G-bands (Saitoh and Laemmli, 1994; Sumner, 1996). Their role in organizing the chromosome scaffold that anchors

Table 7.1. Characteristics of the Different Types of Chromosome Bands

Feature	T-bands	R-bands	G-bands	C-bands
Quinacrine or DAPI	Dull	Dull	Bright	Either
Acridine orange/heat	Bright	Bright	Dull	Dull
Chromomycin A3	Bright	Bright	Dull	Dull
Replication time	Very early	Early	Late	Very late
DNAase I sensitivity	High	High	Low	Low
In or between chromomeres	Between	Between	In	In
Chiasma frequency	Very high	High	Low	Absent
Acetylation of histone H4	High	High	Low	Low
Unmethylated CpG islands	Very many	Many	Few	None
GC-richest (H3) isochores	Many	Some	None	None
GC-rich (H1, H2) isochores	Few	Many	Rare	None
GC-poor (L1, L2) isochores	None	None	Many	Some
Interspersed repeats	SINEs	SINEs	LINEs	Rare
Satellite DNAs	Mini-satellites	Micro-satellites	Micro-satellites	Satellites

each loop domain (Chapter 5) is clearly reflected in their lower concentrations in T- and R-bands, with their longer loops and looser packaging.

Functional Significance of Chromosome Bands

Chromosome bands reflect the functional organization of the genome that is necessary for regulating DNA replication and repair, transcription, genetic recombination, and transposition (Bickmore and Craig, 1997). Table 7.1 summarizes some of the characteristic differences among T-, R-, G-, and C-bands. There are, of course, additional differences already known and possibly many yet to be discovered. T-bands contain only 15% of the DNA but 65% of the mapped genes, most of the meiotic chiasmata, and most of the cancer-associated or X-ray-induced rearrangements (Holmquist, 1992). Genes in T- and R-bands are at or near the periphery of interphase chromosome domains, while genes in G-bands tend to be buried in the interior of the domain (Chapter 5). C-bands, in contrast, tend to clump together at the nuclear envelope or around a nucleolus.

References

Antiqua F, Bird A (1995) Number of CpG islands and genes in human and mouse. Proc Natl Acad Sci USA 90:11995–11999

Bickmore WA, Craig J (1997) Chromosome bands: patterns in the genome. Chapman & Hall, New York

Disney JE, Johnson KR, Magnuson NS, et al. (1989) High-mobility group protein HMG-1 localizes to G/Q- and C-bands of human and murine chromosomes. J Cell Biol 109:1975–1982

Dombroski B, Mathias SL, Nanthakumar E, et al. (1991) Isolation of an active human transposable element. Science 254:1805–1808

Fukagawa T, Nakamura Y, Okumura K, et al. (1996) Human pseudoautosomal boundary-like sequences: expression and involvement in evolutionary formation of the present-day pseudoautosomal boundary of human sex chromosomes. Hum Mol Genet 5:23–32

Furuno N, Nakagawa K, Eguchi U, et al. (1991) Complete nucleotide sequence of the human RCC1 gene involved in coupling between DNA replication and mitosis. Genomics 11:459–461

Glukhova LA, Zoubak SV, Rynditch AV, et al. (1999) Localization of HTLV-1 and HIV-1 proviral sequences in chromosomes of persistently infected cells. Chrom Res 7:177–183

Holmquist GP (1992) Review article: chromosome bands, their chromatin flavors, and their functional features. Am J Hum Genet 51:17–37

Jeppesen P, Turner BM (1993) The inactive X chromosome in female mammals is distinguished by a lack of histone H4 acetylation, a cytogenetic marker for gene expression. Cell 74:281–289

Kazazian HH Jr, Moran JV (1998) The impact of retrotransposons on the human genome. Nat Genet 19:19–24

Kerem B-S, Goitein R, Diamond G, et al. (1984) Mapping of DNAase I sensitive regions in mitotic chromosomes. Cell 38:493–499

Korenberg JR, Rykowski MC (1988) Human genome organization: Alu, Lines, and the molecular structure of metaphase chromosome bands. Cell 53: 391–400

Larsen F, Gunderson G, Lopez R, et al. (1992) CpG islands as gene markers in the human genome. Genomics 13:1095–1107

Muratani K, Hada T, Yamamoto Y, et al. (1991) Inactivation of the cholinesterase gene by Alu insertion: Possible mechanism for human gene transposition. Proc Natl Acad Sci USA 88:11315–11319

Pfeifer D, Kist R, Dewar K, et al. (1999) Campomelic dysplasia translocation breakpoints are scattered over 1 Mb proximal to SOX9: evidence for an extended control region. Am J Hum Genet 65:111–124

Reiter LT, Liehr T, Rautenstrauss B, et al. (1999) Localization of mariner DNA transposons in the human genome by PRINS. Genome Res 9:839–843

Saccone S, Caccio S, Kasuda J, et al. (1996) Identification of the gene-richest bands in human chromosomes. Gene 174:85–94

Saitoh Y, Laemmli UK (1994) Metaphase chromosome structure: bands arise from a differential folding path of the highly AT-rich scaffold. Cell 76: 609–622

Sassaman DM, Dombroski BA, Moran JV, et al. (1997) Many human L1 elements are capable of retrotransposition. Nat Genet 16:37–43

Sumner AT (1996) The distribution of topoisomerase II on mammalian chromosomes. Chrom Res 4:5–14

Tazi J, Bird A (1990) Alternative chromatin structure at CpG islands. Cell 60:909–920

Vansant G, Reynolds WF (1995) The consensus sequence of a major *Alu* subfamily contains a functional retinoic acid response element. Proc Natl Acad Sci USA 92:8229–8233

Zoubak S, Clay O, Bernardi G (1996) The gene distribution of the human genome. Gene 174:95–102

8

In Situ Hybridization

A fascinating property of DNA is the complementarity of the nucleotide bases in its two anti-parallel strands, with G always pairing with C and A always pairing with T. This does not involve strong covalent chemical bonds but weak hydrogen bonds. There are three hydrogen bonds between G-C pairs and two between A-T pairs, so strand separation is easier in AT-rich DNA than in GC-rich DNA. Mild heating breaks these hydrogen bonds and is one way to separate the two strands, called *denaturation* or *dissociation*. Reducing the temperature under the right salt conditions leads to *renaturation* (*reassociation* or *reannealing*) of the two strands by reconstitution of the hydrogen bonds. The rate of renaturation depends on the frequency of collision between complementary sequences, which depends on their concentration. The concentration and time required for renaturation determines the *Cot value* (concentration × time). If a high concentration of labeled probe DNA is used, hybridization to complementary nucleic acid sequences in the target preparation can be achieved in a reasonably short time. These properties of DNA are extremely important, because they make it possible to detect specific DNA sequences (such as genes) on a nitrocellulose filter (*molecular hybridization*) or in cytological preparations (*in situ hybridization*) by using labeled DNA or RNA probes.

In Situ Hybridization of Repetitive and Unique DNA Sequences

Typically, in situ hybridization involves denaturing a cloned DNA fragment, labeling it with a radioactive, fluorescent, or antigenic tag, and hybridizing it to denatured DNA in fixed metaphase chromosome spreads on a slide. Under highly stringent conditions of temperature and salt concentration, the DNA probe will hybridize only to its complementary strands and not to any of the enormously greater number of strands from other parts of the genome. Under these conditions, bonds formed between imperfectly paired sequences will be short-lived, but bonds between perfectly paired sequences will be stable. Sites of in situ hybridization of radioactive probes can then be detected autoradiographically and those of nonradioactive probes by fluorescence or enzyme-mediated staining.

The strength of signal depends on the length of the cloned DNA probe and the density of label, and determines the *sensitivity*. When ^{3}H-thymidine is used to label probe DNA, the sensitivity is limited, because the half-life of tritium (^{3}H) is 12.8 years, so only a tiny fraction of the atoms will decay in a 3- to 4-week autoradiographic exposure period. Furthermore, the electrons produced by the radioactive decay of tritium may travel some distance before they nucleate silver grains in the photographic emulsion overlying the microscope slide. The grains are spread out within a radius of 1 μm or more around the site of hybridization, which yields rather low *resolution*. However, autoradiographic in situ hybridization does have one advantage: It allows comparisons of the relative sizes of two or more targets. De Capoa et al. (1988) used this approach to determine the relative numbers of ribosomal RNA genes on different acrocentric chromosomes.

Despite the limited sensitivity and resolution of autoradiographic detection of sites of in situ hybridization, this method was successfully used to map tandemly repetitive sequences such as satellite DNAs in heterochromatin and several classes of multicopy genes (40 to 200 copies each). These included the 18S and 28S ribosomal RNA genes in the NORs of the acrocentric chromosomes, the 5S ribosomal RNA genes at lq42–q43, and the histone genes at 7q32–q36. Autoradiographic detection of single-copy genes was much harder, though achievable (Harper and Saunders, 1984). A major problem was that fewer than 15% of the autoradiographic silver grains were localized at the two specific allelic sites, with the rest of the grains widely scattered and sometimes also concentrated at a few secondary sites. That is, the background was high and the signal-to-noise ratio was low. An obvious reason for the high background was the presence of extremely abundant repetitive sequences throughout the genome. Because of their high number of copies, the repetitive sequences could hybridize more quickly (at a lower Cot value) than unique sequence genes.

A major advance was the discovery that hybridization of labeled probes to repetitive sequences can be blocked by prehybridization to unlabeled genomic DNA or, preferably, Cot1 DNA, the highly repetitive fraction of the genome (Cremer et al., 1988; Pinkel et al., 1988). As a result, nonradioactive (enzyme or fluorescence) labeling techniques have largely replaced autoradiographic labeling because of their greater speed, resolution, and reliability. Almost any gene or other component of the genome can now be mapped.

Fluorescence In Situ Hybridization

Methods involving fluorescent labels are generally preferred over enzyme labels and are the best developed (Trask, 1991; Lichter and Ward, 1990). Bromo-deoxyuracil, digoxigenin, dinitrophenol, or other side-groups can be used to label DNA probes, and their sites of hybridization can be detected by highly specific fluorescent or enzyme-linked antibodies (Fig. 8.1). Fluorescent side-groups can be covalently bound to the probe DNA itself, thus eliminating the need to add a fluorescent molecule that binds to the probe. Chromosome-specific probe sets have been prepared that "paint" different chromosomes or parts of chromosomes (Cremer et al., 1988; Pinkel et al., 1988). The power and beauty of these new techniques is reflected by their increasingly wide use, and illustrated in Figs. 8.2, 8.3, 8.4 (see color insert), and 26.1.

Very long DNA fragments can be used as hybridization probes after block-ing the hybridization of interspersed repeats, thus markedly reducing the noise level generated by such hybridization. This is useful because long probes yield strong fluorescent signals that are readily detectable by simple fluorescence

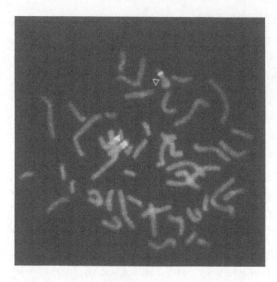

Figure 8.1. Demonstration by FISH of a deletion of the Prader-Willi/Angelman syndrome region, 15q11–q13. Hybridization of a YAC cloned from this region produces a signal on one homologue (filled arrowhead) but none on the other homologue (open arrowhead). The classical satellite probe, D15Z1, produces a large signal in the centromeric region of each chromosome 15 (reproduced from Kuwano et al., Hum Mol Genet 1:417–425, 1992, with permission of Oxford University Press).

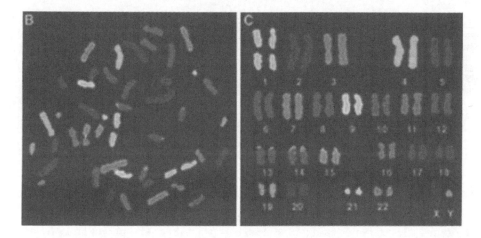

Figure 8.2. Spectral karyotyping after simultaneous hybridization of 24 combinatorially labeled chromosome-specific (painting) probes. B, a metaphase spread; C, the karyotype of this metaphase. Regions rich in repetitive sequences, such as the short arms of numbers 13, 14, 15, 21, and 22, show color variations that are expected after suppression hybridization (reprinted with permission from Schröck et al., Multicolor spectral karyotyping of human chromosomes. Science 273:494–497, copyright 1996, American Association for the Advancement of Science) (See color insert).

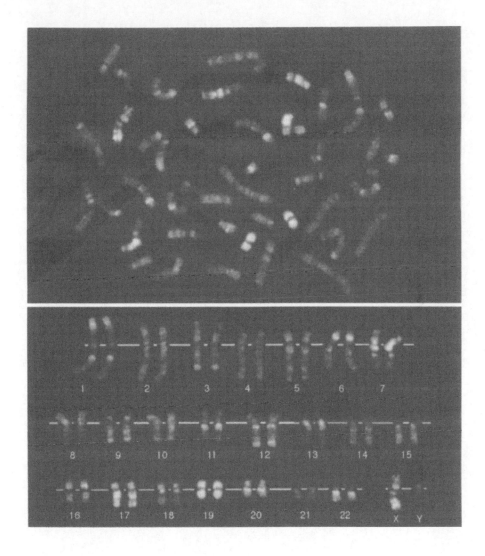

Figure 8.3. A metaphase spread (above) and a karyotype (below) after in situ hybridization and detection of multiplex probes. Red and green bars represent regions that hybridized to fragments in only one probe pool, and yellow bars represent regions that hybridized equally to fragments in both probe pools. Mixed colors are due to overrepresentation of fragments from one pool. Regions that fail to hybridize to fragments in either pool show background DAPI blue fluorescence (reproduced from Müller et al., Toward a multicolor chromosome bar code for the entire human karyotype by fluorescence in situ hybridization, Hum Genet 100, fig. 3, p 273, copyright 1997, Springer-Verlag) (See color insert).

Figure 8.4. High-resolution fiber-FISH using the DAZ G5 probe from the 5′ end of the DAZ gene (green fluorescence) and the DAZ G21 probe from the 3′ end of the gene (red fluorescence) to label extended S phase Y chromatin. Seven DAZ red-green signals (genes or pseudogenes) are seen in a linear array, with at least one gene inverted with respect to the others (Gläser et al., Chromosome Research 6: figure 4, p 484, 1998, with kind permission from Kluwer Academic Publishers) (See color insert).

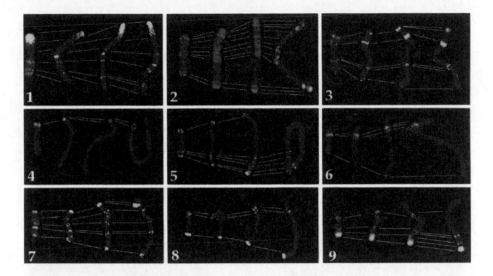

Figure 7.2. Hybridization of GC-rich H3 isochore DNA to chromosomes 1–9 at different stages of contraction. Biotinylated H3 isochore DNA was detected with avidin-FITC (yellow) on propidium iodide-stained chromosomes (red) (Saccone et al., Identification of the gene-richest bands in human prometaphase chromosomes, Chrom Res 7: figure 1, p 382, copyright 1999, with kind permission from Kluwer Academic Publishers).

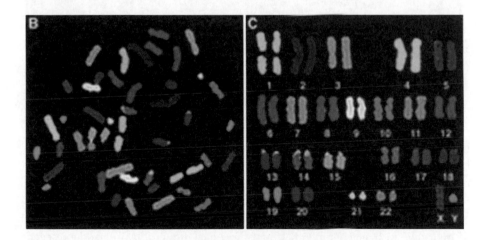

Figure 8.2. Spectral karyotyping after simultaneous hybridization of 24 combinatorially labeled chromosome-specific (painting) probes. B, a metaphase spread; C, the karyotype of this metaphase. Regions rich in repetitive sequences, such as the short arms of numbers 13, 14, 15, 21, and 22, show color variations that are expected after suppression hybridization (reprinted with permission from Schröck et al., Multicolor spectral karyotyping of human chromosomes. Science 273:494–497, copyright 1996, American Association for the Advancement of Science).

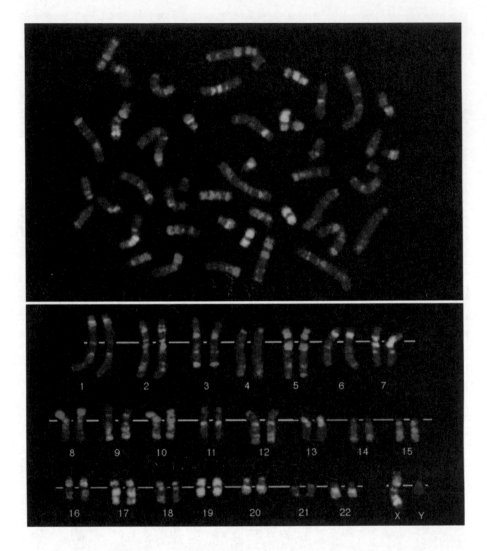

Figure 8.3. A metaphase spread (above) and a karyotype (below) after in situ hybridization and detection of multiplex probes. Red and green bars represent regions that hybridized to fragments in only one probe pool, and yellow bars represent regions that hybridized equally to fragments in both probe pools. Mixed colors are due to overrepresentation of fragments from one pool. Regions that fail to hybridize to fragments in either pool show background DAPI blue fluorescence (reproduced from Müller et al., Toward a multicolor chromosome bar code for the entire human karyotype by fluorescence in situ hybridization, Hum Genet 100, fig. 3, p 273, copyright 1997, Springer-Verlag).

Figure 8.4. High resolution fiber-FISH using the DAZ G5 probe from the 5′ end of the DAZ gene (green fluorescence) and the DAZ G21 probe from the 3′ end of the gene (red fluorescence) to label extended S phase Y chromatin. Seven DAZ red-green signals (genes or pseudogenes) are seen in a linear array, with at least one gene inverted with respect to the others (Gläser et al., Chromosome Research 6: figure 4, p 484, 1998, with kind permission from Kluwer Academic Publishers).

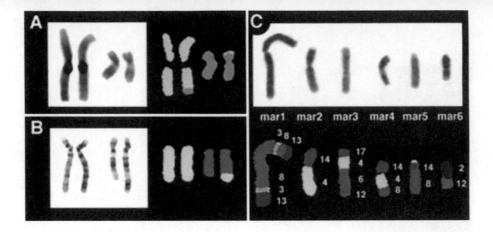

Figure 8.6. (A) Partial C-band and spectral (painting) karyotypes of the father of a mentally retarded child, showing an otherwise cryptic t(1;11) translocation. (B) Partial G-band karyotype of an ataxic patient; note extra material on a chromosome 11; the spectral karyotype shows its origin from a chromosome 4. (C) Six marker chromosomes from a breast cancer cell line; spectral karyotyping shows the chromosomes involved in these complex rearrangements (reprinted with permission from Schröck et al., Multicolor spectral karyotyping of human chromosomes. Science 273:494–497, copyright 1996, American Association for the Advancement of Science).

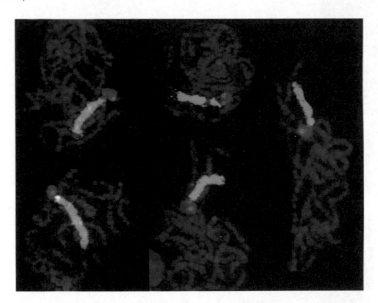

Figure 9.6. Association of bivalent 15 with the sex vesicle in five normal pachytene nuclei stained blue with DAPI. Bivalent 15 is made visible with a 15-specific paint (green) and its centromeric region with a 15-specific α-satellite probe (red). The sex vesicle (blue clump) is marked by an X-specific α-satellite probe (red) (Metzler-Guillemain et al., Chrom Res 7:369, Bivalent 15 regularly associates with the sex vesicle in normal male meiosis, fig. 1, copyright 1999, with kind permission from Kluwer Academic Publishers).

microscopy. The nick translation reaction used to label the probes also yields the short fragments best suited for hybridization. The fluorescence does not spread from the site of hybridization nearly as far as autoradiographic silver grains do, so the resolution, or precision, of mapping is much higher with fluorescence in situ hybridization (FISH) than with autoradiography. The results do not require a statistical analysis of the distribution of signals throughout the entire genome, because the background is so low that virtually every fluorescent signal counts (see Figs. 3.5, 8.4, and 9.6).

Highly sensitive CCD (charge-coupled device) digital cameras and computer-assisted data acquisition and processing methods permit routine detection of even faint fluorescent signals. With them, FISH can detect hybridization to sequences as short as 1 kb. It is even possible to distinguish single RNA transcripts within a cell and to determine the average number of transcripts of a particular gene within a cell. By using three different fluorochromes to label three probes, each complementary to a different segment of a gene, one can determine the rate of elongation of a growing (nascent) RNA transcript and the distance separating the RNA polymerase II transcriptional complexes. These correspond well to the results determined by electron microscopy (Femino et al., 1998).

Probes for highly repetitive (TTAGGG)n telomeric sequences are widely used, and for a variety of purposes. They have been used to show that apparently terminal deletions have been stabilized by the capture of TTAGGG repeats (Figs. 8.1 and 15.3). Pericentric inversions (Chapter 13) can be detected by a fascinating method called Chromosome Orientation and Direction FISH, or COD-FISH (Bailey et al., 1996). This takes advantage of the fact that telomeric repeats (TTAGGG in the G-rich strand, CCCTAA in the C-rich strand) all have the same orientation with respect to the end of the chromosome, just as centromeric α-satellite sequences do. Cells grown for one cycle in the presence of BrdU incorporate BrdU in place of T into each newly replicated strand of DNA. UV light selectively breaks down BrdU-substituted DNA, and exonuclease III can then destroy the entire damaged strand, leaving a single strand of DNA in each sister chromatid. A single-stranded probe complementary to one strand of telomeric or centromeric DNA will thus find a hybridizable (complementary) sequence in only one of the two sister chromatids. The single-stranded telomeric probe will hybridize to the short-arm telomere of one chromatid and the long-arm telomere of the other chromatid; the centromeric probe will hybridize to only one of these chromatids. However, if a pericentric inversion has occurred, the centromeric probe will hybridize to the other chromatid (Fig. 13.6).

Replication Timing by FISH

The time of replication of individual genes can be determined in various ways. The standard method involves incorporating BrdU during part of the S phase in synchronized cell cultures and detecting the BrdU-containing strand with antibodies to BrdU. Alternatively, the BrdU-containing strand can be destroyed by photolysis and its complementary strand destroyed with single-strand-specific nuclease. This procedure eliminates only the DNA segments or genes that were replicated while BrdU was present. Specific gene probes can then be hybridized to the remaining DNA in solution or on Southern blots (Chapter 3). Selig et al. (1992) devised an ingenious cytological method to determine the time at which a particular gene is replicated. DNA probes for the genes in question are hybridized to interphase nuclei in synchronized cultures and the sites of hybridization detected by FISH. Each allele appears as a single dot before replication and as paired dots after replication has occurred and a sufficient degree of separation of sister chromatids has taken place in that region (Fig. 3.5). Torchia et al. (1994) and Boggs and Chinault (1994) applied this technique to several X-linked loci. Those known to undergo X-inactivation, such as *HPRT* and *FRAXA*, showed a high degree of asynchrony in the time of replication of the two alleles, whereas the two alleles of genes that escape X-inactivation, such as *ZFX* and *RPS4X*, replicated in synchrony. The *XIST* gene, which is active only on the inactive X chromosome, showed marked asynchrony, and both research groups suggested that the active XIST allele replicated early and the inactive allele late. However, Gartler et al. (1999), using both the standard method and that of Selig et al. (1992), concluded that the inactive XIST allele replicated early and the active allele late (Fig. 8.5).

Cloned, PCR-Generated, and In Situ-Generated Probes

The first probes for in situ hybridization were fragments of DNA that had been inserted into bacterial viruses (circular plasmids or linear bacteriophages) that

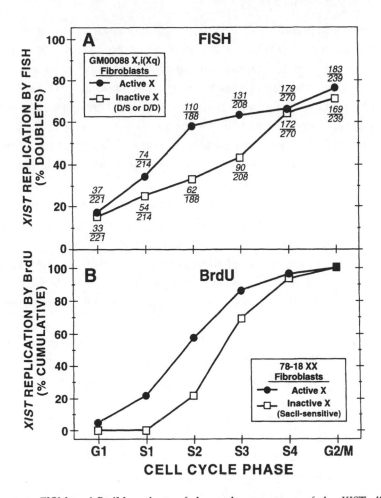

Figure 8.5. FISH and BrdU analysis of the replication time of the *XIST* alleles in X,i(Xq) fibroblasts. (A) Percentage of cells with doublets of the active X (closed circles) and the inactive i(Xq) (open squares) in each fraction of the cell cycle. The inactive i(Xq) was scored as replicated if doublets were present at either or both *XIST* loci. (B) Cumulative replication values for the active X (closed circles) and inactive i(Xq) (open squares) in each fraction of the cycle (reproduced from Gartler et al., Hum Mol Genet 8:1085–1089, 1999, with permission of Oxford University Press).

were then grown in bacteria to develop clones of identical fragments. Larger fragments have been successfully cloned in cosmids (35–55 kb) and in bacterial artificial chromosomes (BACs, 100–250 kb; Shizuya et al., 1992) or yeast artificial chromosomes (YACs, 300–1500 kb; Burke et al., 1987). An example of their use is the creation of a complete set of probes specific for each chromosome end, capable of detecting by FISH deletions, translocations, or other rearrangements (Knight et al., 1997). The polymerase chain reaction (PCR) has been used to generate unique probes or libraries of probes from a whole genome, a whole chromosome, or a chromosome segment (Lüdecke et al., 1989). Some of the most useful probes have been produced using an oligonucleotide complementary to part of an *Alu* sequence as one primer and an arbitrary oligonucleotide sequence for the other. This generates a library of fragments, most of them from the gene-rich R-bands, which are rich in *Alu* sequences. This technique can be used with a variety of chromosome sources: flow-sorted chromosome-specific (or enriched) pools, rodent-human hybrids that contain a specific human chromosome or chromosome fragment, and microdissected fragments from a specific chromosome region. Microdissection (a scrape from a single or several chromosomes on a slide) and formation of a PCR-amplified pool of DNA fragments from a tiny region can be used to yield a chromosome-band-specific painting probe (Lüdecke et al., 1989; Chen et al., 1997).

A method has been developed for the in situ generation of label, called *primed in situ* (PRINS) *labeling*. A labeled oligonucleotide primer is annealed to its complementary sequences in the genome and used to initiate replication in situ. This works quite well with repetitive DNA sequences, such as centromeric α-satellite DNA, and takes no more than an hour. PRINS gives good results with even fairly low copy repeats if the labeled product is at least 1 kb in length. For lower copy repeats, a modification involving PCR in situ sometimes works (Gosden and Lawson, 1994). By using either end of the consensus *Alu* sequence as the oligonucleotide primer (Alu-PRINS), one can selectively label euchromatin (T- and R-bands, mainly) and detect small regions of euchromatin in an aberrant location (Cullen et al., 1997).

Chromosome-, Region-, and Band-Specific Painting Probes

In 1988, two groups reported the successful use of chromosome-specific libraries for the recognition of numerical and structural chromosome abnormalities (Cremer et al., 1988; Pinkel et al., 1988). Flow sorting was used to prepare fractions highly enriched for each chromosome, and the DNA from each of these was fragmented and the pool of fragments ligated into bacterial virus DNA. Bacteria were infected with the mixture of recombinant virus DNA to yield a library of human DNA fragments derived almost entirely from a single chromosome: a *chromosome-specific library*. Although repetitive human sequences common to many chromosomes are abundant in such libraries, their hybridization can be blocked by prehybridization with an excess of unlabeled DNA or the highly repetitive Cot1 fraction of DNA. The nonrepetitive sequences in the labeled library hybridize, producing a fluorescent signal along the length of the chromosome (Fig. 8.2). The number of chromosomes labeled with such a probe is easily identifiable, in both metaphase spreads and interphase nuclei. Translocations involving the painted chromosome are also readily apparent, even when they are very tiny (Popp et al., 1993). Chromosome painting probes can be used in conjunction with locus-specific probes (Fig. 9.6).

In principle, a probe specific for any region or band can be generated by microdissection followed by PCR amplification of sequences from the region or band, and a number of such probes have been developed. Using a more global approach, Rocchi and his collaborators have prepared a panel of subchromosomal painting libraries (paints) representing over 300 regions of the human genome (Antonacci et al., 1995). The presence of ribosomal RNA gene clusters and several repetitive sequence families on all the acrocentric chromosomes reduces the specificity of chromosome painting probes prepared from these chromosomes. A novel way around this problem is to prepare painting probes from gorilla or chimpanzee acrocentric chromosomes, which lack all these repetitive elements but are otherwise highly homologous to the human acrocentrics (Müller et al., 1997a).

Multicolor FISH, Spectral Karyotyping, and Bar Codes

Simultaneous detection of each chromosome in a metaphase spread is now routine, using either multicolor FISH (M-FISH; Speicher et al., 1996) or spectral karyotyping (SKY; Schröck et al., 1996). These methods use 24 chromosome painting probes and five fluorochromes in a combinatorial labeling scheme that labels each of the 22 autosomes and the X and Y differently. M-FISH requires a series of image acquisitions with a change of optical filters, while SKY uses Fourier spectroscopy and an interferometer to evaluate the fluorescence emission patterns in a single, longer image acquisition (Fig. 8.2). Neither method detects intrachromosomal rearrangements such as most deletions, duplications, or inversions, but they are very useful for detecting other structural as well as numerical changes (Fig. 8.6; see color insert). Even minute chromosome rearrangements may be detectable using multicolor FISH with chromosome-specific painting probes (Popp et al., 1993).

Alu-PCR from human-mouse hybrids that contain a limited number of human chromosomes has been used to generate two different human-specific probe sets. These sets were derived pretty much at random from the human genome, and for use one is labeled with red and one with green fluorescence. Chromosome regions that hybridize to probes from both pools appear yellow, while regions that hybridize to probes from only one pool appear either red or green. The 110 distinct signals per haploid set produce a unique multicolor. pattern of variously colored dots along each chromosome, a "bar code" that allows chromosome identification and recognition of rearrangements (Fig. 8.3). Extension of this approach using additional signals or additional colors should make this technique increasingly useful (Müller et al., 1997b).

High-Resolution (Interphase and Fiber) FISH

Techniques have been developed for multicolor fluorescence labeling of chromosomes, which allows two or more specific probes to be identified in a single nucleus. Interphase chromatin in hypotonically swollen nuclei can be stained with these fluorescent probes; the distance between the bands in such fluorescence-labeled chromosomes is directly proportional to the molecular distance (up to about 1 million base pairs) separating the two binding sites (Lichter and Ward, 1990; Trask, 1991). These new techniques will improve the diagnostic power of cytogenetics and even allow chromosome changes to be detected in nondividing differentiated cells. The use of multiple probes and multicolor fluorescence labeling permits much more accurate analysis and detection of subtle changes in structure and copy number (Trask, 1991).

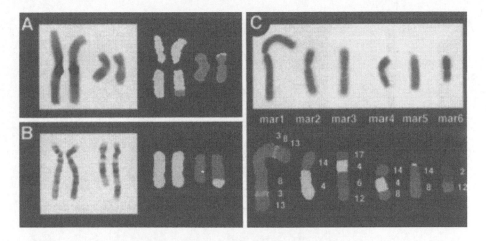

Figure 8.6. (A) Partial C-band and spectral (painting) karyotypes of the father of a mentally retarded child, showing an otherwise cryptic t(1;11) translocation. (B) Partial G-band karyotype of an ataxic patient; note extra material on a chromosome 11; the spectral karyotype shows its origin from a chromosome 4. (C) Six marker chromosomes from a breast cancer cell line; spectral karyotyping shows the chromosomes involved in these complex rearrangements (reprinted with permission from Schröck et al., Multicolor spectral karyotyping of human chromosomes. Science 273:494–497, copyright 1996; American Association for the Advancement of Science) (See color insert).

Chromatin can be released from interphase nuclei and the extended chro matin fibers spread out on lysine-coated slides. These are used for fiber FISH, to determine the order of three or more genes that are close to one another, or to analyze chromosome abnormalities (Mann et al., 1997). Pelizaeus-Merzbacher disease of the central nervous system is frequently caused by a duplication of the *PLP* (proteolipid protein) gene at Xq22. Fiber FISH detected a tiny duplication in three different families. In a fourth family, the two *PLP* signals were widely separated even in metaphase chromosomes, indicating there had been a major structural rearrangement (Woodward et al., 1998). Using differentially fluorescent probes from the 5′ and 3′ ends of the *DAZ* gene, Gläser et al. (1998) demonstrated a linear array of seven *DAZ* genes in Y chromatin fibers, with at least one of them inverted with respect to the others (Fig. 8.4). Shiels et al. (1997) used fiber FISH to estimate the length of ribosomal RNA genes (rDNA) and α-satellite arrays on human chromosome 22 from a human-rodent cell hybrid. There were up to 10 repeats of the 43-kb rRNA gene in a 155-μm cluster, although the full

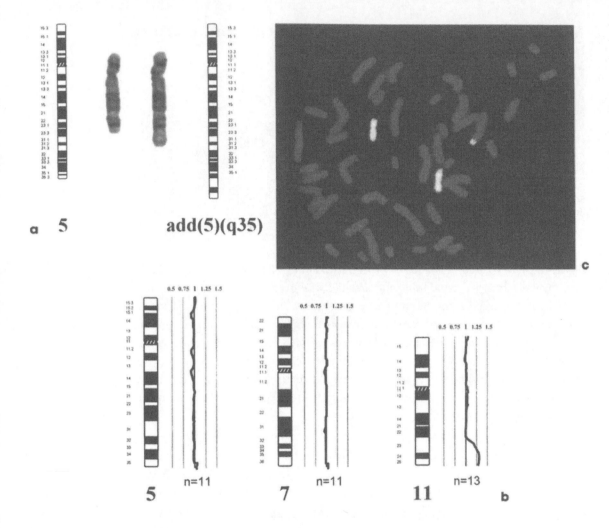

Figure 8.7. (a) Partial karyotype and ideogram of two chromosomes 5, one with additional material on it. (b) Ideograms and CGH (comparative genomic hybridization) profiles indicate that the extra material is from 11q23–qter. (c) Confirmation by FISH using a chromosome 11–specific painting probe (Levy et al., 1997, with permission of S. Karger AG, Basel).

rDNA array on this specific chromosome has nearly 40 copies (Srivastava et al., 1993). The fiber length of the centromeric α-satellite array was 930 μm, providing a better minimal estimate of the size of this array: 2600 kb (2.6 Mb). A modification of this technique uses linearized cosmid DNA fibers attached to a glass slide and aligned in parallel by a process called molecular combing (Weier et al., 1995).

Comparative Genomic Hybridization

Comparative genomic hybridization (CGH) is a molecular cytogenetic approach for genome-wide screening for differences in copy number of any DNA sequence from an individual, and it requires only a few nanograms of DNA as FISH probes. Equal amounts of differently labeled DNA from the test individual (green) and reference DNA (red) are cohybridized to normal metaphase spreads. Copy number differences show up as altered ratios of green-to-red fluorescence intensities (Fig. 26.1). This ratio is calculated along the length of each chromosome by a digital image analysis system. Increased copy number gives an increased ratio, while losses or deletions give a decreased ratio. Levy et al. (1997) used CGH to determine the origin of a de novo translocation in a newborn with multiple congenital anomalies and an unbalanced karyotype (Fig. 8.7). This method has been applied to more than 1500 tumor DNA samples and revealed six previously unknown gene amplifications (Forozan et al., 1997). For a list of publications on CGH, see http://nhgri.nih.gov/dir/ccg/cgh.

References

Antonacci R, Marzella R, Finelli P, et al. (1995) A panel of subchromosomal painting libraries representing over 300 regions of the human genome. Cytogenet Cell Genet 68:25–32

Bailey SM, Meyne J, Cornforth MN, et al. (1996) A new method for detecting pericentric inversions using COD-FISH. Cytogenet Cell Genet 75:248–253

Boggs BA, Chinault AC (1994) Analysis of replication timing of human X-chromosomal loci by fluorescence in situ hybridization. Proc Natl Acad Sci USA 91:6083–6087

Burke DT, Carle GF, Olson MV (1987) Cloning of large segments of exogenous DNA into yeast by means of artificial chromosome vectors. Science 236:806–812

Chen Z, Grebe TA, Guan X-Y, et al. (1997) Maternal balanced translocation leading to partial duplication of 4q and partial deletion of 1p in a son: cytogenetic and FISH studies using band-specific painting probes generated by chromosome microdissection. Am J Med Genet 71: 160–166

Cremer T, Lichter P, Borden J, et al. (1988) Detection of chromosome aberrations in metaphase and interphase tumour cells by in situ hybridization using chromosome specific library probes. Hum Genet 80:235–246

Cullen DF, Yip M-Y, Eyre HJ (1997) Rapid detection of euchromatin by Alu-PRINS: use in clinical cytogenetics. Chrom Res 5:81–85

DeCapoa A, Felli MP, Baldini A, et al. (1988) Relationship between the numbers and function of human ribosomal genes. Hum Genet 79:301–304

Femino AM, Fay FS, Fogarty K, et al. (1998) Visualization of single RNA transcripts in situ. Science 280:585–590

Forozan F, Karhu R, Kononen J, et al. (1997) Genome screening by comparative genomic hybridization. Trends Genet 13:405–409

Gartler SM, Goldstein L, Tyler-Freer SE, et al. (1999) The timing of *XIST* replication: dominance of the domain. Hum Mol Genet 8:1085–1089

Gläser B, Yen PH, Schempp W (1998) Fibre-FISH unravels apparently seven DAZ genes or pseudogenes clustered within a Y-chromosome region frequently deleted in azoospermic males. Chrom Res 6:481–486

Gosden J, Lawson D (1994) Rapid chromosome identification by oligonucleotide-primed in situ DNA synthesis (PRINS). Hum Mol Genet 3: 931–936

Harper ME, Saunders GF (1984) Localization of single-copy genes on human chromosomes by in situ hybridization of ^{3}H-probes and autoradiography. In: Sparkes RS, de la Cruz FF (eds) Research perspectives in cytogenetics. University Park Press, Baltimore, pp 117–133

Knight SJL, Horsley SW, Regan R, et al. (1997) Development and clinical application of an innovative fluorescence in situ hybridization technique which detects submicroscopic rearrangements involving telomeres. Eur J Hum Genet 5:1–8

Kuwano A, Mutirangura A, Dittrich B, et al. (1992) Molecular dissection of the Prader-Willi/Angelman syndrome region (15q11–13) by YAC cloning and FISH analysis. Hum Mol Genet 1:417–425

Levy B, Gershin IF, Desnick RJ, et al. (1997) Characterization of a de novo unbalanced chromosome rearrangement by comparative genomic hybridization and fluorescence in situ hybridization. Cytogenet Cell Genet 76:68–71

Lichter P, Ward DC (1990) Is non-isotopic *in-situ* hybridization finally coming of age? Nature 345:93–94

Lüdecke H-J, Senger G, Claussen U, et al. (1989) Cloning defined regions of the human genome by microdissection of banded chromosomes and enzymatic amplification. Nature 338:348–350

Mann SM, Burkin DJ, Grin DK, et al. (1997) A fast, novel approach for DNA fibre-fluorescence *in situ* hybridization. Chrom Res 5:145–147

Müller S, O'Brien PCM, Ferguson-Smith MA, et al. (1997a) A novel source of highly specific chromosome painting probes for human karyotype analysis derived from primate homologues. Hum Genet 101:149–153

Müller S, Rocchi M, Ferguson-Smith MA, et al. (1997b) Toward a multicolor chromosome bar code for the entire human karyotype by fluorescence in situ hybridization. Hum Genet 100:271–278

Pinkel D, Landegent J, Collins C, et al. (1988) Fluorescent in situ hybridization with human chromosome specific libraries: detection of trisomy 21 and translocation of chromosome 4. Proc Natl Acad Sci USA 85:9138–9142

Popp S, Jauch A, Schindler D, et al. (1993) A strategy for the characterization of minute chromosome rearrangements using multiple color fluorescence in situ hybridization with chromosome-specific DNA libraries and YAC clones. Hum Genet 92:527–532

Schröck E, du Manoir S, Veldman T, et al. (1996) Multicolor spectral karyotyping of human chromosomes. Science 273:494–497

Selig S, Okumura K, Ward DC, et al. (1992) Delineation of DNA replication time zones by fluorescence *in situ* hybridization. EMBO J 11:1217–1225

Shiels C, Coutelle C, Huxley C (1997) Analysis of ribosomal and alphoid repetitive DNA by fiber-FISH. Cytogenet Cell Genet 76:20–22

Shizuya H, Birren B, Kim U-J, et al. (1992) Cloning and stable maintenance of 300-kilobase-pair fragments of human DNA in *Escherischia coli* using an F-factor-based vector. Proc Natl Acad Sci USA 89:8794–8797

Speicher MR, Ballard G, Ward DC (1996) Karyotyping human chromosomes by combinatorial multi-fluor FISH. Nat Genet 12:368–375

Srivastava AK, Hagino Y, Schlessinger D, et al. (1993) Ribosomal DNA clusters in pulsed-field gel electrophoretic analysis of human acrocentric chromosomes. Mammal Genome 4:445–450

Torchia BS, Call LM, Migeon BR (1994) DNA replication analysis of FMR1, XIST, and factor 8C loci by FISH shows nontranscribed X-linked genes replicate late. Am J Hum Genet 55:96–104

Trask BJ (1991) Fluorescence in situ hybridization: applications in cytogenetics and gene mapping. Trends Genet 7:149–154

Weier H-UG, Wang M, Mulliken JC, et al. (1995) Quantitative DNA fiber mapping. Hum Mol Genet 4:1903–1910

Woodward K, Kendall E, Vetrie D, et al. (1998) Pelizaeus-Merzbacher disease: identification of Xq22 proteolipid-protein duplications and characterization of breakpoints by interphase FISH. Am J Hum Genet 63:207–217

9

Main Features of Meiosis

Meiosis is the unique process by which haploid (n) germ cells are produced by two successive cell divisions without an intervening round of DNA replication (Fig. 9.1). This is one of the two key events in the alternation of the haploid and diploid phases of the human life cycle, the other being fusion of haploid egg and sperm (fertilization) to produce a diploid (2n) zygote. Two key features of the first meiotic division are close pairing of homologous chromosomes and their segregation to opposite poles of the meiotic spindle. The first meiotic division, MI, is called the *reduction* division, because it reduces the chromosome number from 2n to n. Sister chromatids separate from each other only at the second, or *equational*, meiotic division, MII. This yields haploid cells that differentiate into ova in females and sperm in males.

The segregation of maternal and paternal homologues of each chromosome at MI is independent of that of all other pairs of homologues, providing the

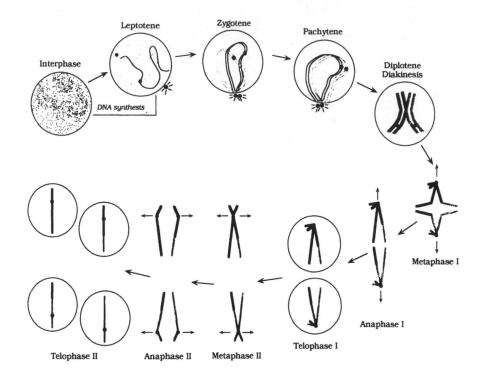

Figure 9.1. Diagram of meiosis with one pair of chromosomes that have a single chiasma. Note: A pachytene bivalent consists of four chromatids.

physical basis for Mendelian segregation of alleles at each locus and independent assortment of alleles at loci on different chromosomes. Alleles at gene loci that are fairly close to one another on a chromosome do not segregate independently of each other and the loci are said to be linked (Chapter 29). Linked gene loci are sometimes separated from each other as a result of another key feature of meiosis called *crossing over*. At prophase of MI, each crossover, or *chiasma*, leads to the exchange of homologous segments between two of the four chromatids of the paired chromosomes. The result is *genetic recombination*, the reciprocal exchange of alleles between two homologues. This can produce new groupings of alleles at different loci and, with mutation, is responsible for the genetic heterogeneity that is the basis of natural selection. The frequency of crossing over increases with the distance separating gene loci, and so does the frequency of two or more crossovers. Independent assortment of genes on different chromosomes is indistinguishable from 50% crossing over between widely separated genes on the same chromosome. Genes on the same chromosome are said to be *syntenic*, even if they are too far apart to show linkage (less than 50% recombination). Crossing over has a second important role: It is generally a prerequisite for orderly meiotic segregation.

The main features of meiosis are homologous pairing and crossing over, segregation of homologues, and independent assortment of nonhomologues. These are universal among eukaryotes, from unicellular organisms such as yeast and fungi to higher organisms throughout both plant and animal kingdoms. Electron microscopic studies have provided key insights into the physical basis of pairing and recombination; immunocytochemical and molecular studies have provided additional insights and shown that the molecular mechanisms underlying meiosis are virtually identical in all eukaryotes (Chapter 10).

Prophase I: Leptotene, Zygotene, Pachytene, Diplotene, and Diakinesis

The durations of some meiotic stages differ markedly from those of their mitotic counterparts. The premeiotic S phase lasts about twice as long as the mitotic S phase. Prophase of MI lasts more than 16 days in males and many years in females, compared to a few hours for mitotic prophase. Cytologically, the early meiotic prophase stages are clearer in oocytes (Fig. 9.2) than in spermatocytes, while from diplotene on they are clearer in spermatocytes than in oocytes. At the beginning of prophase (leptotene or leptonema), the chromosomes become visible as thin threads (Fig. 9.2). Fluorescence in situ hybridization (FISH) analysis in males, using a chromosome 1-specific painting library as a probe (Chapter 8), has shown that the homologous chromosomes, which are invisible by light microscopy prior to meiotic prophase, are in compact, mutually exclusive chromosome territories. FISH analysis with centromeric and telomeric probes has confirmed the general absence of pairing of homologous chromosomes prior to meiosis. As meiosis begins, the compacted chromosomes reorganize into the microscopically visible long, thin leptotene threads. As leptotene progresses, condensed regions appear along the thread-like chromosome strands and become more distinct as the chromosomes gradually contract. These general features have been verified for chromosome 3 by confocal microscopic studies with three-color, five-probe FISH analysis of the individually labeled telomeres and arms (Scherthan et al., 1996, 1998).

Homologous pairing is initiated in the zygotene (zygonema) stage (Fig. 9.2). The earliest detectable event is the movement of centromeres to the nuclear

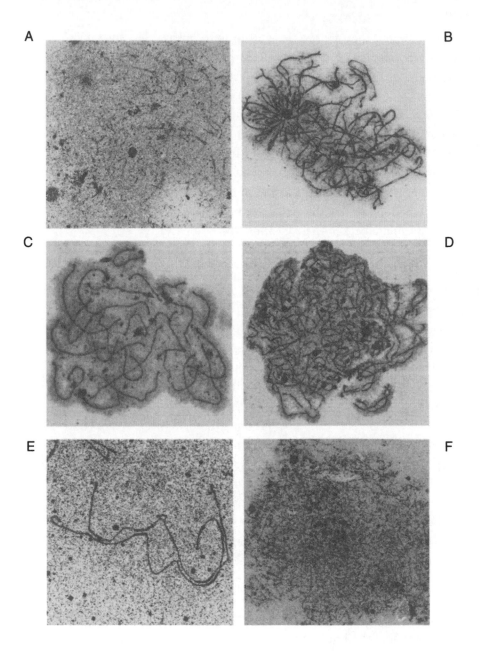

Figure 9.2. Early stages of meiosis in human fetal ovaries by light (LM) and electron (EM) microscopy. (A) Leptotene, LM. (B) Zygotene, LM. (C) Pachytene, LM. (D) Diplotene, LM. (E) Single diplotene bivalent, EM. (F) Dictyotene, LM (reproduced from Speed, The meiotic stages in human foetal oocytes studied by light and electron microscopy, Hum Genet 84, fig 1, p 70, copyright 1985, Springer-Verlag).

envelope, followed shortly by the clustering of the telomeres from their previously dispersed sites of attachment to the nuclear envelope. These movements result in the so-called *bouquet* formation. Visible pairing of homologous chromosomes starts at a number of points along each pair as zygotene progresses. Parts of the chromosomes are thin leptotene threads, whereas the rest have paired (synapsed) to form thicker pachytene-like chromosomes marked by distinct condensed regions called *chromomeres*. Synapsis (close pairing) usually takes place between strictly homologous segments of homologous chromosomes. This is convincingly demonstrated by the pairing configuration of two homologues when one of them has an inversion. To achieve point-by-point pairing in the inverted segment, one of the chromosomes has to form a loop, and such loops are readily visualized (Chapter 16). Similarly, if a translocation has taken place between two chromosomes, the corresponding segments of translocated and nontranslocated chromosomes synapse, leading to a characteristic cross-shaped figure (Chapter 16).

Pairing of homologues begins at a time when the 46 elongated chromosomes are in a rather tangled state. Electron microscopy reveals that interlocking of nonhomologous chromosomes is frequent in zygotene of both male and female meiosis but is very rare in diplotene and metaphase I. Clearly, the interlocked chromosomes break, untangle, and then heal. Consequently, no interlocking of two or more bivalents is visible at later stages (Wettstein et al., 1984). This process is presumably mediated by topoisomerase II (TOPO II), which introduces double-strand breaks, allowing the chromosomes to "walk through" each other, and then re-ligates the broken ends, thus maintaining the integrity of each chromosome.

In pachytene (pachynema), synapsis is complete and each paired chromosome consists of a long series of chromomeres that are arranged like different-sized "beads" along the chromosome thread (Figs. 9.2 and 9.3). Each pair of homologues forms a *bivalent*. The length of a bivalent at pachytene is about one-fifth the length of the same chromosome in leptotene. The chromomere pattern (see Fig. 9 in ISCN 1995) corresponds closely to the bands in G-banded mitotic prophase chromosomes (Jhanwar et al., 1982). Crossing over, which consists of an exchange of homologous segments between two of the four chromatids, takes place in the pachytene stage, although the results of this process cannot be seen until the next meiotic stage, diplotene.

The condensation (shortening and thickening) of chromosomes continues through diplotene (diplonema). The two homologous chromosomes forming each bivalent begin to repel each other by an unknown mechanism until they are held together only at *chiasmata* (singular: *chiasma*) (Figs. 9.2, 9.4, and 9.5).

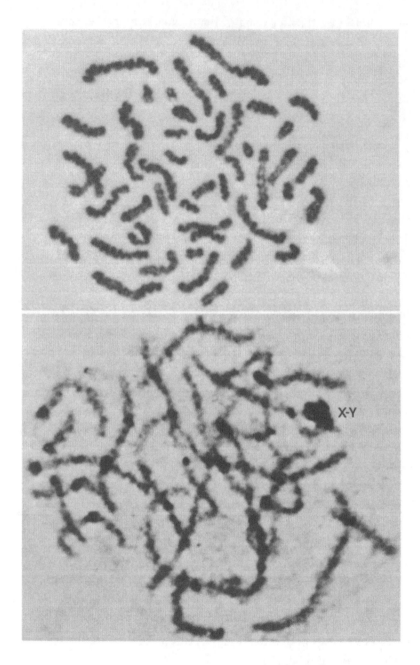

Figure 9.3. (Top) Spermatogonial metaphase. (Bottom) Pachytene spermatocyte with an XY body (sex vesicle). Squashing has destroyed the bouquet arrangement (courtesy of Maj Hultén).

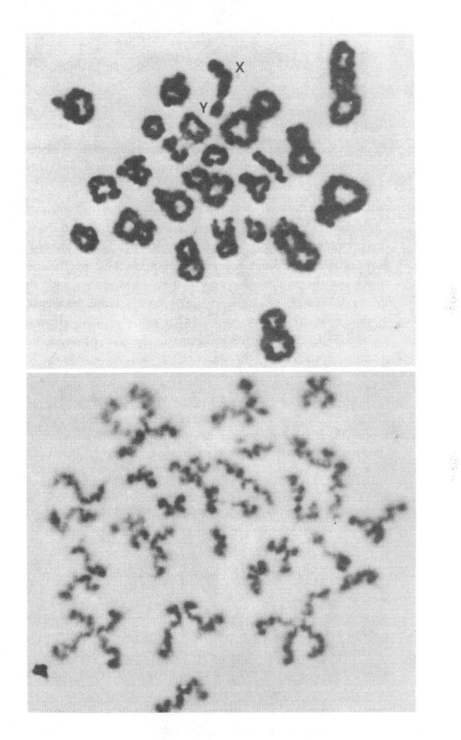

Figure 9.4. (Top) First meiotic metaphase in a spermatocyte; the X and Y are attached at the tips of their short arms. (Bottom) Second meiotic metaphase in a spermatocyte (courtesy of Maj Hultén).

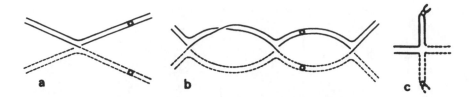

Figure 9.5. Diagrams of three bivalent chromosomes: (a) Diplotene bivalent with one chiasma. (b) Bivalent with three chiasmata and one chromatid overlap. (c) Metaphase bivalent with one chiasma.

Chiasmata are scored more accurately at diplotene than in the more condensed chromosomes of diakinesis and metaphase I. Each chiasma is the result of crossing over (breakage and rejoining) involving one chromatid of each of two homologous chromosomes (Fig. 9.5), in keeping with the chiasmatype theory Johansson proposed in 1909. Chromosome condensation continues in diakinesis. The hairy appearance characteristic of diplotene disappears, and the bivalents are smooth and compact. They may be attached to the nuclear envelope, whereas until this stage they appear free within the nucleus.

Chiasmata and Genetic Recombination

Chiasmata are not randomly distributed. Each pair of chromosomes usually forms at least one chiasma, and each chromosome arm, except the short arms of the acrocentrics, usually shows at least one chiasma. The mean chiasma frequency per cell in males, as determined from diakinesis and metaphase I stages in spermatocytes, is slightly above 50 (Holm and Rasmussen, 1983; Hultén, 1974; Laurie and Hultén, 1985). The largest human bivalents usually have four chiasmata; the medium-sized, two or three; and the smallest, one each. Although direct evidence is limited, the chiasma frequency must be higher in oocytes than in spermatocytes, because the frequencies of recombination are considerably higher. In both male and female meiosis, the greatest density of chiasmata is next to the telomeres and a smaller peak is found in the middle of the longest chromosome arms (Bojko, 1985; Laurie and Hultén, 1985). Chiasmata and crossing over are markedly reduced or absent in heterochromatin (John, 1988).

Statistical studies show that two chiasmata are unlikely to occur close to one another. This phenomenon is called *chiasma interference*. Genetic recombination, which is detected by pedigree analyses using genetic markers, shows a similar interference. This is one of the proofs that chiasmata are the cytological accompaniment of crossing over, or genetic recombination. The total frequency of crossing over, as measured genetically by pedigree analysis, is one-half the frequency of chiasmata, because at each chiasma only two of the four chromatids in the bivalent recombine.

Recombination frequencies vary along the length of each chromosome, differ in males and females, and decrease with maternal age (Chapter 11). Broman et al. (1998) scored 8000 short tandem-repeat polymorphic markers (STRPs) in DNA from individuals in eight large CEPH (Centre d'Etude de Polymorphisme Humain) families. This yielded estimates for the total genetic map length of 44 morgans (4400 cM; 1 cM = 1% recombination) in females and 27 morgans (2700 cM) in males, with some individual variation. Recombination frequencies also varied along each chromosome (Fig. 29.1). At the telomeres, male and female recombination frequencies were about equal, but a marked reduction in male rates occurred near the centromeres. In chromosome 7, the 20-Mb centromeric region showed *no* recombination in males but showed nearly 20% recombination (a 20-cM length) in females. In contrast, Mohrenweiser et al. (1998) used 180 genetic markers spaced an average of 320 kb apart on chromosome 19 to show that males have very high recombination frequencies in the telomeric regions (13 cM/Mb), whereas the corresponding value for females was only 1 cM/Mb. The sex-averaged recombination frequencies varied from a high of 4 cM/Mb across the telomeric bands to a low of 1 cM/Mb for the centromeric bands.

Metaphase I, Anaphase I, Telophase I, Interkinesis, and Meiosis II

The nuclear envelope disappears (*disassembles*) during a short prometaphase stage, and the bivalents collect onto a metaphase plate at the midpoint of the spindle that has formed between the two centrosomes. The two centromeres of each bivalent orient themselves toward the two poles, and only the presence of chiasmata appears to prevent the homologues from separating into univalent chromosomes. A bivalent with one chiasma may appear cross-shaped, or the homologues may be connected end to end by a more or less terminal chiasma. Two chiasmata often result in a ring-shaped bivalent. A bivalent with several chiasmata resembles a chain (Fig. 9.5). At anaphase, each homologue, with its two chromatids, is moved to one pole by its single functional centromere. The cohesion of the centromeres of sister chromatids throughout meiosis I appears to require an evolutionarily conserved cohesion phosphoprotein, pREC8, that is produced throughout meiosis I but disappears at anaphase of meiosis II and is absent from mitotic cells (Watanabe and Nurse, 1999).

In telophase of MI, each of the two chromosome groups, one at each spindle pole, contains the haploid number of centromeres. Telophase of MI is followed by a short interphase stage, *interkinesis*, during which the chromosomes decondense. However, during this stage the chromosomes are not replicated, and thus they maintain the same basic structure in prophase of MII as they had in the preceding telophase.

The second meiotic division is similar to an ordinary mitosis except for the haploid chromosome number. However, the chromosomes in metaphase II differ from mitotic chromosomes by being irregular in shape, being shorter, and having their widely separated chromatids held together only at the centromere (Fig. 9.4). In anaphase II, the centromeres divide and move to the poles, with one chromatid attached to each daughter centromere. The end result of meiosis is four nuclei, each with a haploid chromosome complement. If crossing over has occurred anywhere in the genome (a virtual certainty), no two haploid nuclei will have an identical genetic make-up, in contrast to the genetically identical daughter cells produced by mitosis. In males, the four meiotic products differentiate into four functional gametes (sperm). In females, on the other hand, only one functional ovum is formed, containing one of the four haploid nuclei and most of the cytoplasm of the oocyte. The other three nuclei segregate into tiny cells called polar bodies, which normally degenerate.

Female Meiosis: Dictyotene Arrest, Metaphase II Arrest, and Apoptosis

Meiosis begins in the fifth month of fetal life in females and is preceded by about 33 mitotic divisions, from the fertilized egg to the onset of meiosis, when the two fetal ovaries contain several million oocytes. No mitotic divisions occur in the female germline after the fifth month of fetal life. The first meiotic stage, leptotene, is present only in 12- to 20-week-old embryos, when all the millions of germline stem cells, or oogonia, switch from mitosis to meiosis. After progressing through the pachytene stage, which reaches its peak frequency at 19 weeks, and into diplotene by the sixth month of fetal life, oocytes are arrested by an unknown mechanism in what is called the *dictyotene* stage of prophase I

(Fig. 9.2). The chromosomes are greatly extended at this stage and invisible with most cytological techniques.

The dictyotene stage is very long, lasting many years. During a woman's reproductive years, cohorts of oocytes are released from the dictyotene cell cycle arrest by progesterone-stimulated production of the mitosis-promoting factor, MPF (cyclin B/CDK1), described in Chapter 2. The oocytes continue through the first meiotic division but arrest again in metaphase II until fertilization. This second arrest is mediated by a cytostatic factor (CSF), the product of the c-MOS gene, better known for its role as a protooncogene. This protein kinase activates and stabilizes MPF. Fertilization triggers an increase in intracellular calcium. This activates the calcium-dependent protease calpain which degrades the c-MOS protein and induces cyclin B degradation and MPF inactivation, just as in the mitotic metaphase-to-anaphase transition (Sagata, 1996; Watanabe et al., 1989).

Apoptosis, or programmed cell death, is the fate of most cells of the female germ line. This reduces their number from a peak of about eight million at the onset of meiosis to 1 million–2 million at birth. Each monthly ovulation in the adult is accompanied by a wave of apoptosis in oocytes in partially mature follicles. The product of the TP53 gene, p53, is involved in apoptosis (Chapter 26), and so is the product of a BCL2 gene family member, BAX, whose expression is elevated at the onset of apoptosis in the ovary (Kugu et al., 1998).

Male Meiosis

Mitotic divisions continue to occur in germline stem cell spermatogonia throughout the adult life of males. About 205 cell divisions have occurred from the zygote to meiosis in a 20-year-old male and about 265 in a 25-year-old (the mean age of reproduction in men generally falls between these extremes). Gene mutations usually arise as copying mistakes during replication. Consequently, the six- to nine-fold greater number of cell divisions in the male germline before gamete formation leads to the expectation of a six- to nine-fold higher mutation rate in males than in females, and that appears to be the case (Shimmin et al., 1993). Meiosis in males is initiated by hormonal stimulation at the onset of puberty. Type A spermatogonia form type B spermatogonia. The switch from mitotic divisions in the type B spermatogonia (germline stem cells) to meiosis is restricted to one of the two daughter cells produced at each type B spermatogonial division. This cell differentiates into a primary spermatocyte, which undergoes meiosis. Thus, through adult life, males continue to produce each day many tens of millions of spermatogonia, primary spermatocytes, and sperm. Figure 9.3 (top) illustrates a spermatogonial metaphase in which the spiral structure of the chromosomes is clearly visible.

Both the X and the Y chromosome are heteropycnotic in spermatocytes from zygotene to diplotene. These condensed chromosomes join to form an *XY body*, or *sex vesicle* (Fig. 9.3, bottom). A protein homologous to *Drosophila* heterochromatin protein 1 (HP1) appears to play a role in this condensation (Motzkus et al., 1999). The heteropycnosis eventually disappears, so in diplotene and metaphase I the XY bivalent is distinguishable from any autosomal bivalent only by its characteristic asymmetric shape, with end-to-end pairing of the short arms (Fig. 9.3, top). Pairing of the X and Y appears to be necessary for the completion of meiosis and the formation of sperm, just as pairing of the autosomes is (Speed and Chandley, 1990). The short arm of bivalent chromosome 15 associates rather frequently with the sex vesicle (Fig. 9.6; see color insert), presumably because of sequence similarities in their DNA. This may be responsible for the relatively high frequency of translocations between chromosome 15 and the X and Y chromosomes (Metzler-Guillemain et al., 1999).

Segregation Distortion and Meiotic Drive

Mendelian inheritance is predicated on 1 : 1 *segregation ratios*, that is, equal representation among the gametes of the two alleles at any gene locus. In most cases, that is what one observes. However, a number of significant deviations from a 1 : 1 segregation ratio have been found. Several examples of this are known in diseases that are caused by expansions of tandem trinucleotide repeats, such as (CTG)n (Chapter 20). At the myotonic dystrophy locus, the allele with the larger number of CTG repeats is preferentially transmitted to the offspring (Chakraborty et al., 1996). Some *meiotic drive* mechanism is more likely to account for this than is differential survival of zygotes or embryos. In females, this might involve preferential inclusion of one allele in the egg rather than in a polar body. In males, differential survival or motility of sperm could be responsible. For unknown reasons, such *segregation distortion* usually occurs only in offspring of one sex. For example, only female meiosis is affected in myotonic dystrophy (CTG expansion) and only male meiosis in two disorders of CAG expansion (Chapter 20).

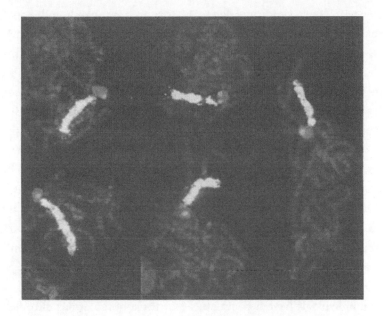

Figure 9.6. Association of bivalent 15 with the sex vesicle in five normal pachytene nuclei stained blue with DAPI. Bivalent 15 is made visible with a 15-specific paint (green) and its centromeric region with a 15-specific α-satellite probe (red). The sex vesicle (blue clump) is marked by an X-specific α-satellite probe (red) (Metzler-Guillemain et al., Chrom Res 7:369, Bivalent 15 regularly associates with the sex vesicle in normal male meiosis, fig. 1, copyright 1999, with kind permission from Kluwer Academic Publishers) (See color insert).

Meiotic Behavior of Three Homologous Chromosomes

Trisomic individuals have three homologous chromosomes. These usually pair to form a *trivalent* in meiosis I. However, at any one point along the trivalent, usually only two of them synapse. Depending on the number and length of the paired segments and the number and location of chiasmata, trivalents come in a variety of shapes. Three homologues may also form a bivalent and a univalent. Of the three homologues, usually two go to one pole and one to the other in anaphase I. This type of segregation is called *secondary nondisjunction*. It can lead to what is called *tertiary trisomy* among the progeny.

Women with Down syndrome due to 21-trisomy are the only individuals with autosomal trisomy who have reproduced. Their offspring consist of children with Down syndrome (tertiary trisomics) and normal children, in a ratio that does not differ significantly from the 1:1 ratio expected with secondary nondisjunction. However, given the considerable embryonic lethality of trisomy 21 (Chapter 12), the ratio is not 1:1 at conception. Individuals with trisomy 18 never reproduce, but it has been possible to analyze the meiotic behavior of 18-trisomic oocytes in two 18- to 20-week-old female fetuses aborted after prenatal diagnosis by using a chromosome 18-specific painting probe and FISH. At the pachytene stage, 75% showed a trivalent configuration, and about half of these had virtually complete triple synapsis; 24% had a bivalent plus a univalent; and 1% had complete asynapsis (Cheng et al., 1995). Bivalent and univalent chromosomes 21 have also been observed in oocytes of a 21-trisomic fetus (Cheng et al., 1998). The segregation of the sex chromosomes during meiosis in XXX, XXY, and XYY individuals is discussed in Chapter 19.

Triploid individuals have a complete set of three homologous chromosomes. Although triploids usually die prenatally, it has been possible to analyze the early meiotic stages in fetuses; the results were consistent with the observations in trisomics (Luciani et al., 1978).

References

Bojko M (1985) Human meiosis. IX. Crossing over and chiasma formation in oocytes. Carlsberg Res Commun 50:43–72

Broman KW, Murray JC, Sheffield VC, et al. (1998) Comprehensive human genetic maps: individual and sex-specific variation. Am J Hum Genet 63:861–869

Chakraborty R, Stivers DN, Daka R, et al. (1996) Segregation distortion of the CTG repeats at the myotonic dystrophy locus. Am J Hum Genet 59:109–118

Cheng EY, Chen Y-J, Gartler SM (1995) Chromosome pairing analysis of early oogenesis in human trisomy 18. Cytogenet Cell Genet 70:205–210

Cheng EY, Chen Y-J, Bonnet G, et al. (1998) An analysis of meiotic pairing in trisomy 21 oocytes using fluorescent in situ hybridization. Cytogenet Cell Genet 80:48–53

Holm PB, Rasmussen SW (1983) Human meiosis. VI. Crossing over in human spermatocytes. Carlsberg Res Commun 48:385–413

Hultén M (1974) Chiasma distribution at diakinesis in the normal human male. Hereditas 76:55–78

ISCN (1995) An international system for human cytogenetic nomenclature. Mitelman F (ed) S Karger, Basel

Jhanwar SC, Burns JP, Alonso ML, et al. (1982) Mid-pachytene chromomere maps of human autosomes. Cytogenet Cell Genet 33:240–248

John B (1988) The biology of heterochromatin. In: Verma RS (ed) Heterochromatin: molecular and structural aspects. Cambridge University Press, Cambridge, pp 1–147

Kugu K, Ratts VS, Piquette GN, et al. (1998) Analysis of apoptosis and expression of bcl-2 gene family members in the human and baboon ovary. Cell Death Differ 5:67–76

Laurie DA, Hultén MA (1985) Further studies on bivalent chiasma frequency in human males with normal karyotypes. Ann Hum Genet 49:189–201

Luciani JM, Devictor M, Boue J, et al. (1978) The meiotic behavior of triploidy in a human 69,XXX fetus. Cytogenet Cell Genet 20:226–231

Metzler-Guillemain C, Mignon C, Depetris D, et al. (1999) Bivalent 15 regularly associates with the sex vesicle in normal male meiosis. Chrom Res 7:369–378

Mohrenweiser HW, Tsujimoto S, Gordon L, et al. (1998) Regions of sex-specific hypo- and hyper-recombination identified through integration of 180 genetic markers with the metric physical map of human chromosome 19. Genomics 47:153–162

Motzkus D, Singh PB, Hoyer-Fender S (1999) M31, a murine homolog of Drosophila HP1, is concentrated in the XY body during spermatogenesis. Cytogenet Cell Genet 86:83–88

Sagata N (1996) Meiotic metaphase arrest in animal oocytes: its mechanism and biological significance. Trends Cell Biol 6:22–28

Scherthan H, Eils R, Trelles-Sticken E, et al. (1998) Aspects of three-dimensional chromosome reorganization during the onset of human male meiotic prophase. J Cell Sci 111:2337–2351

Scherthan H, Weich S, Schwegler H, et al. (1996) Centromere and telomere movements during early meiotic prophase in mouse and man are associated with the onset of chromosome pairing. J Cell Biol 134: 1109–1125

Shimmin LC, Chang BH-J, Li W-H (1993) Male-driven evolution of DNA sequence. Nature 362:745–747

Speed RM (1985) The meiotic stages in human foetal oocytes studied by light and electron microscopy. Hum Genet 69:69–75

Speed RM, Chandley A (1990) Prophase of meiosis in human spermatocytes analysed by EM microspreading in infertile men and their controls and comparisons with human oocytes. Hum Genet 84:547–554

Watanabe N, Vande Woude GF, Ikawa Y, et al. (1989) Specific proteolysis of the c-mos proto-oncogene product by calpain on fertilization of Xenopus eggs. Nature 342:505–511

Watanabe Y, Nurse P (1999) Cohesin Rec8 is required for reductional chromosome segregation at meiosis. Nature 400:461–464

Wettstein D von, Rasmussen SW, Holm PB (1984) The synaptonemal complex in genetic segregation. Annu Rev Genet 18:331–431

10

Details of Meiosis

The Switch from Mitosis to Meiosis

Most of the cell divisions that occur in the germline are mitotic: a series of over 30 in females and over 200 in males (Chapter 9). It is only the final two divisions that are meiotic. What mediates the switch from mitosis to meiosis, and by what mechanisms do all oogonia switch almost simultaneously during fetal life while only one of the two daughter cells from each type B spermatogonial mitosis switches to meiosis throughout the adult life of men? The switch from mitosis to meiosis is the first process of cell differentiation to appear during the evolution of eukaryotes. Consequently, this switch is likely to have an extremely conserved molecular basis, like the other steps in meiosis.

One gene involved in the switch is *OCT-4*, a transcription factor that is expressed specifically in totipotent cells—the fertilized egg, cleavage, morula, blastocyst, inner cell mass, and epiblast stages of the early embryo—and in the germline of both males and females. Methylation of a CpG island in the 5′ upstream region of this gene can inhibit its expression. *OCT-4* expression continues throughout life in type A spermatogonia but declines as these differentiate into spermatocytes. A decline in *OCT-4* expression also accompanies, or initiates, female meiosis, but dictyotene arrest is associated with and perhaps caused by an increase in *OCT-4* expression (Pesce et al., 1998).

Pairing of Homologous Chromosomes

The molecular mechanisms by which homologous chromosomes recognize each other and initiate close pairing (*synapsis*) are under intense study. Homologous chromosomes tend to lie far apart in the interphase nucleus. Pairing is brought about in several stages. The first, long known to cytogeneticists, involves telomere clustering. This brings the ends of homologous chromosomes much closer together (Scherthan et al., 1996) and facilitates the second stage, a genome-wide search for homologous sequences (*homology searching*), the basis for homologous recombination. The human TRF1 protein associates with telomeres; its homology with a fission yeast protein that is necessary for both telomere clustering and efficient recombination suggests it may play a similar role in human meiosis (Nimmo et al., 1998). The RAD51 multiprotein complex appears to be involved in homology searching and in all the later stages of pairing and recombination. In human spermatocytes, RAD51 is a component of early recombination nodules, which may be the sites for initiation of strand invasion and chromosome synapsis, as described later in this chapter and by Barlow et al. (1997). The third stage of pairing involves the synapsis of homologous chromosomes, which usually begins at the telomeres. The synaptonemal complex plays a critical role in synapsis.

The Synaptonemal Complex and Recombination Nodules

Electron microscopic studies led to a major breakthrough in our understanding of how homologous chromosomes pair and recombine in prophase of meiosis I: the discovery of the *synaptonemal complex* (SC). The synaptonemal complex is a protein-rich structure that is involved in pairing of homologous chromosomes in zygotene and maintaining the pairing throughout pachytene, when crossing over takes place. The synaptonemal complex has remained remarkably stable throughout eukaryotic evolution, reflecting its critical importance and the complexity of its multiprotein components, which are themselves extremely highly conserved.

Synaptonemal complexes can be visualized in surface spreads by electron microscopy (Fig. 10.1). Three-dimensional reconstructions from serial sectioned material have clarified the relationship of SCs to paired homologous chromosomes (reviewed by Wettstein et al., 1984). After chromosomes have been surface-spread on slides and silver-stained, the protein-rich SCs can be seen by either light or electron microscopy, although without the high resolution needed for distinguishing critical details (Pathak and Hsu, 1979). Using light microscopy, Barlow and Hultén (1996) combined FISH with immunostaining, using antibodies to the SC. This enabled them to confirm earlier reports that synapsis of homologues usually starts at the telomeres. They showed that the telomeric TTAGGG repeats are tightly associated with the SCs, while other repetitive DNAs, such as α-satellite DNA, are mainly in loops that extend out from the SCs.

Each synaptonemal complex contains two *lateral elements*, with a *central element* between them. The assembly of SCs begins in leptotene, with the formation of an *axial element* along each of the 46 duplicated chromosomes. In early zygotene, as each pair of homologous chromosomes becomes roughly aligned, the axial elements come together at several points to form the lateral elements of the nascent SC, which is stabilized by bridging proteins of the central element. Synaptonemal complex protein 1 (SCP1) may be the major component of the transverse filaments of SCs. The *SCP1* gene was identified by screening a cDNA expression library (thousands of bacteria, each containing a human cDNA that can be translated into a polypeptide in the bacterial cells) with a monoclonal antibody to an SC component. The gene was mapped to region 1p12–p13 by two-color FISH, using a biotin-dUTP-labeled SCP1 cDNA (detected with yellow fluorescein isothiocyanate [FITC] fluorescence) and a digoxigenin-dUTP-labeled chromosome 1–specific subtelomeric repeat probe (detected with Texas red fluorescence).

The predicted SCP1 protein is similar to nuclear lamin proteins. The amino acid sequence, or *domain*, that SCP1 shares with lamins A and C is the target of the cyclin B–dependent kinase CDK1. This kinase activity, which phosphorylates the lamins, is required for the dissolution of the nuclear membrane in cell division. It is likely that phosphorylation of SCP1 is required for the dissolution of SCs (Meuwissen et al., 1997). The separation of homologues at diplotene is mediated by the dissolution of the SCs, which soon disintegrate (Wettstein et al., 1984). Fragments of SCs may temporarily mark the positions of chiasmata

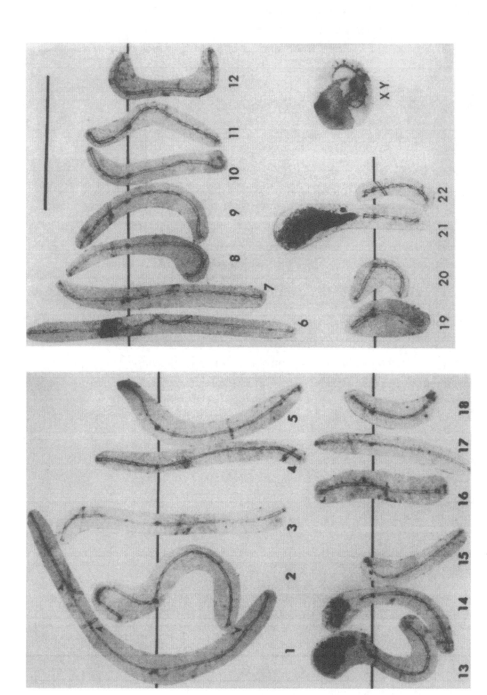

Figure 10.1. Karyotype of a microspread pachytene spermatocyte stained to show the synaptonemal complex by electron microscopy. Note the two lateral elements running the length of each chromosome, the electron-dense material at each centromere, and the widely scattered bars or recombination nodules (some indicated by arrowheads on chromosome 1). The nucleoli vary markedly in size (reproduced from Solari, Synaptonemal complexes and associated structures in microspread human spermatocytes, Chromosoma 81:315–337, copyright 1980, Springer-Verlag).

in diplotene. The disintegration of SCs requires ubiquitin-directed protein degradation, which is mediated by the ubiquitin-conjugating enzyme Ubc9. This enzyme is present in all tissues but has its highest level in the testis, where it is localized to the synaptonemal complexes by its interaction with RAD51 (Kovalenko et al., 1996). Ubc9 also degrades B-type cyclins, a step required for entry into metaphase.

Another protein that is an essential component of SCs is HSPA2, a heat shock protein encoded by a gene on chromosome 14 (Bonnycastle et al., 1994). This protein is 98 per cent identical in amino acid sequence, and functionally equivalent, to the Hsp70–2 protein of the mouse and hamster. Hsp70 proteins are molecular chaperones that assist in folding, unfolding, transport, assembly, and disassembly of proteins. Hsp70–2 is detectable in the axial elements of SCs in leptotene and zygotene stages, increases in amount along the SCs until pachytene, and persists until the SCs are disassembled in diplotene. Gene knockout mice that lack Hsp70–2 still assemble SCs, but these progress only as far as midpachytene and then fragment; there is no desynapsis, and meiosis is blocked at this point (Dix et al., 1996).

Electron-dense structures called *recombination nodules* are associated with the central element of the synaptonemal complex. Recombination nodules are thought to be a prerequisite for crossing over. They arise in zygotene and transform during pachytene into bridgelike structures, called *bars*, or late recombination nodules (Fig. 10.1). Crossing over takes place during this transformation, as inferred from the correlation between the number and distribution of crossovers and that of the nodules and bars (Wettstein et al., 1984). Each synaptonemal complex is the same length as its pachytene chromosome, whereas the DNA molecule of each chromosome is vastly longer. The precision of pairing provided by the synaptonemal complex thus appears to be inadequate to provide a basis for the exact molecular pairing required for homologous recombination. A consideration of the mechanism of double-strand break repair (below) suggests a resolution of this paradox.

Nonrandom Distribution of Meiotic Recombination Sites

Chiasmata and recombination nodules are not distributed at random but show regional and sex differences. So do the sites of recombination, which tend to be concentrated in T-bands and some R-bands (Chapter 7). Recombination frequencies show many differences between different regions and between the two sexes. In a recent example, recombination involving the most telomeric 2.3 Mb of chromosome 21q has been seen only in males (Blouin et al., 1995). Some chromosome regions show a very high frequency of recombination and are called *hotspots*. For example, Orkin et al. (1982) identified a hotspot for recombination in an 11-kb region immediately 5′ of the β-*globin* gene by linkage analysis with 12 molecular markers scattered over a 65-kb region containing the gene.

What determines where meiotic recombination events will be initiated? There is growing evidence that sites containing expressed genes are more likely to be involved than other sites and that chromatin organization plays a role. The most striking evidence for the importance of chromatin organization comes from a comparison of recombination frequencies in a segment of human DNA analyzed in situ and in a yeast artificial chromosome (YAC). This human DNA is packaged 20 times more tightly in human male meiosis at pachytene than it is in yeast meiosis. When the YAC is transferred into yeast, its degree of condensation and the recombination frequencies of its human genes match those of the yeast genome itself (Loidl et al., 1995). In yeast, the double-strand breaks that initiate meiotic recombination occur at or near promoters of transcription, sites that are hypersensitive to cutting by DNAase I (Wu and Lichten, 1994). This provides a possible explanation for the differences in locations and frequencies of chiasmata and of recombination in human male and female meiosis: sex-specific patterns of gene expression in the germline.

Molecular Mechanisms of Meiotic Recombination

The genes and gene products that mediate meiotic recombination (Table 10.1) have been highly conserved throughout evolution. They require a protein complex made up of RAD50, RAD51, RAD52, MRE11, and other components, including DNA polymerase β (Dolganov et al., 1996). RAD51 promotes homologous pairing and DNA strand exchange; RAD52 binds single-stranded DNA and stimulates its reassociation with complementary single-stranded DNA. A change in the amino acid sequence of any one of the proteins in such a large multifunctional complex is likely to disrupt complex formation or function and is thus unlikely to become established; hence the evolutionary conservatism

Table 10.1. Proteins Involved in Meiotic Recombination

Protein	Role
TRF1	Telomere clustering
SPO11	Endonuclease, produces DSBs
ATM	DNA damage checkpoint
ATR	DNA damage checkpoint
p53	DNA damage checkpoint
RAD50	Double-strand break (DSB) repair
RAD51	Homologous pairing, strand exchange
RAD52	Binds single-stranded DNA and RAD51
BRCA1	Binds RAD51; DSB repair
BRCA2	Binds RAD51; DSB repair
MRE11	Binds RAD50; DSB repair
NBS1	Binds RAD50; DSB repair
DMC1	Binds RAD51; DSB repair
MLH1	Mismatch repair
MSH4	Meiosis-specific DNA repair
DNA polymerase β	Fills in short gaps in DNA

of these proteins. The double-strand break (DSB) repair model of recombination is well established (Szostak et al., 1983; Kanaar and Hoeijmakers, 1998). Meiotic recombination begins with the introduction of DSBs in the DNA by a highly conserved protein called SPO11, which was first identified in the yeast *Saccharomyces cerevisiae* and then in both *Drosophila melanogaster* and the worm *Caenorhabditis elegans*. By taking advantage of the extreme conservation of key domains in the molecule, Romanienko and Camerino-Otero (1999) cloned the mouse and human homologues. They mapped the human gene to 20q13.2–q13.3, a region that is amplified in some breast and ovarian cancers and may be responsible for the chromosome instability that leads to these cancers (Chapter 26).

After the introduction of DSBs by SPO11, an exonuclease chews back one strand at each broken end of the DSB, leaving a single-stranded tail at the 3′ end (Fig. 14.1). These tails invade undamaged homologous sequences and bind to the complementary sequence by base pairing, forming an unstable joint molecule whose resolution may result in recombination and a visible crossover. The DNA repair associated with meiotic recombination requires DNA synthesis, which takes place in recombination nodules, as shown by electron microscopic auto-radiographic analysis (Carpenter, 1981). It is mediated by DNA polymerase β, which fills in short gaps such as those produced during DNA excision repair but is not involved in S-phase replication. This enzyme has a high level of expression in testes. Antibodies to DNA polymerase β show foci of the enzyme along meiotic prophase chromosomes in zygotene and early pachytene (Plug et al., 1997).

The mismatch repair protein MLH1 is localized at sites of crossing over, as shown by immunofluorescence localization of anti-MLH1 monoclonal antibodies. A site is seen on 56% of XY bivalents. There are about 51 sites along the autosomes, providing an estimate of 2545 centiMorgans (cM) as the total genetic length in males (Barlow and Hultén, 1998). This is very close to earlier estimates based on chiasma frequencies (Chapter 9) or comprehensive genetic linkage studies (Chapter 29). The mismatch repair gene *MSH5* appears to be required for meiotic pairing in prophase of MI (Edelman et al., 1999). The meiosis-specific DNA mismatch repair protein MSH4 may play a role in recombination similar to that of MLH1 (Paquis-Flucklinger et al., 1997). Another, called DMC1, shows some homology to RAD51 and colocalizes with RAD51 on zygotene chromosomes, supporting its involvement in the repair of double-strand breaks there (Habu et al., 1996). The *BRCA2* breast cancer suppressor gene is also critical for RAD51 binding and DNA repair (Chen et al., 1998).

The premeiotic S phase plays an important role in recombination. The main meiosis-specific event during S is the binding of the RAD51 multiprotein complex. In mitotic cells, RAD51, which is required for DNA double-strand break repair, binds to chromatin only after DNA damage has occurred. In meiotic cells, it binds to chromatin in the absence of prior DNA damage. One of the key features of RAD51 is that it preferentially binds to DNA in a subset of GC-rich T- and R-bands early in the premeiotic S phase (Plug et al., 1996). The RAD51-associated protein RAD52 binds preferentially to single-stranded DNA. RAD52 may thus be responsible for the binding of RAD51 to DNA during the early premeiotic S phase, when short single-stranded regions appear at the replication fork in GC-rich DNA during DNA replication. Why doesn't RAD52 bind as avidly to the single-stranded later-replicating DNA? Either RAD52 is present in limiting amounts during the premeiotic S phase so that only the early-replicating DNA is fully charged with it, or the chromatin conformation of later-replicating sequences hinders binding of this multiprotein complex. RAD52 stimulates DNA strand exchange by targeting the RAD51 protein to a complex of single-stranded DNA and replication protein A (New et al., 1998). Components of the RAD51 multiprotein complex interact with the broken ends, as described above, using an undamaged complementary strand as template for the synthesis of a new DNA strand.

The details of how the search for this complementary strand is conducted are unclear. It presumably takes place during the 36 hours between completion of the S phase and the beginning of visible synapsis. Many individual single-stranded regions (marked by bound RAD51) may come into contact with their complementary strands during this interval. When they reassociate with these strands, the homology search is completed. Using antibodies to human RAD51, Ashley et al. (1995) have shown that the RAD51 protein is present as early as premeiotic prophase in chromatin foci that correspond to the lateral elements of the synaptonemal complex and may be part of early recombination nodules. The RAD51-associated sites are the initial points of contact where pairing of homologues begins and the sites of localized DNA synthesis. As leptotene progresses, the chromatin condenses and the RAD51-DNA foci coalesce into larger foci. By the time homologues begin to synapse in zygotene, each focus is composed of multiple RAD51 complexes bound to many different GC-rich sequences within a single R-band (Fig. 10.2). Crossing over is completed after the formation of pachytene synaptonemal complexes with recombination nodules. R-bands are the preferred sites for synaptic initiation and recombination, as shown by the presence of early recombination nodules at the initial points of contact (Plug et al., 1996). The RAD51 sites along the fully synapsed bivalents of early pachytene correspond to an R-band pattern, emphasizing their critical role in this process.

These considerations provide an explanation for several observations: The most GC-rich R-bands make up a disproportionate fraction of pachytene chromosome and SC length (Luciani et al., 1988), synapsis is delayed or absent and crossing over markedly reduced in heterochromatin (John, 1988), and late recombination nodules are absent from heterochromatin (Carpenter, 1987). They do not fully explain how a single-stranded tail at a DSB finds a complementary DNA strand in meiotic prophase. Further studies are needed to explain how homology searches are conducted.

Meiotic DNA Damage Checkpoint

As in mitosis, cell cycle checkpoints play an essential role in meiosis, slowing the meiotic cycle and thus ensuring that there is time for repair of the double-strand breaks that initiate recombination (Freire et al., 1998). Checkpoint path-

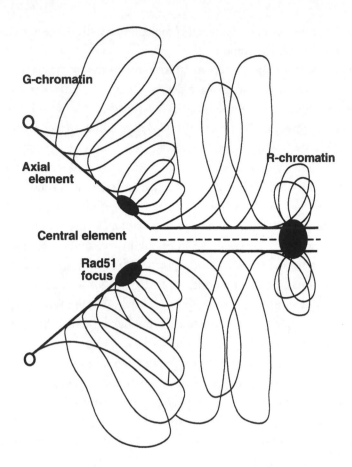

Figure 10.2. A model of chromatin organization and synaptic initiation. Chromatin in R-bands is bound to RAD51 foci (dark ovals) at the bases of its meiotic chromatin loops; chromatin in G-bands is also organized into loops, and the bases of these loops are also associated with the axial elements of each SC, but not with RAD51 foci. Only sequences that are components of RAD51 foci are involved in a search for homology and synaptic initiation (Plug et al., 1996).

ways contain a detector molecule that recognizes DNA damage, a signaling cascade that amplifies and transmits this information, and effector molecules that interact with cell cycle components. Two signaling molecules, the ATR and ATM proteins, appear to play a role in the meiotic, as well as the mitotic, DNA damage checkpoint (Chapters 2 and 26). Antibody studies have shown that ATR is localized to sites along unpaired, or *asynapsed*, chromosomal axes, while ATM is found along synapsed axes. The sites correspond to sites of RAD51 binding in both spermatocytes and oocytes (Keegan et al., 1996). Targeted disruption of the *ATM* gene in a transgenic mouse model leads to abnormal synapsis and chromosome fragmentation, causing arrest at zygotene/pachytene. This indicates the critical nature of the DNA damage checkpoint (Xu et al., 1996). The effector molecule in both mitotic and meiotic DNA damage checkpoints is the p53 protein. In somatic cells, double-strand breaks (DSBs) induce high levels of p53, leading to cell cycle arrest and apoptosis (Chapters 2 and 26). Pachytene spermatocytes also show a high level of p53, presumably in response to the DSBs generated during meiosis. The p53 protein acts by binding directly to RAD51. This repair pathway is so highly conserved that human p53 will even bind to RecA, the bacterial homologue of RAD51 (Stürzbecher et al., 1996).

Absence of a Spindle Assembly Checkpoint in Female Meiosis: Role in Nondisjunction?

While there is an effective DNA damage checkpoint in meiosis, mediated by the same ATM-ATR-p53 proteins as in mitosis, there is no spindle assembly or chromosome misalignment checkpoint in female meiosis. This may account, in part or totally, for the high frequency of meiotic nondisjunction in women, including the maternal age effect. For example, the presence of a univalent arrests meiosis I at metaphase in males but not in females, as shown by studies using immunofluorescence and FISH (Hunt et al., 1995). Is this deficiency related to the make-up of the meiotic spindle? In mitosis, the centrosome at each spindle pole is organized around a core *centriole*. In meiosis, the centrosomes lack a centriole. At the first cleavage division after fusion of sperm and egg (*syngamy*), at least one of the two centrosomes has a centriole, which is paternal in origin (Sathananthan et al., 1991). The absence of a centriole may reflect molecular changes that contribute to the high frequency of nondisjunction, especially in female meiosis.

References

Ashley T, Plug AW, Xu J, et al. (1995) Dynamic changes in Rad51 distribution in chromatin during meiosis in male and female vertebrates. Chromosoma 104:19–28

Barlow AL, Benson FE, West SC, et al. (1997) Distribution of the Rad51 recombinase in human and mouse spermatocytes. EMBO J 16:5207–5215

Barlow AL, Hultén MA (1996) Combined immunocytogenetic and molecular cytogenetic analysis of meiosis I human spermatocytes. Chrom Res 4:562–573

Barlow AL, Hultén MA (1998) Crossing over analysis at pachytene in man. Eur J Hum Genet 6:350–358

Blouin J-L, Christie DH, Gos A, et al. (1995) A new dinucleotide repeat polymorphism at the telomere of chromosome 21q reveals a significant difference between male and female rates of recombination. Am J Hum Genet 57:388–394

Bonnycastle LLC, Yu C-E, Hunt CR, et al. (1994) Cloning, sequencing, and mapping of the human chromosome 14 heat shock protein gene (HSPA2). Genomics 23:85–93

Carpenter ATC (1981) DNA synthesis occurs in recombination nodules. Chromosoma 83:59–80

Carpenter ATC (1987) Gene conversion, recombination nodules, and the initiation of meiotic synapsis. BioEssays 6:232–236

Chen P-L, Chen C-F, Chen Y, et al. (1998) The BRC repeats in BRCA2 are critical for RAD51 binding and resistance to methyl methane-sulfonate treatment. Proc Natl Acad Sci USA 95:5287–5292

Dix DJ, Allen JW, Collins BW, et al. (1996) Targeted gene disruption of HSP70–2 results in failed meiosis, germ cell apoptosis, and male infertility. Proc Natl Acad Sci USA 93:3264–3268

Dolganov GM, Maser RS, Novikov A, et al. (1996) Human Rad50 is physically associated with human Mre11: identification of a conserved multiprotein complex implicated in recombinational DNA repair. Mol Cell Biol 16:4832–4841

Edelman W, Cohen PE, Kneitz B, et al. (1999) Mammalian MutS homologue 5 is required for chromosome pairing in meiosis. Nat Genet 21:123–127

Freire R, Murguia JR, Tarsounas M, et al. (1998) Human and mouse homologues of *Schizosaccharomyces pombe* rad 1+ and *Saccharomyces cerevisiae* RAD 17:

linkage to checkpoint control and mammalian meiosis. Genes Dev 12:2560–2573

Habu T, Taki T, West A, et al. (1996) The mouse and human homologues of DMC1, the yeast meiosis-specific homologous recombination gene, have a common unique form of exon-skipped transcript in meiosis. Nucleic Acids Res 24:470–477

Hunt PA, LeMaire R, Embury P, et al. (1995) Analysis of chromosome behavior in intact mammalian oocytes: monitoring the segregation of a univalent chromosome during female meiosis. Hum Mol Genet 4:2007–2012

John B (1988) The biology of heterochromatin. In: Verma RS (ed) Heterochromatin: molecular and structural aspects. Cambridge University Press, Cambridge, pp 1–147

Kanaar R, Hoeijmakers JHJ (1998) Genetic recombination: from competition to collaboration. Nature 391:335–337

Keegan KS, Holtzman DA, Plug AW, et al. (1996) The Atr and Atm protein kinases associated with different sites along meiotically pairing chromosomes. Genes Dev 10:2423–2437

Kovalenko OV, Plug AW, Haaf T, et al. (1996) Mammalian ubiquitin-conjugating enzyme Ubc9 interacts with Rad51 recombination protein and localizes in synaptonemal complexes. Proc Natl Acad Sci USA 93:2958–2963

Loidl J, Scherthan H, Den Dunnen JT, et al. (1995) Morphology of a human-derived YAC in yeast meiosis. Chromosoma 104:183–188

Luciani JM, Guichaoua MR, Cau P, et al. (1988) Differential elongation of autosomal pachytene bivalents related to their DNA content in human spermatocytes. Chromosoma 97:19–25

Meuwissen RLJ, Meerts I, Hoovers JMN, et al. (1997) Human synaptonemal complex protein 1 (SCP1): isolation and characterization of the cDNA and chromosomal localization of the gene. Genomics 39:377–384

New JH, Sugiyama T, Zaitseva E, et al. (1998) Rad52 protein stimulates DNA strand exchange by Rad51 and replication protein A. Nature 391:407–410

Nimmo ER, Pidoux AL, Perry PE, et al. (1998) Defective meiosis in telomere-silencing mutants of Schizosaccharomyces pombe. Nature 392:825–828

Orkin SH, Kazazian HH Jr, Antonarakis SE, et al. (1982) Linkage of β-thalassaemia mutations and β-globin gene polymorphisms with DNA polymorphisms in human β-globin gene cluster. Nature 296:627–631

Paquis-Flucklinger V, Santucci-Darmanin S, Paul R, et al. (1997) Cloning and expression analysis of a meiosis-specific mut S homologue: the human MSH4 gene. Genomics 44:188–194

Pathak S, Hsu TC (1979) Silver-stained structures in mammalian meiotic prophase. Chromosoma 70:195–203

Pesce M, Gross MK, Schöler HR (1998) In line with our ancestors: Oct-4 and the mammalian germ. BioEssays 20:722–732

Plug AW, Xu J, Reddy G, et al. (1996) Presynaptic association of Rad51 protein with selected sites in meiotic chromosomes. Proc Natl Acad Sci USA 93:5920–5924

Plug AW, Clairmont CA, Sapi E, et al. (1997) Evidence for a role for DNA polymerase beta in mammalian meiosis. Proc Natl Acad Sci USA 94:1327–1331

Romanienko PJ, Camerino-Otero RD (1999) Cloning, characterization, and localization of mouse and human SPO11. Genomics 61:156–169

Sathananthan AH, Kola I, Osborne J, et al. (1991) Centrioles in the beginning of human development. Proc Natl Acad Sci USA 88:4806–4810

Scherthan H, Weich S, Schwegler H, et al. (1996) Centromere and telomere movements during early meiotic prophase in mouse and man are associated with the onset of chromosome pairing. J Cell Biol 134:1109–1125

Solari AJ (1980) Synaptonemal complexes and associated structures in microspread human spermatocytes. Chromosoma 81:315–337

Stürzbecher H-W, Donzelmann B, Henning W, et al. (1996) p53 is linked directly to homologous recombination processes via RAD51/RecA protein interaction. EMBO J 15:1992–2002

Szostak JW, Orr-Weaver TL, Rothstein RJ, et al. (1983) The double-strand-break repair model for recombination. Cell 33:25–35

Wettstein D von, Rasmussen SW, Holm PB (1984) The synaptonemal complex in genetic segregation. Annu Rev Genet 18:331–431

Wu T-C, Lichten M (1994) Meiosis-induced double-strand break sites determined by yeast chromatin structure. Science 263:515–518

Xu Y, Ashley T, Brainerd EE, et al. (1996) Targeted disruption of ATM leads to growth retardation, chromosomal fragmentation during meiosis, immune defects, and thymic lymphoma. Genes Dev 10:2411–2422

11

Meiotic Abnormalities: Abnormal Numbers of Chromosomes

A neuploidy is the loss or gain of individual chromosomes. It can be the result of nondisjunction in a premeiotic mitotic division in the germline of either parent, a first or second meiotic division in either parent, or an early embryonic mitotic (*postzygotic*) division in the affected individual. Nondisjunction refers to any process that causes two homologous chromosomes to go to the same pole instead of segregating to opposite poles. Some meiotic aberrations leading to nondisjunction are described in Table 11.1 and illustrated in Fig. 11.1. When homologous chromosomes fail to pair or fail to form chiasmata the homologues fall apart and appear as univalents in diplotene. Univalents may drift at random to the two poles in the first division and divide regularly in the second. Alternatively, they may divide mitotically in anaphase I and in anaphase II drift at random to opposite poles or fail to go to either pole, rarely, one may misdivide at the centromere, just as univalents might in the first meiotic division. Only a small

Table 11.1. Principal Meiotic Events and Outcomes of Their Failures

Stage	Meiotic events	Results of unsuccessful completion of meiotic events
Leptotene	Chromosomes become visible; lateral elements begin to form	Germ cell degeneration; sometimes nondisjunction
Zygotene	Chromosomes form bouquet; each chromosome pairs with its lateral element; homologous lateral elements unite into a synaptonemal complex, which completes the pairing	Germ cell degeneration; sometimes nondisjunction
Pachytene, early	Recombination nodules attach to the central elements	No crossing over; chromosomes remain univalent
Pachytene, late	During crossing over, recombination nodules change into bars	Because of lack of chiasmata, bivalents fall into univalents
Diplotene	Homologues repel each other until they are held together only at the chiasmata	More univalents visible than in earlier stages
Metaphase I, Anaphase I, Metaphase II, Anaphase II	Orderly segregation of chromosomes is prerequisite for regular gametogenesis	Univalents may undergo nondisjunction, loss, or misdivision; spindle abnormalities interfere with chromosome segregation

segment of the XY bivalent forms a synaptonemal complex in which crossing over takes place (Fig. 17.2). Thus, the X and Y remain as univalents much more often than even the smallest autosome pair. Frequencies of univalents vary among different individuals, but the mean frequency of unpaired sex chromosomes in the male is about 11 per cent (Laurie and Hultén, 1985). Multiple aneuploidy of one or several chromosomes is very uncommon, except for the sex chromosomes. Polyploidy (triploidy or tetraploidy) is the gain of whole sets of chromosomes.

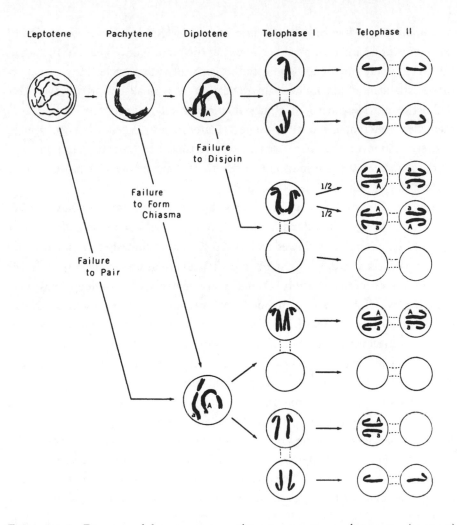

Figure 11.1. Diagram of the processes resulting in meiotic nondisjunction (see text) (Patau, 1963).

Incidence of Nondisjunction in Meiosis and Gametes

Direct analysis of male and female meiosis is difficult and limited by the availability of material. All stages of male meiosis can be studied in the adult testis, but this provides only a small amount of information about the incidence of nondisjunction in a particular individual. Corresponding information in the female is much more limited. The early prophase stages in the female are observable only in the fifth month of fetal life. The later stages of female meiosis I and meiosis II occur during the reproductive years of the adult, but usually only one ovum is released in each menstrual cycle. However, hormonal stimulation can induce superovulation of a small cohort of ova. This technique is widely used in clinics that carry out in vitro fertilization as a means of overcoming infertility. Superovulated ova in excess of those needed for fertilization or that have remained unfertilized by added sperm may be used to analyze the chromosomes of both the metaphase II oocyte and its corresponding polar body.

Angell (1995) studied oocytes removed by aspiration from the ovaries of 26- to 37-year-old women and matured in vitro. Germinal vesicle breakdown took about 12 hours and was followed by a 10- to 12-hour diakinesis. Chiasmata could not be well visualized at this stage, but their distribution was clearly different from that seen in males. Only 22 analyzable metaphase I configurations were seen, but two of these each contained a pair of univalent chromosomes and two each contained two pairs of univalents. In contrast, Hultén (1974) saw no univalents in 2168 metaphase I configurations in males, including 500 from an 86-year-old man. Clearly, female meiosis is much more error-prone than male meiosis, and increasingly so at advancing maternal ages.

The haploid products of meiosis must successfully complete gametogenesis if they are to become sperm or ova, and this might be less likely to occur if a particular chromosome is missing or present in excess, as in some nullisomics or disomics. Examination of meiosis is essential for working out the mechanisms involved in nondisjunction, but it cannot explain the widely different frequencies of the different trisomies or other numerical abnormalities in abortuses or liveborns.

Two methods have been developed to analyze the chromosome content of male gametes. The first involves fusion of sperm with hamster oocytes from which the zona pellucida has been removed enzymatically and observation of the first cleavage division several hours later, when the male and female metaphase plates are distinguishable and every human chromosome identifiable by banding (Rudak et al., 1978). Surprisingly high frequencies of chromosome abnormalities are seen using this technique, with considerable variation among studies. In a comprehensive review, Guttenbach et al. (1997) summarized data on 18,000 sperm karyotypes from normal men. Hyperhaploidy (usually disomy) occurred, on average, in 1.7% and hypohaploidy (usually monosomy) in 3.3% of the sperm. The chromosomes most frequently present in an extra copy were 21, X or Y, 16, 9, and 1. Structural abnormalities were seen in 6–7% of sperm.

This may be relevant to the observation that 84% of de novo translocations have a paternal origin (Olson and Magenis, 1988).

The second method is to examine chromosomes in chemically decondensed sperm using molecular techniques, such as multicolor FISH with two to five chromosome-specific α-satellite probes. This method has the advantage that very large numbers of sperm from an individual can be scored quickly. Consistent results can be obtained from person to person if the same technique of sperm decondensation is used. Again, the frequency of disomic sperm is quite high. Guttenbach et al. (1997) examined chromosomes 1, 7, 10, 17, X, and Y in both fertile and infertile men; 43 of 45 infertile men showed no differences from the fertile men. The frequencies of disomic sperm ranged from 0.1% for the Y to 0.14% for chromosome 10 and the X. The frequency of diploid sperm was 0.1%. Two of the 45 infertile men had 7- and 23-fold higher frequencies of diploid sperm: 0.35 and 1.6% vs. 0.06% in the other 43.

Guttenbach et al. (1997) reviewed results of more than 40 studies, involving every chromosome. In general, no chromosome showed a higher incidence of hyperhaploidy than any other chromosome. Pellestor et al. (1996) used primed in situ labeling (PRINS; Chapter 8) to determine the incidence of disomy for chromosomes 8, 9, 13, 16, and 21 in 96,292 sperm and found them all to be roughly the same, 0.26–0.32%. This suggests that nondisjunction for each of these chromosomes is equally likely in male meiosis and that the extra chromosome does not affect spermatogenesis. They extended this to chromosome 1 the following year, with a similar result: approximately 0.2% disomy for chromosomes 1 and 16. Since approximately 1 in 500 sperm is disomic for chromosome 1, what accounts for the extreme rarity of trisomy 1, even in early abortuses? Lethality before implantation or soon after seems the most likely explanation. This explanation is supported by the finding of trisomy 1 in an eight-cell pre-implantation embryo (Watt et al., 1987).

Incidence of Nondisjunction in Spontaneous Abortions, Stillborns, and Liveborns

There is no direct evidence that any type of chromosome imbalance interferes with the formation of a gamete that is able to participate in fertilization, but analysis of embryos indicates markedly skewed representations of the various abnormal karyotypes. Large numbers of embryos from presumably normal pregnancies that were terminated for nongenetic reasons have shown a low frequency of chromosome abnormalities, only a bit higher than that found in newborns. In contrast, a very high frequency of karyotypic abnormalities is seen in embryos that spontaneously abort (Boue et al., 1985). A comparison of the incidence of the various karyotypes in embryos that spontaneously abort at various stages of pregnancy with that in stillborn and liveborn infants provides a measure of the relative lethality of each karyotypic abnormality.

The wealth of information on the incidence of numerical and structural chromosome abnormalities in spontaneous abortuses, stillbirths, and live births is illustrated in Table 11.2. Almost half (47.9%) of all spontaneous abortuses are chromosomally abnormal, with 9.8% polyploid (mostly triploid), 8.6% 45,X, and 26.8% trisomic for one or another chromosome (virtually all trisomies have been seen in abortuses). Nearly 6% of stillbirths are chromosomally abnormal, with 0.6% polyploid, 0.25% 45,X, and 3.8% trisomic. Among live births, about 0.3% are aneuploid: most of these are trisomic, with less than one in a thousand 45,X or polyploid. Jacobs and Hassold (1995) also estimated the frequency of chromosome abnormalities in all recognized pregnancies, based on well-established estimates·that 15% of pregnancies are spontaneously aborted and 1% end in stillbirths. Almost 1 in 12 pregnancies involves a karyotypic abnormality: polyploidy in 1.5%, 45,X in 1.3%, and trisomy in 4.3%.

The incidence of abnormal karyotypes is higher the earlier in pregnancy the spontaneous abortion occurs. Delhanty et al. (1997) used multicolor FISH to

Table 11.2. Frequencies of Numerical Chromosome Abnormalities in Various Populations

Population	Frequency (%)					
	Trisomy	Polyploidy	XO	XXY	XXX	XYY
Spontaneous abortions	26.8	9.8	8.6	0.2	0.1	
Stillbirths	3.8	0.6	0.25	0.4	0.3	
Live births	0.3		0.01	0.05	0.05	0.05
All recognized pregnancies*	4.3	1.5	1.3	0.08	0.06	0.04
Survival probability	5.8	0	0.3	55	70	100

*Assumes 15% spontaneous abortions and 1% stillbirths
Source: Adapted from Jacobs and Hassold (1995), with permission, Academic Press

analyze preimplantation embryos that were left over from in vitro fertilization cases. They noted that the chance of a successful pregnancy for an embryo transferred into the uterus after in vitro fertilization (IVF) was 25% (the reason for transferring multiple embryos). This is the same as the chance of conception in any one menstrual cycle in fertile women. Delhanty et al. (1997) studied 93 embryos and found that only half were karyotypically normal. Two were fully aneuploid, 30% were mosaics (usually triploid/diploid), and a few showed chaotic divisions. If IVF embryos are representative of the normal population, these findings offer one explanation for the low pregnancy rate in normal women.

First polar bodies have been used for preimplantation diagnosis of aneuploidy, an appropriate approach because of the preponderance of maternal meiotic aberrations (Munné et al., 1995). An interesting approach to the analysis of female meiosis is the use of FISH with chromosome-specific α-satellite centromeric probes to look for abnormal segregation of chromosomes 13, 18, 21, and X in metaphase II oocytes and the associated polar bodies. Using this method, Dailey et al. (1996) found evidence for an additional or missing univalent chromosome in 15 of 168 oocytes (9%) and aneuploidy in 34 (20%). This is surprisingly high, since they were scoring only four chromosomes. Further studies are needed, because nondisjunction of these four chromosomes is responsible for almost all the aneuploidy seen in liveborn populations.

Different chromosomes show markedly different frequencies of trisomy. Trisomy 16 is present in more than 1% of conceptuses, accounting for nearly a fourth of all trisomies in recognized pregnancies. It is unclear to what extent this overrepresentation is due to a higher frequency of nondisjunction of number 16 and to what extent it is due to differential timing of embryonic lethality. The lethal effect of trisomy 16 usually occurs within a limited period, from 22 to 31 days of gestation, when chromosome studies are feasible (Boue et al., 1985). Trisomy for many of the autosomes may lead to degeneration of the embryo at a stage before chromosome studies are possible, leading to low estimates of their frequency. Trisomy 1 has never been seen in abortuses, but was observed in an eight-cell preimplantation embryo (Watt et al., 1987). This is unlikely to be the case for trisomies 21, 18, and 13, which are seen even in liveborns. Furthermore, a recent study of metaphase II oocytes found a preponderance of chromosome 16 among prematurely separated chromatids (Angell, 1997), suggesting that this chromosome indeed has a preferentially high rate of nondisjunction.

Causes of Meiotic Nondisjunction

Do mutant genes play any role in meiotic nondisjunction? None has yet been identified, although one mutation leading to mitotic nondisjunction is described later in this chapter. In principle, any genetic or acquired defect in centromeric function could lead to aneuploidy. Bernat et al. (1991) showed that microinjection of antibodies to kinetochore proteins disrupted the assembly of kinetochores and led to aneuploidy and micronucleated cells. Mitotic cells from scleroderma patients who produce antibodies to the kinetochore protein CENP-C have a slight increase in aneuploidy (Jabs et al., 1993). The alkaloid colchicine specifically destroys the spindle if given in sufficient dosage and at lower dosage can lead to nondisjunction. However, such agents have not been implicated as a cause of human trisomy. Preconception exposure to ionizing radiation has been suggested as a minor cause of nondisjunction, but studies in human populations have not supported this claim (Tease, 1988). No increase in nondisjunction has been found in the survivors of the atomic bombing of Hiroshima and Nagasaki (Awa et al., 1987).

The most important (cyto)genetic cause of nondisjunction is the presence of an abnormal karyotype in either parent. Aberrant meiotic segregation in translocation heterozygotes is discussed in Chapter 16. Here we need only note that 2:1 segregation of a chromosome in oocytes or spermatocytes trisomic for that chromosome is a well-established cause of trisomy 21. Nearly half the progeny of 21-trisomic women are themselves 21-trisomic, although 21-trisomic women rarely become pregnant. Much more common is germline (gonadal) mosaicism in the father or mother of a 21-trisomic individual. Penrose pointed out, more than 30 years ago, that as many as 15% of the mothers or fathers of 21-trisomics have minimal phenotypic features of Down syndrome, suggesting the presence of mosaicism in such parents. Mosaicism has indeed been demonstrated in some of these parents, and this may account for the fact that couples with one trisomic child have a 10-fold increase in the risk of having another.

Maternal Age, Recombination, and Mechanisms of Nondisjunction

The most important predisposing factor for nondisjunction is increased maternal age. Paternal age has a very small effect on the incidence of nondisjunction.

The incidence of disomic sperm increases by a factor of about 2.5 from 20–29 years of age to over 50 years of age (Griffin et al., 1995), whereas the incidence of trisomy increases exponentially with rising maternal age after age 30 or so. At a maternal age of 20 years, the incidence of 21-trisomic children is 0.4 per 1000 newborns; for women age 45 years and over, the risk has increased over 42-fold, to 17 per 1000 newborns (Hassold and Jacobs, 1984). A similar maternal age effect is found for 18-trisomy, 13-trisomy, and most of the other autosomal trisomies found in spontaneous abortuses. The 10–15% of pregnancies in women over 35 years of age account for one-third to one-half of aneuploid offspring. A smaller maternal age effect is seen for XXX and XXY. The incidence of 45,X, on the other hand, is virtually independent of maternal age. Studies of the Xg blood group gene, which is located on the X chromosome, show that in 78% of XO cases the gamete without a sex chromosome came from the father (Sanger et al., 1971). Similar results have been obtained with molecular methods (Mathur et al., 1991).

In some families, two or more sibs with aneuploidy of different chromosomes have been seen, such as trisomy 13 and 45,X, trisomy 21 and 49,XXXXY, or trisomy 18 and trisomy 21. One karyotypically normal couple had three offspring, one each with trisomy 21, 18, and 13. They were born at maternal ages of 40 to 43 years (Fitzpatrick and Boyd, 1989). No cause other than advanced maternal age has been established in this or similar families. The same is true for cases of double aneuploidy, such as XXY and 21 trisomy. Reddy (1997) observed only 22 cases of double trisomy and one of triple trisomy among 3024 spontaneous abortuses, a frequency even less than that expected if the two aneuploid events were independent of each other. They also occurred at an older mean maternal age.

The generally exponential increase in the frequency of nondisjunction with increasing maternal age is correlated with a decline in genetic recombination frequency, for example, for chromosomes 21, 18, 16, and 15 (Robinson et al., 1993; Sherman et al., 1994; Hassold et al., 1995). Chromosome 16 has a genetic length of 172 cM in controls but only 120 cM in trisomy 16 pedigrees. More important is the region of reduced recombination: the normally 60-cM central region of chromosome 16 in controls is 0 cM long in trisomy 16 pedigrees! That is, there is *no recombination* (and *no chiasmata*) around the centromere in female meioses leading to 16 trisomics. Similar results are seen for chromosome 21, although this appears to be true only for meiosis I errors (Fig. 11.2).

Chiasmata stabilize bivalents in late prophase I, when homologous centromeres repel each other and homologues appear to be held together only by

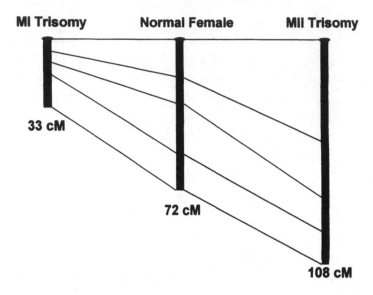

Figure 11.2. Meiotic recombination is reduced in maternal age-related trisomy 21 due to meiosis I errors but increased in that due to meiosis II errors (courtesy of Stephanie Sherman).

their chiasmata. Without a chiasma, the bivalent is likely to fall apart, yielding two univalents. Such univalents have been seen in metaphase II oocytes, as expected in the classical model of nondisjunction (Fig. 11.3; Kamiguchi et al., 1993; Lim et al., 1995). However, Angell (1991; 1997) has marshalled impressive evidence that nondisjunction usually arises in another way (Fig. 11.4). She adapted an improved method for studying the metaphase II oocytes obtained as a by-product of in vitro fertilization. She noted that univalents were much less common than extra or missing chromatids and proposed a novel mechanism of nondisjunction (Angell, 1991). Recently, she presented confirmatory evidence from the analysis of 200 metaphase II oocytes (Angell, 1997). These contained no univalents, but one-third contained prematurely separated chromatids (half-univalents) of three types: 23 + 1/2 (extra chromosome), 22 + 1/2 (missing chromosome), and 22 + 1/2 + 1/2 (balanced, possibly an artifact). Her hypothesis is that the cause of this type of aberration is the premature separation of the bivalent in meiosis I, leading to premature equational division (separation of the chromatids) in MI (Fig. 11.4). Very few MI oocytes have been analyzed, but these have shown some prematurely separated bivalents, particularly in older women.

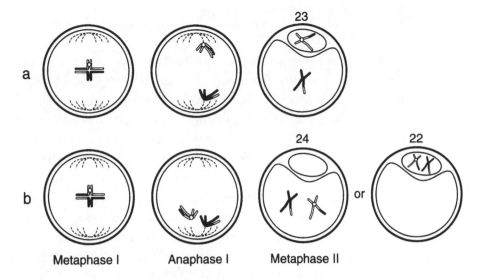

Figure 11.3. Behavior of chromosomes at MI division in oocytes, according to the classic model: (a) Normal disjunction. (b) Nondisjunction (reproduced from Angell, the Am J Hum Genet 61:23, Fig. 1, copyright 1997, the American Society of Human Genetics, with permission of the University of Chicago Press).

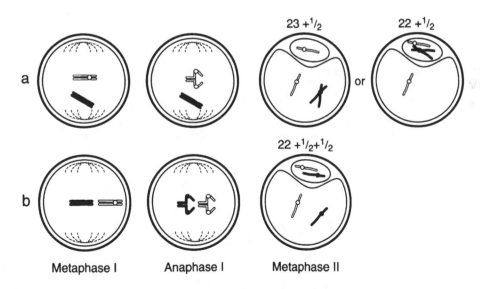

Figure 11.4. (a) Precocious MI division of one univalent leads to 23 + 1/2 and 22 + 1/2 MII oocytes. (b) Precocious MI division of both univalents leads to 22 + 1/2 + 1/2 MII oocytes (reproduced from Angell, the Am J Hum Genet 61:23, Fig. 7, copyright 1997, the American Society of Human Genetics, with permission of the University of Chicago Press).

Parental Origin of Aneuploid Gametes

The marked maternal age effect and minimal paternal age effect on the incidence of aneuploidy indicates that most aneuploidy in older women originates in a maternal meiotic error. This is, in fact, also true for younger women. The development of a large number of highly polymorphic molecular markers for loci scattered throughout the genome (Chapter 29) has provided a means of determining not only the parent of origin of the extra chromosome but whether it arose from an error in meiosis I or meiosis II. Genetic markers close to the centromere are the most useful because they provide the most reliable information on the segregation of the various centromeres; markers farther away are more likely to have been separated from the centromere by recombination.

In individuals with extra sex chromosomes, the extra chromosome was maternal in origin in 90% (45 of 50) XXX females and in 54% (76 of 142) XXY males. In those of maternal origin, the error arose in MI in over two-thirds of cases but in MII in nearly one-fourth and in a postzygotic mitosis in about one-tenth (Table 11.3). The maternal MI errors were associated with the absence of chiasmata in 30%, abnormally distributed chiasmata in 24%, and normal chiasmata in 45%. In all cases with four or five X chromosomes, the

Table 11.3. Percentage of Aneuploidy Attributable to Parent and Stage (MI, MII, or Postzygotic) of Origin, and the Role of Chiasmata, as Reflected by Recombination Frequency

			Maternal			
Karyotype	Paternal	Maternal	Meiosis I	Meiosis II	Postzygotic	Recombination
45,X	80	20				
47,XXX	10	90	66	18	16	Decreased
47,XXY	46	54	70	25	5	Decreased
Trisomy 2	44	56				
Trisomy 13, 14, 15, 22	12	88	68	32		Decreased
Trisomy 16	0	100	100	0		Decreased
Trisomy 18	3	97	29	65	6	Decreased
Trisomy 21	10	90	73	25	2	Decreased

Source: Modified from Jacobs and Hassold (1995), with permission, Academic Press, Fisher et al. (1995), and Savage et al. (1998), with permission, Oxford University Press

extra X chromosomes came from the mother, presumably by nondisjunction in meiosis I followed by nondisjunction again in meiosis II (Hassold et al., 1990).

In trisomy 16, the extra chromosome is always maternal in origin and results from an error in meiosis I (Hassold et al., 1995). Why this is so is unclear, since more than 1 in 1000 sperm is 16-disomic (Pellestor et al., 1996). Trisomy 18 is maternal in origin in 97% (61 of 63) individuals, with 29% due to a meiosis I error, 65% to a meiosis II error, and 6% to a postzygotic mitotic error (Table 11.3). The high density of genetic markers for this chromosome enabled Fisher et al. (1995) to demonstrate the absence of recombination in one-third of those due to meiosis I errors and normal recombination in the other two-thirds due to meiosis I errors and in all 35 due to meiosis II errors. The extra chromosome in trisomies 13, 14, 15, and 22 was maternal in 88% and paternal in 12% (Table 11.3). About two-thirds of the former arose from a meiosis I error and one-third from a meiosis II error. For chromosome 21, nearly one-fourth arose from an error in maternal meiosis II, as shown by homozygosity for markers near the centromere for which the mother is heterozygous. However, there is evidence that in some cases scored as having a meiosis II error, the error was really a failure of segregation of homologues at meiosis I. The evidence for this is the marked maternal age effect in the cases scored as having a meiosis II error. This is not expected, because maternal age does not affect the length of meiosis II but does that of meiosis I (which lasts 10–40 years, roughly). Lamb et al. (1996) proposed that the presence of a decreased number of chiasmata in the pericentromeric region tended to produce premature separation of bivalents at meiosis I. Angell (1997) has confirmed the premature separation of bivalents in meiosis I in older women. However, persistent bivalents have not been observed in any of the hundreds of metaphase II oocytes that have been observed (Kamiguchi et al., 1993; Lim et al., 1995; Angell, 1997).

The Origin of Diploid Gametes and Polyploidy

A triploid zygote may arise in various ways. The egg or sperm may be diploid as a result of restitution in either the first or the second meiotic division. The second polar body may reunite with the egg nucleus. Two sperm may fertilize the same egg (a rare event, because sperm penetration of the zona pellucida usually triggers a barrier to penetration by other sperm). Jacobs et al. (1978) obtained maximum-likelihood estimates that 66.4% of 24 triploid abortuses arose by dispermy, 23.6% by fertilization of a haploid ovum by a diploid sperm (failure of male meiosis I), and 10% from a diploid egg (failure of female meiosis I). How diploid gametes arise is unclear. Incomplete meiosis I might lead to a restitution nucleus' containing the unreduced diploid complement. Alternatively, endoreduplication in a gonial cell would lead to tetraploid meiocytes in which meiosis would produce diploid gametes.

Tetraploidy is rarer than triploidy among both spontaneous abortuses and live-born infants. This is probably due to the production of fewer tetraploid zygotes rather than to their greater lethality, because there are fewer mechanisms (and these rare) for producing tetraploid zygotes than for producing triploid ones. The most probable origin of tetraploidy is chromosome duplication in a somatic cell in an early-cleavage-stage embryo, a postzygotic event. Fertilization of a rare diploid ovum by an equally rare unreduced sperm may be possible. Another rare event, fertilization of one egg by three sperm, has produced a few tetraploids (Surti et al., 1986), but these have developed as hydatidiform moles rather than fetuses, because of genomic imprinting effects (Chapter 21).

Aneuploidy of Somatic (Mitotic) Origin: Mosaicism

Chromosomal mosaicism is the presence of two or more cell lines with different karyotypes that have arisen from a single fertilized egg. Chromosomal mosaics arise from postzygotic events in somatic cells. Sex chromosome mosaics, the most common, include such types as $n - 1/n$ (XO/XX or XO/XY), $n - 1/n + 1$ (XO/XXX or XO/XXY), and $n - 1/n/n + 1$ (XO/XX/XXX). The first two types arise at the first cleavage division, while the third type must arise later, since a normal cell line is also present. Trisomy 21/disomy 21 mosaicism is fairly common, while mosaicism involving chromosome 18, 13, or 8 is much less common and mosaicism for any other autosome is quite rare. Mosaicism usually arises by loss of a chromosome from one daughter cell, but it can also arise by mitotic nondisjunction, producing $n - 1$ and $n + 1$ daughter cells. Examples of this are common for the sex chromosomes but are almost never seen for the autosomes. This probably reflects the lethality of autosomal monosomy, even at the cellular level, since there is abundant evidence that autosomal mitotic nondisjunction does occur.

The use of highly polymorphic genetic markers has shown that a significant proportion of autosomal trisomies originate from postzygotic nondisjunction, including 4.5% of cases of trisomy 21 (Antonarakis et al., 1993). In a study of 63 cases of trisomy 18, Fisher et al. (1995) found a postzygotic origin in both of the cases in which the extra chromosome was of paternal origin and in three of the 61 cases in which the extra chromosome was of maternal origin. Robinson et al. (1998) found mitotic errors to be responsible for nearly 10% (12 of 128) of cases with an extra chromosome 15 of maternal origin.

An unusual genetic cause of mitotic nondisjunction has recently been identified. Two unrelated infants had multiple mosaic trisomies and monosomies (multiple variegated aneuploidy). In repeated leukocyte cultures, premature chromatid separation (PCS) of all mitotic chromosomes was observed in 67–87.5% of the cells. Both parents of each infant showed total PCS in 2.5–47% of their cultured leukocytes but did not have mosaic variegated aneuploidy. The phenotype of the presumptive homozygous infants was one of severe growth retardation, microcephaly, seizures, and multiple malformations (Kajii et al., 1998).

Minimal mosaicism is seen in some older people. The frequency of cells with a 45,X karyotype tends to increase with age in normal XX women, and so does that of XX cells in which the centromere of one X is inactivated (Fig. 22.2). The absence of a functional centromere leads to random segregation of this X chromosome, with some cells lacking it and other cells accumulating various numbers of copies.

References

Angell RR (1991) Predivision in human oocytes at meiosis I: a mechanism for trisomy formation in man. Hum Genet 86:383–387

Angell RR (1995) Meiosis I in human oocytes. Cytogenet Cell Genet 69:266–272

Angell RR (1997) First-meiotic-division nondisjunction in human oocytes. Am J Hum Genet 61:23–32

Antonarakis SE, Avramopolous D, Blouin J-L, et al. (1993) Mitotic errors in somatic cells cause trisomy 21 in 4.5% of cases and are not associated with advanced maternal age. Nat Genet 3:146–150

Awa AA, Honda T, Neriishi S, et al. (1987) Cytogenetic study of the offspring of atomic bomb survivors, Hiroshima and Nagasaki. In: Obe G, Basler A (eds) Cytogenetics. Springer, Berlin/Heidelberg, pp 166–183

Bernat RL, Delannoy MR, Rothfield NF, et al. (1991) Disruption of centromere assembly during interphase inhibits kinetochore morphogenesis and function in mitosis. Cell 66:1229–1236

Boue A, Boue J, Gropp A (1985) Cytogenetics of pregnancy wastage. In: Harris H, Hirschhorn K (eds) Advances in human genetics, Vol 14. Plenum, New York, pp 1–57

Dailey T, Dale B, Cohen J, et al. (1996) Association between nondisjunction and maternal age in meiosis II oocytes. Am J Hum Genet 59:176–184

Delhanty JDA, Harper JC, Ao A, et al. (1997) Multicolor FISH detects frequent chromosomal mosaicism and chaotic division in normal preimplantation embryos from fertile patients. Hum Genet 99:755–760

Fisher JM, Harvey JF, Morton NE, et al. (1995) Trisomy 18: studies of the parent and cell division of origin and the effect of aberrant recombination on nondisjunction. Am J Hum Genet 56:669–675

Fitzpatrick DR, Boyd E (1989) Recurrences of trisomy 18 and trisomy 13 after trisomy 21. Hum Genet 82:301

Griffin DK, Abruzzo MA, Millie EA, et al. (1995) Non-disjunction in human sperm: evidence for an effect of increasing paternal age. Hum Mol Genet 4:2227–2232

Guttenbach M, Engel W, Schmid M (1997) Analysis of structural and numerical abnormalities in sperm of normal men and carriers of constitutional chromosome aberrations. A review. Hum Genet 100:1–21

Hassold TJ, Pettay D, May K, et al. (1990) Analysis of non-disjunction in sex chromosome tetrasomy and pentasomy. Hum Genet 85:648–650

Hassold TJ, Merrill M, Adkins K, et al. (1995) Recombination and maternal age-dependent nondisjunction: molecular studies of trisomy 16. Am J Hum Genet 57:867–874

Hassold TJ, Jacobs PA (1984) Trisomy in man. Annu Rev Genet 18:69–97

Hassold TJ, Jacobs PA, Kline J, et al. (1980) Effect of maternal age on autosomal trisomies. Ann Hum Genet 44:29–36

Hultén M (1974) Chiasma distribution at diakinesis in the normal human male. Hereditas 76:55–78

Jabs EW, Tuck-Muller CM, Anhalt GJ, et al. (1993) Cytogenetic survey in systemic sclerosis: correlation of aneuploidy with the presence of anticentromeric antibodies. Cytogenet Cell Genet 63:169–175

Jacobs PA, Hassold TJ (1995) The origin of numerical chromosome abnormalities. Adv Genet 33:101–133

Jacobs PA, Angell RR, Buchanan IM, et al. (1978) The origin of human triploids. Ann Hum Genet 42:49–57

Kajii T, Kawai T, Takumi T, et al. (1998) Mosaic variegated aneuploidy with multiple congenital abnormalities: homozygosity for total premature chromatid separation trait. Am J Med Genet 78:245–249

Kamiguchi Y, Rosenbusch B, Sterizk K, et al. (1993) Chromosome analysis of unfertilized human oocytes prepared by a gradual fixation-air drying method. Hum Genet 90:533–541

Lamb NE, Freeman SB, Savage-Austin A, et al. (1996) Susceptible chiasma configurations of chromosome 21 predispose to non-disjunction in both maternal meiosis I and meiosis II. Nat Genet 14:400–405

Laurie DA, Hultén MA (1985) Further studies on bivalent chiasma frequency in human males with normal karyotypes. Ann Hum Genet 49:189–201

Lim AST, Ho ATN, Tsakok MFH (1995) Chromosomes of oocytes failing in-vitro fertilization. Hum Reprod 10:2570–2575

Mathur A, Stekol L, Schatz D, et al. (1991) The parental origin of the single X chromosome in Turner syndrome: lack of correlation with parental age or clinical phenotype. Am J Hum Genet 48:682–686

Munné S, Dailey T, Sultan KM, et al. (1995) The use of first polar bodies for preimplantation diagnosis of aneuploidy. Hum Reprod 10:1015–1021

Olson SB, Magenis RE (1988) Preferred paternal origin of de novo structural chromosome rearrangements. In: Daniel A (ed) The cytogenetics of mammalian autosomal rearrangements. Liss, New York, pp 583–599

Patau K (1963) The origin of chromosomal abnormalities. Pathol Biol 11:1163–1170

Pellestor F, Girardet A, Coignet L, et al. (1996) Assessment of aneuploidy for chromosomes 8, 9, 13, 16, and 21 in human sperm by using primed in situ labeling techniques. Am J Hum Genet 58:797–802

Reddy KS (1997) Double trisomy and spontaneous abortions. Hum Genet 101:339–345

Robinson WP, Bernasconi F, Mutirangura A, et al. (1993) Nondisjunction of chromosome 15: origin and recombination. Am J Hum Genet 53: 740–751

Robinson WP, Kuchinka BD, Bernasconi F, et al. (1998) Maternal meiosis I non-disjunction of chromosome 15: dependence of the maternal age effect on level of recombination. Hum Mol Genet 7:1011–1019

Rudak E, Jacobs PA, Yanagimachi R (1978) Direct analysis of the chromosome constitution of human spermatozoa. Nature 274:911–913

Sanger R, Tippett P, Gavin J (1971) Xg groups and sex abnormalities in people of northern European ancestry. J Med Genet 8:417–426

Savage AR, Petersen MB, Pettay D, et al. (1998) Elucidating the mechanisms of paternal non-disjunction of chromosome 21 in humans. Hum Mol Genet 7:1221–1227

Sherman SL, Petersen MB, Freeman SB (1994) Nondisjunction of chromosome 21 in maternal meiosis I: evidence for a maternal age-dependent mechanism involving reduced recombination. Hum Mol Genet 3:1529–1535

Surti U, Szulman AK, Wagner K, et al. (1986) Tetraploid partial hydatidiform moles: two cases with a triple paternal contribution and a 92,XXXY karyotype. Hum Genet 72:15–21

Tease C (1988) Radiation-induced aneuploidy in germ cells of female mammals. In: Vig BK, Sandberg AA (eds) Aneuploidy, Part B: Induction and test systems. Liss, New York, pp 141–157

Watt JL, Templeton AA, Messinis SI, et al. (1987) Trisomy 1 in an eight cell human preembryo. J Med Genet 24:60–64

12

Abnormal Phenotypes Due to Autosomal Aneuploidy or Polyploidy

Only three autosomal trisomies, those for chromosomes 13, 18, and 21, occur with an appreciable frequency in liveborn infants, and only one, trisomy 21, is compatible with long-term survival in a large proportion of cases. Trisomy 21, trisomy 18, and trisomy 13 produce three different and characteristic syndromes despite some degree of overlap. Traits they share include mental retardation, congenital heart defects, malformations in other organ systems, seizures, and growth retardation. These features are also common with other chromosome imbalances. For example, a ventricular septal defect is found in 47% of infants with trisomy 13, in 34% of 18-trisomics, and in 20% of 21-trisomics (Lewandowski and Yunis, 1977). In general, no single trait is exclusive to a particular chromosome imbalance, but a few tend to be associated with a particular chromosome imbalance, such as the persistence of fetal hemoglobin and abnormal neutrophil projections in trisomy 13. The increasingly routine use of ultrasound sometimes identifies fetal abnormalities at a stage of pregnancy at which a rapid decision is needed if termination of pregnancy is an option. Interphase FISH with DNA probes for chromosomes 13, 18, 21, X, and Y permits rapid diagnosis of aneuploidy of any of these chromosomes, or of polyploidy (Gersen et al., 1995).

Trisomy 21 and Down Syndrome

Down syndrome is the least severe of the autosomal trisomy syndromes. It is described in detail in many books and reviews, because it is the most common genetic cause of mental retardation and multiple congenital anomalies (for example, Epstein, 1995; Gardner and Sutherland, 1996). Down syndrome is characterized by moderate mental retardation, hypotonia, short stature, short extremities, oblique palpebral fissures, epicanthic folds, Brushfield spots in the irises of the eye, flat nasal bridge, open mouth, protruding tongue, small head, flat occiput, transverse palmar creases, and abnormal dermatoglyphic (palm- and fingerprint) patterns. Congenital heart disease (especially an endocardial cushion defect) is seen in about 40% and duodenal atresia, tracheo-esophageal fistula, or other gastrointestinal abnormality in about 5%. A clinical diagnosis can be made with over 95% accuracy in expert hands but only 75% overall. Ultrasound findings of a short femur and increased nuchal transparency (due to an abnormal accumulation of fluid at the nape of the neck), especially if accompanied by elevated maternal serum α-fetoprotein (of fetal origin), has led to the prenatal diagnosis of Down syndrome by chromosome studies (Verdin et al., 1997).

Mental retardation is almost universal in Down syndrome. There is usually a steady and fairly rapid decline in intellectual performance, with average IQs falling from over 75 at 1 year of age to less than 30 in those over 11 years of age. The IQ rarely stays above 80, even with early intervention and environmental enrichment. Neuronal degeneration identical to that in Alzheimer disease is frequent in middle-aged adults; cataracts and keratoconus are fairly common and may affect vision. Life expectancy now exceeds 60 years in those without congenital heart disease. The major causes of death in Down syndrome are severe congenital heart disease, cancer, and infections. There is a 20-fold increased risk of developing leukemia, and an excess of certain other tumors, particularly male germ cell tumors. Males with Down syndrome are more at risk for tumor development than females, for unknown reasons (Satge et al., 1998).

The chromosomal cause of Down syndrome, trisomy 21, was discovered by Lejeune et al. (1959). It is by far the most frequent of the autosomal trisomy syndromes, the usual estimate of its incidence being about 1 in 750 newborns. Trisomy 21 is twice as prevalent at 10 weeks of gestation as in newborns, indicating that half of those who survive the first 10 weeks of fetal life die between 10 weeks and birth (Snijders et al., 1995). In 93–96% of individuals with Down syndrome, the chromosome constitution is 47,+21. An additional 2–3% of subjects are mosaics, with a disomic and a trisomic cell line. Trisomy 21 mosaicism with minimal effects on the phenotype is sometimes found, particularly in parents of children with trisomy 21. Even gonadal mosaicism (limited to the germline) in a parent may account for the birth of a child, or more than one, with Down syndrome. In the remaining 2–5% of subjects, the extra chromosome 21 is part of a Robertsonian translocation chromosome or, rarely, a reciprocal translocation. Aberrant meiotic segregation in reciprocal translocation carriers (Chapter 16) can lead to duplication of a segment of chromosome 21 and to Down syndrome.

Trisomies corresponding to trisomy 21, in both phenotype and karyotype, are found only in the great apes (chimpanzee, gorilla, and orangutan). However, duplication (partial trisomy) of the segment of mouse chromosome 16 that is homologous to a particular segment of human chromosome 21 produces some phenotypic features of Down syndrome (Chapter 15; Reeves et al., 1995; Sago et al., 1998). Attempts to define a critical region of chromosome 21 containing genes whose extra dosage is responsible for the characteristic phenotype are described in Chapter 15.

How does the presence of a third copy of a chromosome lead to phenotypic abnormalities? The most likely reason is that the extra dosage of one or more genes on the chromosome is responsible. Some of the likely candidates are discussed in Chapter 15. It is important to remember that a gene may have an enhancing effect or an inhibitory effect. There is evidence that trisomy 21 has important inhibitory effects on genes scattered throughout the genome. Amiel et al. (1998) used FISH to analyze the replication timing of four genes: *TP53* and *HER2* on chromosome 17, *RB1* on chromosome 13, and *MYC* on chromosome 8. The two alleles of each gene replicated synchronously in normal cells but very asynchronously in 21-trisomic cells, suggesting that one allele may have been inactivated. If the two tumor suppressor genes *TP53* and *RB1* are indeed inactivated in some cells, this may be responsible for the increased risk of leukemia and certain other tumors (Chapter 28). Alleles at some loci on other chromosomes show asynchronous replication in trisomy 13 and 18, just as they do in trisomy 21. This suggests that the inactivation of the same genes on other chromosomes is responsible for some of the phenotypic overlaps of these three trisomies (Amiel et al., 1999).

Trisomy 18 and Edwards Syndrome

Edwards et al. (1960) reported the first case of what we now call trisomy 18 syndrome. This syndrome is marked by severe neurological and cardiac anomalies, plus defects in many other organ systems. Dysmorphic features include abnormal facies; small jaw; prominent occiput; clenched, overlapping fingers; muscle rigidity; and rocker-bottom (calcaneovalgus) feet (Hodes et al., 1978). Nuchal translucency can be detected by ultrasound in 90% of 11- to 14-week-old trisomy 18 fetuses. Cytogenetic diagnosis can then be made on chorionic villus samples. The parents in these cases usually decide to terminate the affected pregnancy (Hyett et al., 1995).

The incidence of trisomy 18 is about 1 per 6000 in the newborn. It is 85 times as prevalent at 10 weeks of gestation as in newborns, indicating that over 98% of those alive at 10 weeks of gestation die before birth (Snijders et al., 1995). Of those born alive, 30% die within one month and only 10% survive 1 year (Gorlin, 1977). About 80% of the patients have straight trisomy, another 10% are mosaics, and the rest have a translocation. In one case, the karyotype was 46,XY,−18,+psu dic(18)(qter→cen→p11.31::p11.31→psu cen→qter), as verified by multicolor PRINS analysis using chromosome 18–specific α-satellite and telomeric oligonucleotide primers. The subject was monosomic for two maternal 18p12 markers, indicating that the pseudodicentric chromosome 18 was paternal in origin. Both parents had normal karyotypes, so the rearranged chromosome arose either in the father's germline or in an early cleavage division (Graveholt et al., 1997).

Trisomy 13 and Patau Syndrome

Patau et al. (1960) first reported a case of what we now call trisomy 13 syndrome (Fig. 12.1). The frequency of trisomy 13, usually an embryonic lethal condition, is 100 times greater in spontaneously aborted embryos or fetuses than in liveborns. The incidence of trisomy 13 (47,+13) in liveborns is about 1 in 12,000 (Hook, 1980). Increased maternal age is a factor in trisomy 13, as it is in other

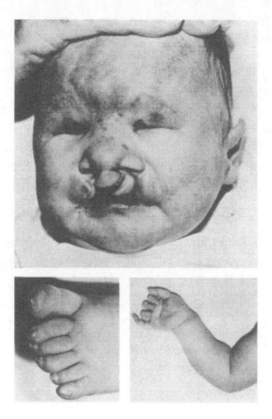

Figure 12.1. An infant with trisomy 13 (Patau) syndrome. Note the anophthalmia, cleft lip, and six toes.

trisomies. Even those infants with trisomy 13 who survive birth have a limited life expectancy; about 45% die within the first month, 90% are dead before 6 months, and fewer than 5% reach the age of 3 years. Two exceptional 13-trisomics lived for 11 and 19 years (Redheendran et al., 1981).

Infants with trisomy 13 show severe neurological impairment, with holoprosencephaly (forebrain defects, absence of the corpus callosum or olfactory bulbs), eye anomalies ranging from anophthalmia (absence of eyes) to microphthalmia, coloboma (fissure) of the iris, cleft lip, cleft palate, capillary hemangiomata, scalp defects, postaxial polydactyly, and rocker-bottom feet. Several types of complex heart anomalies are common (Hodes et al., 1978). Individuals with trisomy 13 show considerable phenotypic variation even when mosaicism is absent; so do patients with other chromosomal abnormalities.

Other Autosomal Aneuploidy Syndromes

In general, trisomy for any autosome except 21, 18, or 13 is an embryonic lethal condition. Survival until after birth is rare, with the possible exception of number 22. Despite numerous reports of 22-trisomic survivors (for instance, Kukolich et al., 1989), the extra chromosome usually appears to have arisen by 3 : 1 segregation in a t(11;22)(q23;q11) translocation carrier or heterozygote and is not a normal chromosome 22 (Zackai and Emanuel, 1980). No carrier of a Robertsonian translocation involving chromosome 22 (Chapter 27) has had offspring with translocation trisomy 22, and this speaks against the viability of trisomy 22.

Monosomy in liveborn infants has been established for only one autosome, chromosome 21, and even that is extremely rare, with six cases reported by 1983 (Wisniewski et al., 1983). It is also infrequent in spontaneous abortions. Monosomy 21 has never occurred among the progeny of a Robertsonian translocation heterozygote (Chapter 16). Haploid newborns or abortuses have never been seen. If haploid zygotes are ever produced, they presumably die before the time of implantation, just like most monosomic and some trisomic zygotes, so there is no recognized pregnancy.

Trisomy/Disomy Mosaicism

Mosaics with trisomic and disomic cell lines usually arise from a trisomic zygote. Subsequent loss of one of the three homologues from one cell gives rise to the disomic line. The disomic line is usually normal; the exceptions, which involve uniparental disomy and imprinting effects, are discussed in Chapter 21. The most common type of autosomal mosaicism involves chromosome 21. About 2–3% or more of individuals with Down syndrome are trisomy 21/disomy 21 mosaics. The proportion of disomic cells varies greatly, and so does the phenotype, which ranges from severe Down syndrome to complete normality. An unknown number of mosaics go undetected, either because the phenotype is normal or because disomic cells are rare in the tissues examined. Parental mosaicism plays an important role in the origin of 21-trisomic individuals: 2.7% have a mosaic parent (Uchida and Freeman, 1985). Mosaics have a much increased risk of producing 21-trisomic offspring and are responsible for some of the 21-trisomics born to young mothers (James et al., 1998).

Mosaicism for chromosome 18 or 13 is found in perhaps 2–10% of those with the corresponding syndromes and has the same moderating effect on the phenotype as noted for chromosome 21. Such mosaicism probably accounts for the rare individuals with these syndromes who survive well past infancy. Does this same moderating effect enable trisomy/disomy mosaics for other autosomes or triploid/diploid (mixoploid) mosaics to survive instead of ending in early abortion? Yes, but not very often. Rarely, an individual case report of such mosaicism has been published. About half the autosomes have been involved, in addition to triploidy/diploidy (Lin et al., 1998). Mosaicism involving chromosome 8 occurs considerably more frequently. Riccardi (1977) was able to describe over 60 cases and to characterize the phenotype. The origin of trisomy 8 mosaics appears to be different from that of most mosaics or trisomies: 20 of 26 arose postzygotically, during early embryonic mitoses, and only two of the 26 arose by maternal meiotic nondisjunction, in sharp contrast to most trisomics and mosaics (Karadina et al., 1998). More complex mosaics rarely occur, such as a severely affected individual with 13-trisomic, 18-trisomic, and normal cell lines (Wilson et al., 1983).

Mosaicism is sometimes detected in cultured second-trimester amniotic fluid cells, and even more frequently in cells cultured earlier in pregnancy from first-trimester chorionic villi, despite the absence of the abnormal cell line in the fetus or newborn. This is called *confined placental mosaicism* and is sometimes associated with abnormalities due to uniparental disomy (Chapter 21). Clinical cytogeneticists must be very careful in evaluating such cases, taking the method of ascertainment into account. If advanced maternal age is the indication, then the prognosis (expected outcome) for pregnancies with confined placental mosaicism is good. That is not the case if amniocentesis is performed in a woman because of abnormal maternal serum α-fetoprotein or maternal serum human chorionic gonadotropin values. A series of 11 women with these findings, nine under age 35, had trisomy 16 mosaicism. At least nine of the fetuses had serious complications, including intrauterine growth retardation and congenital malformations, two dying neonatally (Hsu et al., 1998).

Triploidy and Tetraploidy

The only types of polyploidy found in humans are triploidy (3n) and tetraploidy (4n). Most polyploid embryos die early in pregnancy and are spontaneously aborted. Triploidy is one of the major causes of spontaneous abortions (Chapter

11), accounting for about 17% of cases (Carr and Gedeon, 1977). Only about one in 10,000 triploid zygotes leads to a live birth. Most of these liveborns die within a day, although some have survived for a few months. In such rare cases, one tends to suspect undetected mosaicism, especially since most presumed polyploid infants have been shown to have a diploid cell line as well; they are thus *mixoploid*. They may have malformations discernible prenatally (Lin et al., 1998). The phenotype of triploids is not characteristic but includes growth retardation and multiple malformations. Rapid prenatal diagnosis of triploid fetuses is readily achieved using FISH with multiple chromosome-specific probes (Gersen et al., 1995).

Tetraploidy is rarer than triploidy among both spontaneous abortuses and liveborn infants, due either to the production of fewer triploid zygotes or to their greater lethality. Nonmosaic tetraploidy has been reported in a few newborns and in a 22-month-old girl (Lafer and Neu, 1988). In addition, a few liveborn infants have been diploid/tetraploid mosaics, or mixoploids. Malformations of multiple organ systems may be seen, but there is not a clearly defined syndrome.

Spontaneous Abortions, Fetal Deaths, and Stillbirths

Spontaneous abortions represent pregnancies that have ended prematurely because of death of the embryo or fetus, placental inadequacy, or uterine pathology. Spontaneous abortions have many causes, including hormonal, infectious, vascular, and genetic, including chromosomal. It is therefore rather surprising that almost half of first-trimester spontaneous abortions are caused by chromosome abnormalities (Chapter 11). Early abortions have a higher frequency of chromosome abnormalities than later ones. The incidence of trisomic abortions increases with maternal age, while the frequencies of polyploid and of 45,X abortions are independent of maternal age (Carr and Gedeon, 1977). Less than 1% of 45,X zygotes lead to 45,X liveborns. The generally accepted explanation for this is that 45,X zygotes are lethal but 45,X/46,XX and 45,X/46,XY mosaics are not, and liveborn 45,X individuals are presumed to have started out as mosaics, whether or not they still harbor some XX or XY cells.

Since monosomy may, in principle, arise through either chromosome loss or nondisjunction whereas trisomy results only from the latter process, monosomy should be more frequent among spontaneous abortuses than trisomy. Instead, autosomal monosomy is almost nonexistent (Chapter 11). Monosomic zygotes presumably die so early that there is either no implantation (thus no recognized pregnancy) or no detectable products of conception to study.

References

Amiel A, Avivi L, Gaber E, et al. (1998) Asynchronous replication of allelic loci in Down syndrome. Eur J Hum Genet 6:359–364

Amiel A, Korenstein A, Gaber E, et al. (1999) Asynchronous replication of alleles in genomes carrying an extra chromosome. Eur J Hum Genet 7:223–230

Carr DH, Gedeon M (1977) Population cytogenetics of human abortuses. In: Hook EB, Porter IH (eds) Population cytogenetics. Academic, New York, pp 1–9

Edwards JH, Harnden DG, Cameron AH, et al. (1960) A new trisomic syndrome. Lancet i:787–790

Epstein CJ (1995) Down syndrome (trisomy 21). In: Scriver CR, Beaudet AL, Sly WS, Valle D (eds) The metabolic and molecular bases of inherited disease, 7th edn. McGraw-Hill, New York, pp 749–794

Gardner RJM, Sutherland GR (1996) Chromosome abnormalities and genetic counseling, 2nd edn. Oxford, New York

Gersen SL, Carelli MP, Klinger KW, et al. (1995) Rapid prenatal diagnosis of 14 cases of triploidy using FISH with multiple probes. Prenatal Diag 15:1–5

Gorlin RJ (1977) Classical chromosome disorders. In: Yunis JJ (ed) New chromosomal syndromes. Academic, New York, pp 59–117

Graveholt CH, Bugge M, Stromkjaer H, et al. (1997) A patient with Edwards syndrome caused by a rare pseudodicentric chromosome 18 of paternal origin. Clin Genet 52:56–60

Hodes ME, Cole J, Palmer CG, et al. (1978) Clinical experience with trisomies 18 and 13. J Med Genet 15:48–60

Hook EB (1980) Rates of 47,+13 and 46 translocation D/13 Patau syndrome in live births and comparison with rates in fetal deaths and at amniocentesis. Am J Hum Genet 32:849–858

Hsu W-T, Shchepin DA, Mao R, et al. (1998) Mosaic trisomy 16 ascertained through amniocentesis: evaluation of 11 new cases. Am J Med Genet 80:473–480

Hyett JA, Moscoso G, Nicolaides KH (1995) Cardiac defects in 1st-trimester fetuses with trisomy 18. Fetal Diagn Ther 10:381–386

James RS, Ellis K, Pettay D, et al. (1998) Cytogenetic and molecular study of four couples with multiple trisomy 21 pregnancies. Eur J Hum Genet 6:207–212

Karadina G, Bugge M, Nicolaides P, et al. (1998) Origin of nondisjunction in trisomy 8 and trisomy 8 mosaicism. Eur J Hum Genet 6:432–438

Kukolich MK, Kulharya A, Jalal SM, et al. (1989) Trisomy 22: no longer an enigma. Am J Med Genet 34:541–544

Lafer CZ, Neu RL (1988) A liveborn infant with tetraploidy. Am J Med Genet 31:375–378

Lejeune J, Turpin R, Gautier M (1959) Le mongolisme, premier example d'aberration autosomique humaine. Ann Génét 1:41–49

Lewandowski RC, Yunis JJ (1977) Phenotypic mapping in man. In: Yunis JJ (ed) New chromosomal syndromes. Academic, New York, pp 369–394

Lin HJ, Schaber B, Hashimoto CH, et al. (1998) Omphalocoele with absent radial ray (ORR): a case with diploid-triploid mixoploidy. Am J Med Genet 75:235–239

Patau K, Smith DW, Therman E, et al. (1960) Multiple congenital anomaly caused by an extra autosome. Lancet i:790–793

Redheendran R, Neu RL, Bannerman RM (1981) Long survival in trisomy-13 syndrome: 21 cases including prolonged survival in two patients 11 and 19 years old. Am J Med Genet 8:167–172

Reeves RH, Irving NG, Moran TH, et al. (1995) A mouse model for Down syndrome exhibits learning and behavior deficits. Nat Genet 11:177–184

Riccardi VM (1977) Trisomy 8: an international study of 70 patients. In: Birth defects: original article series, XIII, 3C. The National Foundation, New York, pp 171–184

Sago H, Carlson EJ, Smith DJ, et al. (1998) Ts1Cje, a partial trisomy 16 mouse model for Down syndrome, exhibiting learning and behavioral abnormalities. Proc Natl Acad Sci USA 95:6256–6261

Satge D, Sommelet D, Geneix A, et al. (1998) A tumor profile in Down syndrome. Am J Med Genet 78:207–216

Snijders RJM, Sebire NJ, Nicolaides KH (1995) Maternal age- and gestational age-specific risk for chromosome defects. Fetal Diagn Ther 10:356–367

Uchida IA, Freeman VCP (1985) Trisomy 21 Down syndrome: parental mosaicism. Hum Genet 70:246–248

Verdin SM, Braithwaite JM, Spencer K, et al. (1997) Prenatal diagnosis of trisomy 21 in monozygotic twins with increased nuchal transparency and abnormal serum biochemistry. Fetal Diagn Ther 12:153–155

Wilson WG, Shires MA, Wilson KA, et al. (1983) Trisomy 18/trisomy 13 mosaicism in an adult with profound mental retardation and multiple malformations. Am J Med Genet 16:131–136

Wisniewski K, Dambska M, Jenkins EC, et al. (1983) Monosomy 21 syndrome: Further delineation including clinical, neuropathological, cytogenetic and biochemical studies. Clin Genet 23:102–110

Zackai EH, Emanuel BS (1980) Site-specific reciprocal translocation, t(11;22)(q23;q11), in several unrelated families with 3:1 meiotic disjunction. Am J Med Genet 7:507–521

13

Chromosome Structural Aberrations

S tructural chromosome abnormalities are relatively frequent in human populations. They are the result of breaks that disrupt the continuity of one or more chromosomes. Chromosome breaks in the germline can lead to heritable structural abnormalities; those occurring in somatic cells may increase the risk of cancer. Chromosomes may break at almost any point, but there are sites of preferred breakage, called *hotspots*. The breaks may be repaired, but because any two broken ends that are sufficiently close together in the nucleus may rejoin, an extremely wide variety of structurally altered chromosomes occur. The main classes of structural abnormalities are described in this and the next two chapters, and some of the resulting phenotypes, or clinical syndromes, in Chapters 15–20.

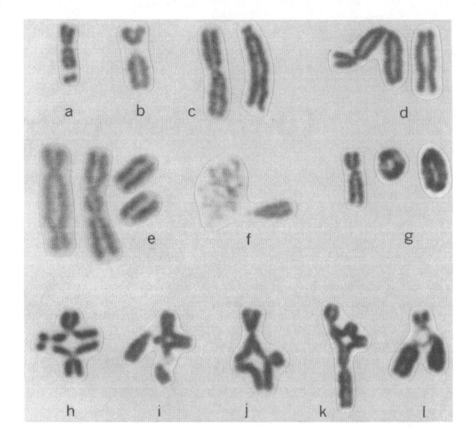

Figure 13.1. Structural aberrations: (a) Gap. (b) Gap at centromere. (c) Normal chromosome 1 and a pericentric inversion of its homologue. (d) Dicentric chromosome and acentric fragment. (e) Two dicentrics and two acentrics from one cell (courtesy of EM Kuhn). (f) A D group chromosome in satellite association with an interphase-like acrocentric. (g) Chromosome 9, ring (9) and double-sized ring (9). (Courtesy of ML Motl). (h) Mitotic chiasma between heteromorphic homologues. (i) Class II quadriradial between two D group chromosomes in satellite association with a D and a G. (j–k) Class IVb chromatid translocations. (l) Hexaradial chromatid translocation, or a satellite association between two D and one G (all, except as noted, courtesy of E Therman).

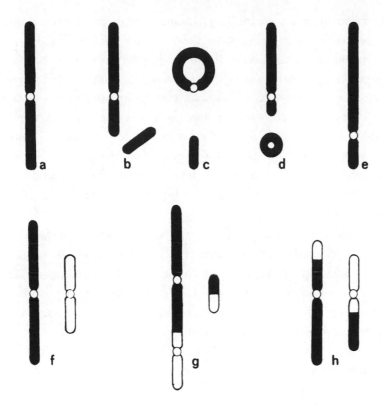

Figure 13.2. Results of G1 breaks in one chromosome (a–e), and in two chromosomes (f–h). (b) Broken chromosome. (c) Centric ring and acentric fragment. (d) Acentric ring and centric fragment. (e) chromosome with pericentric inversion. (g) Dicentric chromosome and acentric fragment. (h) Reciprocal translocation.

Chromosome and Chromatid Breaks and Rearrangements

A chromosome may break at any stage of the cell cycle. If it breaks during the G1 stage and is unrepaired through S, it will be visible in both chromatids (a *chromosome break*) in the next metaphase. If there is no displacement of the fragments, one sees only a gap (Fig. 13.1a and b). If one chromosome (Fig. 13.2a) undergoes a single break, this may lead to a terminally deleted chromosome and an acentric fragment (Fig. 13.2b) that can be lost in a subsequent mitosis. Alternatively, the acentric fragment may be included in a daughter nucleus and replicate, with double fragments visible at the next metaphase.

Two breaks in the same chromosome may result in the formation of either an acentric fragment plus a centric ring chromosome or an acentric ring plus a centric chromosome with an interstitial deletion (Figs. 13.2c, d; Fig. 13.1g). Very small fragments are called minutes. If two breaks take place in the same arm and the deleted segment reunites with the chromosome in an inverted orientation, a *paracentric inversion* is produced, with no change in the position of the centromere. If one break occurs in each arm and the segment between the breaks is reunited in an inverted orientation, a pericentric inversion is produced (Fig. 13.2e; Fig. 13.1c). The latter rearrangement sometimes shifts the position of the centromere, but many pericentric inversions and all paracentric inversions would be undetectable without banding techniques or in situ hybridization with molecular probes.

If two chromosomes (Fig. 13.2f) undergo an interchange following a single break in each, this may produce a dicentric chromosome and an acentric fragment (Fig. 13.2g; Figs. 13.1d, e) or a reciprocal translocation (Fig. 13.2h). The acentric fragment will eventually be lost, and the dicentric one as well, if the two centromeres are some distance apart, because the centromeres may be pulled towards opposite poles, forming an anaphase bridge. This bridge may undergo breakage at a random point; each broken end is prone to fusion, forming another dicentric chromosome that is subject to the same *bridge-breakage-fusion-bridge cycle* and eventual loss. A three-break interchange may produce either an insertion of an interstitial segment from one chromosome into the same or another chromosome or interchanges among three chromosomes. Multiple breaks in a cell may lead to more complex rearrangements, including chromosomes with several centromeres.

When a break takes place during G2, it involves only one of the two chromatids and is therefore called a *chromatid break*. A single break yields a deleted chromatid and an acentric fragment. A chromatid break in each of two chromosomes can lead to chromatid exchanges and result in *quadriradial* configurations (Fig. 13.1i). These are of two types, one in which alternate chromatids will segregate to opposite poles (I, IIIa, and IVa in Fig. 13.3) and one in which adjacent chromatids will segregate to opposite poles (II, IIIb, and IVb in Fig. 13.3). Alternate segregation leads to a reciprocal translocation involving two of the chromatids, and two unchanged chromatids. Adjacent segregation gives rise to a dicentric chromatid, an acentric chromatid, and two unchanged chromatids. Mitotic chiasmata can occur as a result of crossing over between two homologous mitotic chromosomes (Fig. 13.1h) and form a special subgroup of the alternate type of quadriradial segregation.

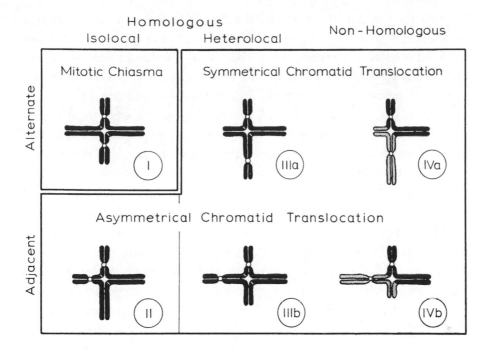

Figure 13.3. The classification of quadriradials. Breaks at homologous points are called isolocal, those at nonhomologous points heterolocal (Therman and Kuhn, 1976).

Deletions (Partial Monosomies), Including Ring Chromosomes

Chromosome deletions, or deficiencies, range in size from just a few base pairs to many megabase pairs in length. Their size can be evaluated by bivariate cytophotometric flow karyotyping, even with deletions as small as 9–26 Mb in size (Silverman et al., 1995). Deletions may arise from a single break (terminal deletion) or from two breaks, which can lead to an interstitial deletion if both breaks are in the same arm or to a ring chromosome if the breaks are in different arms. Large deletions are rare; some of the more important ones are 4p−, 5p−, 9p−, 11p−, 11q−, 13q−, 18p−, and 18q−. Deletions that appear to be terminal, as well as many interstitial deletions, have been found for every chromosome arm in liveborn children (Borgaonkar, 1997). They are common in deletion syndromes despite the known instability (stickiness) of broken ends.

Molecular cytogenetic methods have resolved this paradox: Many of these apparently terminal deletions are capped by a telomeric (TTAGGG)n repeat as a result of a cryptic translocation or the action of telomerase. The first clue to this was the demonstration that a terminal deletion of the short arm of chromosome 16 had been stabilized by the addition of a normal telomere at the point of breakage (Wilkie et al., 1990). Meltzer et al. (1993) showed, using chromosome-specific subtelomeric sequences, that translocations with one breakpoint in the subtelomeric region are very common. These provide a source of telomeres to cap broken chromosome ends, a process they called *telomere capture*. Such telomere capture would stabilize apparently terminal deletions.

This explanation could account for several seeming inconsistencies that have been observed. When human chromosomes are broken with x-rays, the sister chromatids rarely fuse to yield dicentrics that go through bridge-breakage-fusion-bridge cycles, as expected of chromosomes with broken ends. Ring chromosomes sometimes open up and act like normal biarmed chromosomes despite having what should be broken ends. Terminally deleted chromosomes occur too frequently, and have breakpoints that are too consistent, to be the result of two breaks. Niebuhr (1978) reported that in 35 cri du chat patients the 5p– deletion appeared to be terminal in 27, interstitial in four, and capped by a reciprocal translocation in four others. Vermeesch et al. (1998) showed that the distal end of all four 5p– and the one 4p– deletions they studied had telomeric repeats (see Fig. 15.3). Using chromosome microdissection to construct a probe from the area around the apparent deletion, they showed the absence of DNA from any other chromosome, thus ruling out a cryptic translocation.

Ring chromosomes are a class of chromosomes with deletions in which material is lost from the ends of both arms. Rings of almost every chromosome have been observed, with r(21) the most common (Melnyk et al., 1995). Ring chromosomes tend to generate new variants. A sister chromatid exchange may lead to a continuous double ring with one centromere. When the centromere divides in anaphase, the daughter centromeres may go to the same pole. This leads to a double-sized dicentric ring in one daughter cell and no ring in the other. If, on the other hand, the centromeres move to opposite poles, the ring may break at random and the broken ends rejoin, so daughter cells may have rings of unequal size. Thus, rings can generate mosaics in which different cells contain different derivatives of the original ring.

Ring chromosomes are unstable in both mitosis and meiosis and are frequently lost, probably as a result of dicentric formation and subsequent bridge-breakage-fusion-bridge cycles. The result is a monosomic cell, which may or may not be viable. Ring-X chromosomes are frequently associated with Turner syndrome, due to the generation of an XO cell line. An unusual example of this is a woman who was diagnosed as having Turner syndrome at age 14 but went on to have three pregnancies. One of them produced a daughter who, like the mother, was a 45,X/46,X,r(X) mosaic and had signs of Turner syndrome (Blumenthal and Allanson, 1997). A different kind of mosaicism involves the presence of cells in which the ring chromosome has apparently recurred in two or three successive generations of a family, replacing one of the two normal homologues present in other cells of the individual. Examples involving several autosomes are known, and one involving chromosome 21 has been diagnosed prenatally (Melnyk et al., 1995).

Duplications (Partial Trisomies)

Duplications are seldom seen, in contrast to the more common duplication/deficiencies generated by aberrant meiotic segregation in translocation and inversion heterozygotes (Chapter 16). However, small accessory or supernumerary chromosomes are fairly frequent, occurring in about 1.5 per 1000 live births. At least half of these small chromosomes are derived from chromosome 15 and are inverted duplications, inv dup(15), of the pericentromeric region (Schreck et al., 1977), making this one of the most common structural aberrations. These are sometimes associated with phenotypic abnormalities and sometimes not, depending upon whether the deleted segment of 15q11 contains the Prader-Willi/Angelman syndrome critical region (Chapter 15; Wandstrat et al., 1998). The analysis of these small chromosomes has been aided by FISH analysis, especially using probes generated by PCR amplification of the DNA from flow-sorted or microdissected supernumerary chromosomes, a technique called *reverse chromosome painting* (Viersbach et al., 1998).

Isodicentric chromosomes are symmetric structures consisting of segments of two homologous chromosomes that have broken at identical points. Each segment contains the same whole arm, a centromere, and part of the other arm. Usually one centromere is inactivated. An isodicentric chromosome represents an inverted duplication (partial trisomy) of part of a chromosome, combined with a deletion of the remainder. Most isodicentrics involve two X chromosomes (Fig. 22.2). This probably reflects the more severe phenotypic effects of autosomal trisomies or partial trisomies rather than preferential origin of isodicentric X chromosomes. Isodicentrics consisting of two Y chromosomes also occur, and so do Robertsonian translocation chromosomes with two centromeres (Chapter 27).

Isochromosomes are inverted duplications: palindromic structures in which the breakpoints are very close to or in the centromere. Isochromosomes are thus metacentric and have two homologous arms that are either genetically identical (homozygous at all gene loci) or non-identical (heterozygous at some loci). Isochromosomes for acrocentric chromosomes, including the Y, are fairly common, because the lack of the short arm does not affect viability. Isochromosomes for Xq, designated i(Xq), are also fairly common, because preferential inactivation of the i(Xq) prevents abnormal dosage effects (Chapter 18). The presence of an isochromosome in addition to a normal chromosome complement has been described for very short arms, such as 9p (Jalal et al., 1991). In such cases, the individual is tetrasomic for the arm concerned. Even tetrasomy for major parts of chromosome 9 seems to be compatible with a limited life span. The largest partial tetrasomy has been described in a patient with an extra isodicentric, consisting of two chromosomes 9 attached long arm to long arm (both breakpoints in q22), with the second centromere inactivated. The infant was highly abnormal and lived for only a couple of hours (Wisniewski et al., 1978).

The most probable origin of isodicentric chromosomes, including dicentric isochromosomes, is segregation of an adjacent quadriradial (type II in Fig. 13.3), with the centromeres of the dicentric chromatid going to the same pole (Fig. 13.4b). Therman and Kuhn (1985) have shown that this is the usual mechanism for creating symmetric dicentrics in Bloom syndrome. Dicentrics between two non-homologous chromosomes arise through segregation of a type IVb quadri-radial (Fig. 13.3) or through a G1 break in two nonhomologous chromosomes. Molecular studies have shown that in monocentric isochromosomes for Xq and 21q, the two arms may be homozygous or heterozygous (Lorda-Sanchez et al., 1991). The former may arise by misdivision of the centromere and the latter through segregation in an adjacent quadriradial between two homologous chromosomes. A quadriradial may segregate in various ways. However, common to all of them is that the daughter cells are different from each other, whereas descendants of a cell in which a G1 aberration has taken place are identical. A number of human mosaics display cell lines with different chromosome constitutions that may owe their origin to segregation in a quadriradial (Daly et al., 1977). Good examples of such mosaics are provided by persons having one cell line with an isodicentric t(X;X) chromosome and another with a 45,X karyotype.

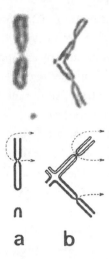

n

a   b

Figure 13.4. Possible origins of isodicentric chromosomes: (a) Isochromatid break and rejoining of the broken chromatids may result in a dicentric. (b) Segregation of an adjacent quadriradial is probably the main mechanism creating isodicentrics.

The simplest explanation for such mosaics, segregation in a quadriradial, has been largely ignored.

Misdivision of the Centromere: Centric Fusion and Centric Fission

Another mechanism (rare) that can lead to isochromosome formation is misdivision of the centromere. This involves transverse fission of centromeric elements rather than the usual longitudinal separation. It segregates the two chromosome arms instead of the two sister chromatids. This is possible because the centromere is a repetitive structure whose subunits are capable of centromeric function (Zinkowski et al., 1991; Chapter 4). Misdivision may occur during either mitosis or meiosis and can lead to isochromosomes by centric fusion. If misdivision takes place between S phase and anaphase, when the chromosomes have already been duplicated, the result is two isochromosomes. The most frequently seen isochromosome involves the long arm of the X chromosome, i(Xq); i(Xp) is not seen, presumably because 46,X,i(Xp) is not viable. Xp has no inactivation center, and trisomy for an active Xp is lethal (Chapter 19).

Most isochromosomes arise by translocation rather than misdivision (Chapter 14). Meiotic misdivision is ruled out because it would produce 46, i(Xq),i(Xp), and this is never seen. If misdivision of an unduplicated chromosome were to occur between anaphase and S phase, two telocentric chromosomes would be produced, a process called centric fission. Janke (1982) observed centric fission of chromosome 7 in three generations of one family. Reports of two telocentrics are rare, involving only chromosomes 4, 7, and 10 (Therman et al., 1981; Rivera and Cantu, 1986). Individuals with two cell lines, one with a telocentric chromosome 13, 21, or X and the other with an isochromosome 13q, 21q, or Xq, have also been described (Therman et al., 1981). In these cases, either the telocentric chromosome or the isochromosome arose through misdivision of a normal chromosome, and the other through further misdivision of the resulting abnormal chromosome. Once a centromere misdivides, it often continues to be unstable.

Pericentric and Paracentric Inversions

Pericentric inversions (Fig. 13.2e) have been described for all chromosomes except number 20, but with quite different frequencies; for instance, the C-band heteromorphism, inv(9), comprises nearly 40% of all pericentric inversions, and inversions of chromosome 7 comprise about 20%. Breakpoints are also nonrandom: for example, those in bands 2p13, 2q21, 5q31, 6q21, 10q22, and 12q13 are seen most often (Kleczkowska et al., 1987). Some pericentric inversions are obvious because of a change in the position of the centromere or the banding pattern, but many are not. Bailey et al. (1996) developed a method capable of detecting any pericentric inversion, using FISH with strand-specific probes for centromeric and telomeric repeats (Fig. 13.5) Microscopically observable paracentric inversions are far less common than pericentric inversions, although they have involved almost every chromosome. Submicroscopic inversions are relatively common, especially the very short ones, and are a major cause of disease (Chapter 16).

The most frequent cytologically visible paracentric inversions have breakpoints in chromosome arms 11q, 7q, and 3p. Of the 184 inversions cited by Madan (1995), 38 had a breakpoint in 11q and in 31 of these it was in 11q21–q23. An additional 24 had a breakpoint in 7q, although these were not so tightly clustered. The rarity of microscopically observed paracentric inversions is probably due to the difficulty of detecting them, since they do not

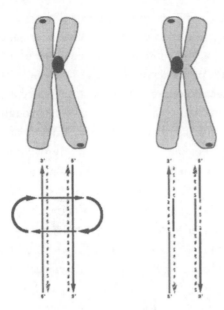

Figure 13.5. COD-FISH, using strand-specific telomeric and centromeric probes that hybridize to only one of the two chromatids. The chromatid to which the centromeric marker hybridizes is switched by a pericentric inversion, as explained in the DNA diagram below each chromosome. (Left) Normal arrangement. (Right) After inversion (modified by Julianne Meyne, from Bailey et al., 1996, reproduced with permission of S. Karger AG, Basel).

change the arm ratio and only very rarely lead to abnormal offspring. The reason for this is that crossing over in a paracentric inversion results in a dicentric and an acentric chromosome that usually segregate to different gametes and produce inviable zygotes. Any family in which a presumed paracentric inversion carrier has produced chromosomally abnormal progeny is generally segregating an intrachromosomal insertion instead; this carries a 15% risk of chromosomally unbalanced progeny (Allderdice et al., 1983; Madan and Menko, 1992).

Reciprocal Translocations

Many different reciprocal translocations (Fig. 13.2h) have been identified and characterized by chromosome banding techniques and, increasingly, by molecular methods, including FISH analysis with chromosome paints (Chapter 8). Reciprocal translocations involving all human chromosome arms have been observed (Borgaonkar, 1997). Translocations usually come to attention (are ascertained) through infertility, a phenotypically abnormal child, or by chance, as in population surveys. The inferred spontaneous rate of interchanges is at least 1–10 per 1000 gametes per generation. The population frequency is so low that matings between heterozygous carriers almost never occur.

The breakpoints in translocations ascertained through a chromosomally unbalanced individual are distributed differently from those detected by chance, because the former reflect the differential viability of individuals who are partially trisomic or monosomic for various chromosome regions. When reciprocal translocations are ascertained by chance, the breaks are distributed at random (at a cytological level; but see Chapter 14). When they are ascertained through chromosomally unbalanced descendants, the breakpoints are most frequent in some parts of the genome (for example, telomeres), with 65% located in R-positive bands, especially those that contain a fragile site and those that are rich in CpG islands and poor in *Alu* repeats. Breakpoint distribution depends to some extent on the agent that induced the chromosome breakage (Cohen et al., 1996).

Translocations involving no more than 2–5 Mb of DNA often can be detected by high-resolution banding methods, but determining the origin of such small translocated segments usually requires molecular cytogenetic techniques. Chromosomal microdissection of the affected region, PCR amplification of the microdissected DNA, and its use as a painting probe for FISH can frequently pinpoint the origin of the rearranged or extra material (Stone et al., 1996). Comparative genomic hybridization (CGH) can detect gains or losses of an extremely short chromosome segment (Chapter 8). For example, CGH, FISH, and genotyping with molecular markers were combined to recognize a complex abnormal karyotype, 46,XY,−13,+der(13)t(6;13)(q23;q34)de novo mat, in an infant with congenital malformations (Erdel et al., 1997).

Analysis of polymorphic molecular markers can clarify the origin of reciprocal translocations and their unbalanced segregation products. A newborn female with multiple anomalies had the karyotype 46,XX,der(18)(18pter–18q23::13q14.3–13qter), with a duplication of 13q14–qter and a deficiency of 18q23–qter. Both parents had normal karyotypes, but could one of them be a germline (gonosomal) mosaic, with an increased risk of further abnormal children? Molecular genotyping showed that the der(18) chromosome was of combined maternal and paternal origin, indicating that the translocation occurred during early embryogenesis in the proband, with the subsequent loss of the der(13) chromosome and duplication of the normal 13 (Eggerman et al., 1997).

Robertsonian Translocations (RTs)

Whole-arm transfers constitute a special class of reciprocal translocations. They almost always involve two acrocentric chromosomes (*Robertsonian translocations*, or RTs). Whole-arm transfers between nonacrocentric chromosomes are extremely rare. RTs are the most commonly observed chromosome aberrations, with a frequency of nearly 4 per 10,000 gametes per generation. This reflects their virtual lack of phenotypic effect and consequent high rate of retention; for example, 85–95% of DqDq translocations are familial (Nielsen and Rasmussen, 1976). Nearly 75% of RTs are 13q14q, and 10% are 14q21q. The eight other possible types involving acrocentric chromosomes account for the rest, with 21q21q the most common. Nucleolar fusion and satellite association can bring acrocentric chromosomes together but play no role in the nonrandom participation of chromosomes in RTs (Therman et al., 1989).

Robertsonian translocations were originally thought to arise by fusion of two chromosomes that had each broken at the centromere. However, banding and molecular studies have shown that this is rarely the case. Most Robertsonian translocations are dicentric but contain no rRNA genes, so the breakpoints are in the short arm, proximal to the NOR (Fig. 13.6 left). FISH analysis enabled

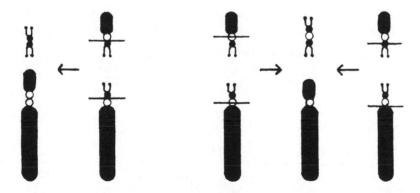

Figure 13.6. Two mechanisms of Robertsonian translocation formation between a G- and a D-group chromosome. (left) (common): Breaks in the short arms produce a dicentric chromosome (bottom) and an acentric fragment (top). (right) (rare): Breaks through the centromeres (open circles) or one break in a short arm and one in a long arm produce two monocentric chromosomes (center).

Han et al. (1994) to narrow the region of the breakpoints in 17 rob(13q14q) translocations to the satellite III region of 14p11 and adjacent to the retained satellite I region of 13p11. Page and Schaffer (1997) isolated rob(13q14q) and rob(14q21q) translocations in human-rodent somatic cell hybrids to facilitate molecular genotyping. They showed that both of the contributing chromosomes are almost always from the same parent, that 90% arise in female meiosis, and that the breakpoints are highly consistent. In contrast, analysis of genetic markers has demonstrated a different (somatic) origin of most isochromosomes and RTs between homologous chromosomes, such as t(14q;14q) and t(21q;21q). The isochromosomes are homozygous at all loci and associated with uniparental disomy (Chapter 21), suggesting a postmeiotic origin. The RTs involving homologous chromosomes contain both a maternal and a paternal contribution and can have arisen only postmeiotically, during early mitotic divisions (Robinson et al., 1994).

The rare formation of a Robertsonian translocation by centric fusion should lead to reciprocal chromosomes, one with the short arms from two acrocentrics plus a centromere (Fig. 13.5). Abeliovich et al. (1985) reported a family in which the reciprocal RT chromosomes were present in both a balanced carrier mother and her daughter with Down syndrome. Such cases are very rare, because most RTs arise from breaks in the short arms, so only the long-arm RT has a centromere. In fact, most RTs have two centromeres, but this does not lead to anaphase bridges and breaks. Is the mitotic and meiotic stability of these chromosomes due to inactivation of one of the two centromeres, or are they so close to each other that there is never a twist between them? Studies with the centromeric proteins C and E (CENP-C and CENP-E), markers for active centromeres, support both explanations.

Intrachromosomal and Interchromosomal Insertions

Insertions require three breaks. Consequently, they are considerably rarer (about one in 5000 live births) than two-break abnormalities (one in 500 live births). An insertion may occur either within one chromosome (intrachromosomal), by a shift in position of a chromosome segment, or between two chromosomes (interchromosomal), by translocation of a segment of one chromosome to an interstitial location in the other. The segment may be inserted either in the same orientation as its original one or in an inverted orientation. If the segment stays in the same arm of a chromosome, it is called a paracentric insertion; if it is inserted into the other arm of the same chromosome, it is a pericentric insertion (Fig. 16.5). Paracentric insertions, like paracentric inversions, do not change the position of the centromere. They became detectable only with the development of precise banding methods (reviewed by Madan and Menko, 1992).

Some insertions are large enough to be detected by chromosome banding or FISH analysis. Tandemly repetitive rRNA genes are sometimes moved from the short arm of an acrocentric chromosome to an interstitial site on another chromosome, presumably as a result of having two breakpoints within a single rRNA gene cluster and one elsewhere (Guttenbach et al., 1998). Many insertions are submicroscopic, requiring molecular cytogenetic methods to detect. They usually involve a transposable element (Chapter 5). One mechanism by which this occurs is that a cDNA copy of a cytoplasmic mRNA is made, using reverse transcriptase, and is inserted back into the genome. These genes or pseudogenes are easily recognized because they have no introns; these are removed from the RNA transcript before it leaves the nucleus. Since integration occurs at random and may disrupt the gene into which the element integrates, this is one cause of mutation.

Complex and Multiple Rearrangements

The occurrence of three or more chromosome breaks can lead to complex rearrangements involving multiple chromosomes. Q-banding enabled Allderdice et al. (1971) to recognize one of the first of these, involving chromosomes 6, 14, and 20 in the mother of a mentally retarded, chromosomally unbalanced girl; without banding, the girl appeared to have a novel D-group trisomy. More than 100 such complex rearrangements, involving two to six chromosomes, have been reported (reviewed by Batanian and Eswara, 1998). A phenotypically abnormal child with five structurally aberrant chromosomes, 1, 4, 7, 12, and 15, was born to a woman who developed malignant melanoma during pregnancy but was not treated during the pregnancy (Fitzgerald et al., 1977). The same unknown agent may have been responsible for both the malignancy in the mother and the chromosome aberrations in the child, because chromosome aberrations predispose to malignancy (Chapter 26).

A specific chromosome segment can be involved in nonreciprocal rearrangements via a common breakpoint as a result of instability of the repaired site. This can lead to multiple rearrangements and complex mosaicism (Lejeune et al., 1979) or to jumping Robertsonian translocations that produce a succession of different Robertsonian translocations in an individual (Gross et al., 1996). The nomenclature of ISCN (1995) is useful for providing an unequivocal description of all these complex karyotypes.

References

Abeliovich D, Katz M, Karplus M, et al. (1985) A de novo translocation, 14q21q, with a microchromosome-14p21p. Am J Med Genet 22:29–33

Allderdice PW, Eales B, Onyett H, et al. (1983) Duplication 9q34 syndrome. Am J Hum Genet 35:1005–1019

Allderdice PW, Miller OJ, Miller DA, et al. (1971) Familial translocation involving chromosomes 6, 14, and 20 identified by quinacrine fluorescence. Humangenetik 13:205–209

Bailey SM, Meyne J, Cornforth MN, et al. (1996) A new method for detecting pericentric inversions using COD-FISH. Cytogenet Cell Genet 75:248–253

Batanian JR, Eswara MS (1998) De novo apparently balanced complex chromosome rearrangement (CCR) involving chromosomes 4, 18, and 21 in a girl with mental retardation: report and review. Am J Med Genet 78:44–51

Blumenthal AL, Allanson JE (1997) Turner syndrome in a mother and daughter: r(X) and fertility. Clin Genet 52:187–191

Borgaonkar DS (1997) Chromosomal variation in man: a catalog of chromosomal variants and anomalies, 8th edn. Wiley, New York

Cohen O, Cans C, Cuillel M, et al. (1996) Cartographic study: breakpoints in 1574 families carrying human reciprocal translocations. Hum Genet 97:659–667

Daly RF, Patau K, Therman E, et al. (1977) Structure and Barr body formation of an Xp+ chromosome with two inactivation centers. Am J Hum Genet 29:83–93

Eggerman T, Engels H, Heidrich-Kaul C, et al. (1997) Molecular investigation of the parental origin of a de novo unbalanced translocation 13/18. Hum Genet 99:521–522

Erdel M, Duba H-C, Verdorfer I, et al. (1997) Comparative genomic hybridization reveals a partial de novo trisomy 6q23–qter in an infant with congenital malformations: delineation of the phenotype. Hum Genet 99:596–601

Fitzgerald PH, Miethke P, Caseley RT (1977) Major karyotypic abnormality in a child born to a woman with untreated malignant melanoma. Clin Genet 12:155–161

Gross SJ, Tharapel AT, Phillips OP, et al. (1996) A jumping Robertsonian translocation: a molecular and cytogenetic study. Hum Genet 98:291–296

Guttenbach M, Nassar N, Feichtinger W, et al. (1998) An interstitial nucleolus organizer region in the long arm of human chromosome 7: cytogenetic characterization and familial segregation. Cytogenet Cell Genet 80:104–112

Han J-Y, Choo KHA, Shaffer LG (1994) Molecular cytogenetic characterization of 17 rob(13q14q) Robertsonian translocations by FISH, narrowing the region containing the breakpoints. Am J Hum Genet 55:960–967

ISCN (1995) An international system for human cytogenetic nomenclature. Mitelman F (ed) Karger, Basel

Jalal SM, Kukolich MK, Garcia M, et al. (1991) Tetrasomy 9p: an emerging syndrome. Clin Genet 39:60–64

Janke D (1982) Centric fission of chromosome No. 7 in three generations. Hum Genet 60:200–201

Kleczkowska A, Fryns JP, Van den Berghe H (1987) Pericentric inversions in man: personal experience and review of the literature. Hum Genet 75:333–338

Lejeune J, Manoury C, Prieur M, et al. (1979) Translocation sauteuse (5p;15q), (5q;15q), (12q;15q). Ann Génét 22:210–213

Lorda-Sanchez I, Binkert F, Maechler M, et al. (1991) A molecular study of X isochromosomes: parental origin, centromeric structure, and mechanisms of formation. Am J Hum Genet 49:1034–1040

Madan K (1995) Paracentric inversions: a review. Hum Genet 96:503–515

Madan K, Menko FH (1992) Intrachromosomal insertions: a case report and a review. Hum Genet 89:1–9

Melnyk AR, Ahmed I, Taylor JC (1995) Prenatal diagnosis of familial ring 21 chromosome. Prenatal Diagn 15:269–273

Meltzer PS, Guan X-Y, Trent JM (1993) Telomere capture stabilizes chromosome breakage. Nat Genet 4:252–255

Niebuhr E (1978) Cytologic observations in 35 individuals with a 5p– karyotype. Hum Genet 42:143–156

Nielsen J, Rasmussen K (1976) Autosomal reciprocal translocations and 13/14 translocations: a population study. Clin Genet 10:161–177

Page SL, Shaffer LG (1997) Nonhomologous Robertsonian translocations form predominantly during female meiosis. Nat Genet 15:231–232

Rivera H, Cantu JM (1986) Centric fission consequences in man. Ann Génét 29:223–225

Robinson WP, Bernasconi F, Basaran S, et al. (1994) A somatic origin of homologous Robertsonian translocations and isochromosomes. Am J Hum Genet 54:290–302

Schreck RR, Breg WR, Erlanger BF, et al. (1977) Preferential derivation of abnormal human G-group-like chromosomes from chromosome 15. Hum Genet 36:1–12

Silverman GA, Schneider SS, Massa HF, et al. (1995) The 18q– syndrome: analysis of chromosomes by bivariate flow karyotyping and the PCR reveals a successive set of deletion breakpoints within 18q21.2–q22.2. Am J Hum Genet 56:926–937

Stone D, Ning Y, Guan X-Y, et al. (1996) Characterization of familial partial 10p trisomy by chromosomal microdissection, FISH, and microsatellite dosage analysis. Hum Genet 98:396–402

Therman E, Kuhn EM (1985) Incidence and origin of symmetric and asymmetric dicentrics in Bloom's syndrome. Cancer Genet Cytogenet 15:293–301

Therman E, Sarto GE, DeMars RI (1981) The origin of telocentric chromosomes in man: A girl with tel(Xq). Hum Genet 57:104–107

Therman E, Susman B, Denniston C (1989) The nonrandom participation of human acrocentric chromosomes in Robertsonian translocations. Ann Hum Genet 53:49–65

Vermeesch JR, Falzetti D, van Buggenhout G, et al. (1998) Chromosome healing of constitutional chromosome deletions studied by microdissection. Cytogenet Cell Genet 81:68–72

Viersbach R, Engels H, Gamerdinger U, et al. (1998) Delineation of supernumerary marker chromosomes in 38 patients. Am J Med Genet 76:351–358

Wandstrat AG, Leana-Cox J, Jenkins L, et al. (1998) Molecular cytogenetic evidence for a common breakpoint in the largest inverted duplications of chromosome 15. Am J Hum Genet 62:925–936

Wilkie AOM, Lamb J, Harris PC, et al. (1990) A truncated human chromosome 16 associated with α-thalassemia is stabilized by addition of telomeric repeat (TTAGGG)n. Nature 346:868–871

Wisniewski L, Politis GD, Higgins JV (1978) Partial tetrasomy 9 in a liveborn infant. Clin Genet 14:147–153

Zinkowski RP, Meyne J, Brinkley BR (1991) The centromere-kinetochore complex: a repeat subunit model. J Cell Biol 113:1091–1110

14

The Causes of
Structural Aberrations

Ionizing radiation and many other exogenous or endogenous chromosome-breaking agents (*clastogens*) produce double-strand breaks (DSBs) in DNA. A DSB may be repaired, reuniting at its original breakpoint, or it may interact with another DSB to produce a chromosome rearrangement. According to the classical model of DSB resolution, the two broken ends of an unrepaired DSB separate from each other, whereas the exchange model postulates that the chromatin of the two broken ends is held together. Lucas and Sachs (1993) used three-color chromosome painting with FISH to examine interchanges involving chromosomes 1, 2, and 4. They found an excess of three-color triplets representing broken and rejoined chromosomes 1, 2, and 4, which is inconsistent with the exchange model and favors the classical model.

The sites of DSBs are referred to as "sticky ends" because of their tendency to rejoin with other broken ends. This is achieved by either of two mechanisms;

both involve repair replication, which requires a DNA template. Meiotic repair (the basis of genetic recombination) uses a segment of the homologous chromosome as a template. Mitotic repair in G2 can use a segment of the sister chromatid as a template, which can lead to sister chromatid exchanges. The genetic makeup of a cell is not altered if a sister chromatid exchange (SCE) takes place at identical sites on the sister chromatids. However, if the breakpoints in the two chromatids are different, the resultant unequal (ectopic) SCE leads to a duplication of the intervening segment on one chromatid and its deletion from the other chromatid. At other stages of the cell cycle, mitotic repair can sometimes use an interspersed repeat as the source of a homologous sequence.

When there is no homologous sequence nearby, repair is achieved using short (1–5 bp) complementary sequences near the nonhomologous ends as a starting template. Using a new experimental approach, Liang et al. (1998) showed that DSBs are repaired 30–50% of the time by homologous processes (Fig. 14.1) and 50–70% of the time by nonhomologous processes. The latter generally result in small deletions, insertions, or larger rearrangements at the break site. The Ku70 and Ku86 proteins are required for nonhomologous DSB repair. They form a complex with DNA protein kinase, which is activated by binding to broken ends. This provides the signal that DNA damage has occurred, and initiates DSB repair by recruiting the additional proteins required for end joining.

The migration to sites of damage by some of these proteins, such as MRE11 and RAD50, has been observed by immunofluorescence with labeled antibodies to the proteins (Nelms et al., 1998). MRE11 has endonuclease activity (cutting at sites along the DNA) and also exonuclease activity (destroying DNA from a broken end). The latter is increased when MRE11 is in a complex with RAD50. These activities promote the joining of noncomplementary ends by making it possible to use 1 to 5-bp complementary sequences near broken ends (Paull and Gellert, 1998). Two other proteins that are an essential part of the DSB repair complex are nibrin (p95) and BRCA1 (Zhong et al., 1999). Nibrin is the product of the *Nijmegen breakage syndrome* (*NBS1*) gene (Chapter 24), and BRCA1 is the product of the *BRCA1* tumor suppressor gene (Chapter 28). XRCC2, a RAD51 homologue, is essential for efficient repair of DSBs by homologous, but not nonhomologous, recombination (Johnson et al., 1999).

Chromosomes may break at any stage of the cell cycle: Gl, S, G2, mitosis, or meiosis. Different cell types and stages show different responses to chromosome-breaking agents. Most mutagens induce both chromosome breaks and gene mutations, changes that can lead to cancer (Chapter 26). Patients with ataxia telangiectasia are especially sensitive to the effects of ionizing

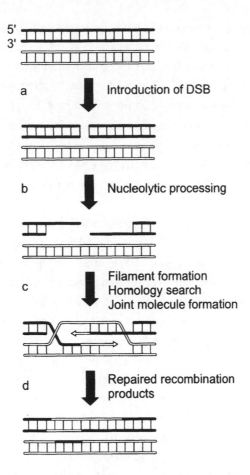

5'
3'

a   Introduction of DSB

b   Nucleolytic processing

c   Filament formation
      Homology search
      Joint molecule formation

d   Repaired recombination
      products

Figure 14.1. Double-strand break (DSB) repair model (modified and reprinted with permission from Kanaar and Hoeijmakers, Nature 391, p 335, copyright 1998, Macmillan Magazines Limited).

radiation and radiomimetic drugs (Chapter 24). This is important because the usual anticancer doses of these agents are lethal for them. Ionizing radiation and anticancer drugs can also cause secondary malignancies in any patient. Various aspects of chromosome breakage are reviewed in Mendelsohn and Albertini (1990).

Chromosome damage has been studied in various ways, such as by scoring dicentric bridges and acentric fragments in anaphase or by counting the micronuclei formed by damaged or lagging chromosomes (Prosser et al., 1988). The induction of breaks, and their healing, can be followed through interphase by fusing the clastogen-treated cells with untreated cells blocked in metaphase; this induces prematurely condensed chromosomes (Chapter 23; Hittelman, 1990). An alternative approach allows the determination of chromosome abnormalities in both metaphase and interphase: chromosome painting by FISH with biotinylated chromosome-specific probes (Cremer et al., 1988). One can enhance the sensitivity of clastogen detection, and also get some idea of the effectiveness of DNA repair, by suppressing the repair process with high doses of caffeine (Puck et al., 1993). Alternatively, one can use cultured cells that have genetic defects in specific signaling and repair pathways (Wright et al., 1998; Chapter 24).

Two-break aberrations, such as quadriradials, dicentrics, and ring chromosomes, are easier to distinguish than one-break aberrations but are much rarer. Many studies on clastogens have used the more sensitive system of SCEs. Since bromodeoxyuridine (BrdU) itself induces SCEs, Pinkel et al. (1985) developed a method for measuring SCEs at very low levels of BrdU substitution in DNA, using a monoclonal antibody to BrdU. Alternatively, one can use a range of doses of BrdU, plot the SCE frequency against dosage, and extrapolate back to zero dosage to determine the spontaneous rate of SCEs or the rate due to particular exogenous agents. The spontaneous rate is about 2–5 per cell per two generations (the time required for the BrdU technique). In Bloom syndrome, the rate is 10–20 times as high (Chapter 24). The highest frequencies of SCEs are produced by bifunctional alkylating agents, such as methylmethane sulfonate, mitomycin C, and nitrogen mustards (Latt, 1981). UV light is also a powerful inducer, but ionizing radiation is not. A dose of X-rays that increases chromosome breakage 20-fold only doubles SCE frequency (Kato, 1977). SCEs frequently take place at common fragile sites (Glover and Stein, 1987; Chapter 20).

Exogenous Causes of Structural Aberrations

X-rays, γ-rays, α-particles, and other forms of ionizing radiation produce oxidants, such as superoxide, hydrogen peroxide, and the hydroxyl radical. These are powerful clastogens, while ultraviolet light is a much weaker clastogen. The effect of a given dose of radiation depends upon whether it is applied within a short time span (for example, an atomic bomb) or over a longer period, which allows antioxidants to neutralize the oxidants more effectively and allows more time for DNA repair. Atomic bomb survivors of Hiroshima and Nagasaki showed a significantly higher incidence of chromosome abnormalities in their lymphocytes than did nonexposed controls. Those who received heavy doses of radiation still exhibited structural abnormalities almost 30 years later (Awa, 1974).

Chromosome breaks can also be induced by radiomimetic and other chemicals, including alkylating agents, purine and pyrimidine analogs, alkyl epoxides, aromatic amines, nitroso compounds, and heavy metals. Most chemicals induce breaks of the G2 type (present in only one chromatid), but they have to be present during the preceding S period. Some virus infections induce chromosome damage, which can vary from single chromosome and chromatid breaks to multiple rearrangements or total pulverization of the chromosome complement. This is found especially in lymphocytes of persons with an acute viral infection (Nichols, 1983). The "rogue" cells seen in a tiny fraction of cultured lymphocytes from different populations around the world may be another example of this. The occurrence of these cells is highly correlated with previous, and probably recent, exposure to the ubiquitous JC human polyoma virus (Neel et al., 1996).

The nature of the DNA damage, or lesion, depends on the clastogen involved. Ionizing radiation and bleomycin cause DNA strand breakage. Short-wave UV induces pyrimidine dimers. Alkylating agents induce base alkylation. Bifunctional alkylating agents, such as psoralen, or long-wave UV induces inter- and intrastrand crosslinks, and acriflavine, proflavin, or similar molecules intercalate (are inserted between base pairs) in the double helix. The lesions may undergo repair or misrepair by a wide range of DNA repair systems (Friedberg et al., 1995). The frequency of visible breaks decreases through G2, indicating that the majority of breaks that have not led to cell death eventually heal (Hittelman, 1990). Actinomycin D and cytosine arabinoside mimic the effect of adenovirus 12, which induces breaks at specific sites: the *RNU1*, *RNU2*, *PSU1*, and *RNS5* genes (Yu et al., 1998). Other fragile sites, hotspots for chromosome breakage, are discussed in Chapter 20. Ionizing radiation and most clastogenic chemicals do not increase the incidence of Robertsonian translocations. Some exogenous clastogens produce a highly nonrandom distribution of chromosome breaks, and the hotspots induced by different agents are not the same (Koskull and Aula, 1977). Chromosome breaks from any cause are usually seen in the gene-rich, GC-rich bands, but even in these regions they are unevenly distributed, with various sites emerging as hotspots. Certain substances, such as mitomycin C, preferentially cause whole-arm interchanges (Hsu et al., 1978).

Spontaneous breaks and rearrangements of indeterminate origin occur infrequently in almost every cell culture and person, especially older people. The incidence of inducible structural changes also increases with age. The additive effects of known breaking agents, such as background cosmic rays and medical or occupational radiation, drugs, viral infections, or even high fevers, probably account for some of the increase in chromosome breakage at older ages. Progressive telomere shortening with age may also play a role (Chapter 4). Clastogenic oxidants are produced in great numbers during normal aerobic respiration and even more during phagocytosis. Most of these are destroyed by the various antioxidants found especially in fruit and vegetables, including vitamins C and E, but those remaining are a potential source of "spontaneous" chromosome breakage (Ames, 1989).

Endogenous Causes of Structural Aberrations

Genetic factors are a major cause of chromosome breaks. This is seen most clearly in the rare autosomal recessive chromosome breakage syndromes with their defective DNA repair enzymes (Chapter 24) and in the much more common somatic mutations in various mismatch repair genes, an important cause of tumor progression (Chapter 26). Inherent features of the genome, such as transposable elements (Chapter 7), interspersed repeats (Chapter 7), gene (including pseudogene) (Chapter 30) duplications, and fragile sites (Chapter 20), also predispose to structural aberrations. So does telomere shortening, from whatever cause. In the Thiberge-Weissenbach syndrome of scleroderma and telangiectasia, unbroken chromosome ends tend to unite with each other, leading to the formation of chains and rings, sometimes involving all the chromosomes (Dutrillaux et al., 1977). Could this be due to premature shutdown of the telomerase gene, which is normally active throughout early embryogenesis, leading to widespread premature telomere shortening and loss? Similar unstable terminal attachments of chromosomes have been described in cells of patients with ataxia telangiectasia (Chapter 24) and in some cancer cells (Fitzgerald and Morris, 1984; Chapter 26).

Transposable Elements and Other Interspersed Repeats

The presence of transposable elements in the genome is a major predisposing factor for structural aberrations, because they encode the enzymatic machinery (endonuclease and reverse transcriptase) needed for their own insertion. The inserted sequence may be stably integrated, becoming a permanent part of the genome. However, insertion of a DNA segment into a chromosome produces a region of nonhomology, whose presence may trigger a repair process that eliminates the insert or leads to chromosome breakage, a common occurrence with transposable elements. This may be how the introduction of telomeric TTAGGG repeats into an interstitial location destabilizes the site. Such introduction has been used for targeted disruption of a gene and even for the construction of an X-derived minichromosome less than 10 Mb in size (Farr et al., 1995).

Over millennia, hundreds of thousands of copies of a few short pieces of DNA (short and long interspersed elements, or SINEs and LINEs) have, by retrotransposition, been integrated into the genome and become fixed, that is, present at a specific site of integration on virtually all copies of the chromosome. The 300-bp *Alu* repeats make up 4–8% of the genome and occur, on average, about every 4 kb, with even closer spacing in the gene-rich R-bands. The longer LINE elements make up about 15% of the genome but are less common in the gene-rich regions than in the gene-poor ones. Most LINE elements are truncated and immobile. However, complete LINE-1 elements can still retrotranspose and cause insertion mutagenesis by disrupting a gene. Several examples involve the inactivation of tumor suppressor genes (Chapter 28), and others, to be described, have interesting features.

The extreme prevalence of interspersed repeated DNA sequences in the genome is a major cause of some structural aberrations, such as deletions, duplications, and inversions, because they predispose to unequal (ectopic or illegitimate) recombination (Fig. 14.2), a common cause of such events. These are usually interchromosomal, as seen in the 22q11.2 deletions responsible for the CATCH22 (Di George) syndrome and most of the 7q11.23 deletions in Williams-Beuren syndrome (Baumer et al., 1998). Sometimes they are intrachromosomal, as has been shown for the 15q11–q13 deletions in Prader-Willi syndrome (Carrozzo et al., 1997) and some of the 7q11.23 deletions in Williams-

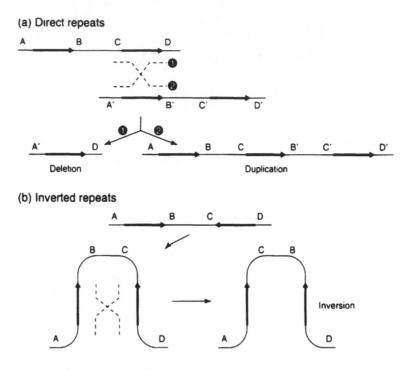

Figure 14.2. Genomic rearrangements resulting from ectopic recombination between repeated sequences (black arrows) on homologous chromosomes. Letters refer to unique sequences flanking the repeats, with A, B, C, and D on one of the homologues and A′, B′, C′, and D′ on the other (reprinted from Trends Genet vol 14, Lupski, Genome disorders: structural features of the genome can lead to DNA rearrangements and human disease traits, pp 417–422, copyright 1998, with permission of Elsevier Science).

Beuren syndrome. Mutations of the *TP53* gene can cause a 5 to 20-fold increase in the rate of spontaneous ectopic homologous recombination between intra-chromosomal direct repeats (Bertrand et al., 1997). Deletions range from just a few base pairs to many megabases in size. The very short deletions, up to about 20 bp, are usually the result of ectopic recombination between short repeats, with the frequency of deletion greater if the repeat is longer or the distance between the repeats is shorter (Krawczak and Cooper, 1991). Unequal (ectopic) homologous recombination between neighboring LINE-1 elements led to a 7.6-kb deletion in the *phosphorylase kinase subunit B* (*PHKB*) gene and to resultant glycogen storage disease (Burwinkel and Kilimann, 1998). Several copies of the *MN7* gene are present near the common breakpoints seen in most individuals with either Prader–Willi or Angelman syndrome, suggesting that ectopic recombination among multiple copies of the *MN7* gene may be responsible for the 3 to 4-Mb deletion commonly seen in these individuals (Buiting et al., 1998).

Double-strand breaks in DNA appear to initiate such ectopic homologous exchange events. This has been shown for the inversion mutation that is so common in the mucopolysaccharidosis called Hunter syndrome. DSBs also lead to the duplication that causes the Charcot-Marie-Tooth type 1A (CMT1A) neurological disorder and to the deletion of the same region that causes hereditary neuropathy with liability to pressure palsies (HNPP) (Lagerstedt et al., 1997). The most intriguing aspect of this is that almost 80% of the breakpoints involved in the 1.5-Mb *CMT1A* duplication/*HNPP* deletion at 17p11.2–p12 occur within a 1.7-kb hotspot for recombination. Reiter et al. (1996) have proposed that the hotspot is due to the initiation of DSBs at a nearby *mariner* transposon-like element by a transposase enzyme. There are about 1000 *mariner* transposon-like elements in the human genome, and numerous other transposable elements, so this may be an extremely important cause of hotspots for chromosome breakage, analogous to the chromosome breakage Barbara McClintock's classic studies in maize showed accompanied the movement of transposable elements.

Interspersed Repeats as Hotspots for Double-Strand Breaks and Rearrangements

Interstitial deletions of short regions can arise by a process called *slipped mispairing* when there is homology (close sequence similarity) between the sequences flanking the region. This generates a single-stranded loop that is recognized by a DNA repair system and excised, producing the deletion (Krawczak and Cooper, 1991). In contrast to the wide distribution of these very short deletions, longer deletions tend to be clustered in particular areas of the genome and produce characteristic clinical syndromes. This may reflect embryonic lethality of deletions in other regions of the genome, but there is growing evidence that some sites are preferentially susceptible to the breaks leading to deletions and duplications, reflecting a particular origin. Thus, most patients with the Smith-Magenis syndrome (Chapter 15) have the same genetic markers deleted, cover-

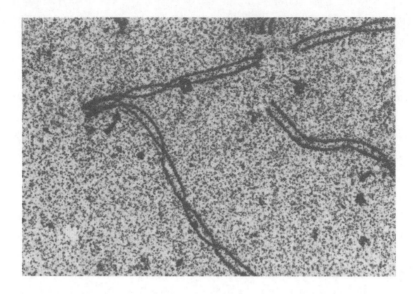

Figure 14.3. Visualization of a de novo translocation as an interchange (arrow) between two pachytene synaptonemal complexes (Speed and Chandley, 1990).

ing a 5-Mb region of 17p11.2. This region is flanked by low-copy repeats of a gene, and unequal homologous recombination between these repeats produces the common deletion (Chen et al., 1997).

Translocations and inversions also show breakpoint hotspots. Reciprocal translocations involving Xp and Yq are the result of rare recombination events between homologous sequences (Yen et al., 1991). That is also the case for some autosomal reciprocal translocations and for most Robertsonian translocations. Nearly 90% of translocations involving 11q also involve 22, with breakpoints at 11q23 and 22q11 (Fraccaro et al., 1980). This suggests that these two bands have a homologous sequence that tends to pair ectopically and lead to a translocation. Figure 14.3 shows the origin of a translocation during meiosis, perhaps generated by this mechanism. Other combinations that occur significantly more frequently than expected are t(9;22) and t(9;15).

Submicroscopic inversions are relatively common. The most abundant ones are very short, and the breakpoints are in inverted repeat sequences. These are fairly abundant throughout the genome. Inversions in the *factor 8C (hemophilia A)* gene in Xq28 provide an example. The underlying predisposing factor is the presence of a short region of homology in intron 22 of the gene and a region in the opposite orientation 500 kb closer to the Xq telomere. Intrachromosomal homologous recombination between the mispaired regions produces an inversion of the specific segment. This inversion arises only in male meiosis, presumably because the presence of the correct pairing partners on the second X chromosome in females inhibits illegitimate pairing and recombination (Rossiter et al., 1994). Hunter syndrome, an X-linked mucopolysaccharidosis caused by a deficiency of iduronate sulfatase (IDS), is frequently the result of an inversion brought about by recombination between sequences in the *IDS* gene at Xq27.3–q28 and sequences in its closely linked pseudogene, *IDS-2*. All recombination appears to take place within a 1-kb region in which the sequence identity is over 98% (Lagerstedt et al., 1997).

The common inverted duplication of chromosome 15, inv dup(15) arises by illegitimate recombination between homologues by a U-type exchange rather than the usual X-type (Schreck et al., 1977). There is a common breakpoint in most of these cases (Wandstrat et al., 1998). This hotspot for breakage may account for the high frequency of this abnormality. Inverted duplications of the short arm of various chromosomes are well known, though not numerous except in the case of those leading to tiny supernumerary chromosomes like inv dup(15). Mono- and dicentric 8p duplications are produced by the same U-type exchange mechanism in maternal meioses (Floridia et al., 1996). Molecular studies have shown that inverted duplications of the short arms of chromosomes 3, 7, 8, and 9 are often accompanied by a deletion (deficiency) distal to the duplicated segment and frequently telomeric (Jenderny et al., 1998). The underlying mechanism for these duplication/deficiencies remains to be determined. Inverted duplications can arise by aberrant segregation in the carrier of an inversion involving an acrocentric chromosome in the rare case when one of the two breaks is in the short arm (Fig. 16.5; Trunca and Opitz, 1977). As a rule, crossing over in an inversion loop leads to duplication/deficiency chromosomes (Chapter 16).

References

Ames B (1989) Endogenous DNA damage as related to cancer and aging. Mutat Res 250:3–16

Awa AA (1974) Cytogenetic and oncogenic effects of the ionizing radiations of the atomic bombs. In: German J (ed) Chromosomes and cancer. Wiley, New York, pp 637–674

Baumer A, Dutley F, Balmer D, et al. (1998) High level of unequal meiotic crossovers at the origin of the 22q11.2 and 7q11.23 deletions. Hum Mol Genet 7:887–894

Bertrand P, Rouillard D, Boulet A, et al. (1997) Increase of spontaneous intrachromosomal homologous recombination in mammalian cells expressing a mutant p53 protein. Oncogene 14:1117–1122

Buiting K, Gross S, Ji Y, et al. (1998) Expressed copies of the MN7 (D15F37) gene family map close to the common deletion breakpoints in the Prader Willi/Angelman syndromes. Cytogenet Cell Genet 81: 247–253

Burwinkel B, Kilimann MW (1998) Unequal homologous recombination between LINE-1 elements as a mutational mechanism in human genetic disease. J Mol Biol 277:513–517

Carrozzo R, Ross E, Christian SL, et al. (1997) Inter- and intrachromosomal rearrangements are both involved in the origin of 15q11–q13 deletions in Prader-Willi syndrome. Am J Hum Genet 61:228–231

Chen KS, Manian P, Koeuth T, et al. (1997) Homologous recombination of a flanking repeat gene cluster is a mechanism for a common contiguous gene syndrome. Nat Genet 17:154–163

Cremer T, Lichter P, Borden J, et al. (1988) Detection of chromosome aberrations in metaphase and interphase tumor cells by in situ hybridization using chromosome-specific library probes. Hum Genet 80:235–246

Dutrillaux B, Aurias A, Couturier J, et al. (1977) Multiple telomeric fusions and chain configurations in human somatic chromosomes. In: Chapelle A de la, Sorsa M (eds) Chromosomes Today, Vol 6. Elsevier/North Holland, Amsterdam, pp 37–44

Farr CJ, Bayne RAL, Kipling D, et al. (1995) Generation of a human X-derived minichromosome using telomere-associated chromosome fragmentation. EMBO J 14:5444–5454

Fitzgerald PH, Morris CM (1984) Telomeric association of chromosomes in B-cell lymphoid leukemia. Hum Genet 67:385–390

Floridia G, Piantanidu M, Minelli A, et al. (1996) The same molecular mechanism at the maternal meiosis produces mono- and dicentric 8p duplications. Am J Hum Genet 58:785–796

Fraccaro M, Lindsten J, Ford CE, et al. (1980) The 11q;22q translocation: a European collaborative analysis of 43 cases. Hum Genet 56:21–51

Friedberg EC, Walker GC, Siede W (1995) DNA repair and mutagenesis. ASM, Washington, DC

Glover TW, Stein CK (1987) Induction of sister chromatid exchanges at common fragile sites. Am J Hum Genet 41:882–890

Han J-Y, Choo KHA, Shaffer LG (1994) Molecular cytogenetic characterization of 17 rob(13q14q) Robertsonian translocations by FISH, narrowing the region containing the breakpoints. Am J Hum Genet 55: 960–967

Hittelman WN (1990) Direct measurement of chromosome repair by premature chromosome condensation. In: Mendelsohn ML, Albertini RJ (eds) Mutation and the environment, Part B. Wiley-Liss, New York, pp 337–346

Hsu TC, Pathak S, Basen BM, et al. (1978) Induced Robertsonian fusions and tandem translocations in mammalian cell cultures. Cytogenet Cell Genet 21:86–98

Jenderny J, Poetsch M, Hoeltzenbein M, et al. (1998) Detection of a concomitant distal deletion in an inverted duplication of chromosome 3. Is there an overall mechanism for the origin of such duplication/deficiencies? Eur J Hum Genet 6:439–444

Johnson RD, Liu N, Jasin M (1999) Mammalian XRCC2 promotes the repair of DNA double-strand breaks by homologous recombination. Nature 401:397–399

Kanaar R, Hoeijmakers JHJ (1998) Genetic recombination: from competition to collaboration. Nature 391:335–337

Kato H (1977) Spontaneous and induced sister chromatid exchanges as revealed by the BUdR-labeling method. Int Rev Cytol 49:55–97

Koskull H von, Aula P (1977) Distribution of chromosome breaks in measles, Fanconi's anemia and controls. Hereditas 87:1–10

Krawczak M, Cooper DN (1991) Gene deletions causing human disease: mechanisms of mutagenesis and the role of the local DNA sequence environment. Hum Genet 86:425–441

Lagerstedt K, Karsten SL, Carlberg B-M, et al. (1997) Double-strand breaks may initiate the inversion mutation causing the Hunter syndrome. Hum Mol Genet 6:627–633

Latt SA (1981) Sister chromatid exchange formation. Annu Rev Genet 15:11–55

Liang F, Han M, Romanienko PJ, et al. (1998) Homology-directed repair is a major double-strand break pathway in mammalian cells. Proc Natl Acad Sci USA 95:5172–5177

Lucas JN, Sachs RK (1993) Using three-colour chromosome painting to test chromosome aberration models. Proc Natl Acad Sci USA 90:1484–1487

Lupski UR (1998) Genomic disorders: structural features of the genome can lead to DNA rearrangements and human disease. Trends Genet 14:417–422

Mendelsohn ML, Albertini RJ (eds) (1990) Mutation and the environment, Part B. Wiley-Liss, New York

Neel JV, Major EO, Awa AA, et al. (1996) Hypothesis: "rogue-cell-" type chromosomal damage in lymphocytes is associated with infection with the JC human polyoma virus and has implications for carcinogenesis. Proc Natl Acad Sci USA 93:2690–2695

Nelms BE, Maser RS, Mackay JF, et al. (1998) In situ visualization of DNA double-strand break repair in human fibroblasts. Science 280:590–592

Nichols WW (1983) Viral interactions with the mammalian genome relevant to neoplasia. In: German J (ed) Chromosome mutation and neoplasia. Liss, New York, pp 317–332

Page SL, Shaffer LG (1997) Nonhomologous Robertsonian translocations form predominantly during female meiosis. Nat Genet 15:231–232

Paull T, Gellert M (1998) The 3′ to 5′ exonuclease activity of Mre11 facilitates repair of DNA double-strand breaks. Mol Cell 1:969–979

Pinkel D, Thompson LH, Gray JW, et al. (1985) Measurement of sister chromatid exchanges at very low bromodeoxyuridine substitution levels using a monoclonal antibody in Chinese hamster ovary cells. Cancer Res 45:5795–5798

Prosser JS, Moquet JE, Lloyd DC, et al. (1988) Radiation induction of micronuclei in human lymphocytes. Mutat Res 199:37–45

Puck TT, Morse H, Johnson R, et al. (1993) Caffeine-enhanced measurement of mutagenesis by low levels of γ-irradiation in human lymphocytes. Somat Cell Mol Genet 19:423–429

Reiter LT, Murakami T, Koeuth T, et al. (1996) A recombination hotspot responsible for two inherited peripheral neuropathies is located near a *mariner* transposon-like element. Nat Genet 12:288–297

Rossiter JP, Young M, Kimberland ML, et al. (1994) Factor VIII gene inversions causing severe hemophilia A originate almost exclusively in male germ cells. Hum Mol Genet 3:1035–1039

Schreck RR, Breg WR, Erlanger BF, et al. (1977) Preferential derivation of abnormal human G-group-like chromosomes from chromosome 15. Hum Genet 36:1–12

Speed, Chandley (1990) Prophase of meiosis in human spermatocytes. Hum Genet 84:551

Trunca C, Opitz JM (1977) Pericentric inversion of chromosome 14 and the risk of partial duplication of 14q (14q31 ~ 14qter). Am J Med Genet 1:217–228

Wandstrat AG, Leana-Cox J, Jenkins L, et al. (1998) Molecular cytogenetic evidence for a common breakpoint in the largest inverted duplications of chromosome 15. Am J Hum Genet 62:925–936

Wright JA, Keegan KS, Herendeen DR, et al. (1998) Protein kinase mutants of human ATR increase sensitivity to UV and ionizing radiation and abrogate cell cycle checkpoint control. Proc Natl Acad Sci USA 95:7445–7450

Yen PH, Tsai S-P, Wenger SL, et al. (1991) X/Y translocations resulting from recombination between homologous sequences on Xp and Yq. Proc Natl Acad Sci USA 88:8944–8948

Yu A, Bailey AD, Weiner AM (1998) Metaphase fragility of the human *RNU1* and *RNU2* loci is induced by actinomycin D through a p53-dependent pathway. Hum Mol Genet 7:609–617

Zhong Q, Chen C-F, Li S, et al. (1999) Association of BRCA1 with the hRad50-hMre11-p95 complex and the DNA damage response. Science 285:747–750

15

Syndromes Due to Autosomal Deletions and Duplications

M ost cytologically visible autosomal deletions cause multiple malformations and mental retardation. A small proportion of deletions produce recognizable phenotypes; these are called *segmental aneusomy* or *contiguous gene syndromes*. The same malformations, and even the same combination of malformations, or syndrome, may be caused by deletions involving two or more different chromosomes. Brewer et al. (1998) have worked out a chromosomal deletion map for 47 human malformations, based on the analysis of the phenotypes associated with 1753 deletions involving 89% of the 289 autosomal bands (based on the ISCN 400 band nomenclature). A series of specific malformation-associated bands were identified in this way.

Molecular analyses can refine such studies. A less-than-1-Mb region of band 13q32 has been identified whose deletion leads to major malformations, the 13q32 deletion syndrome. A constant feature is holoprosencephaly (midline cleft lip and palate and associated brain defects, such as absence of the corpus callosum and olfactory bulb). The *ZIC2* gene maps to this region, and heterozygous mutations of *ZIC2* are associated with holoprosencephaly (Brown et al., 1998). It is intriguing that holoprosencephaly is also seen in trisomy 13 syndrome, suggesting that a triple dose of *ZIC2* may have the same developmental effects as a single dose. The phenotypes associated with deletions are generally more severe than those associated with duplications of the same regions, with an average maximal tolerance of 3% of the genome for deletions and 10% for duplications (Hecht and Hecht, 1987). Ledbetter and Ballabio (1995) and Budarf and Emanuel (1997) have reviewed the clinically important segmental aneusomy syndromes.

Submicroscopic deletions are a much more common cause of diseases that show Mendelian patterns of inheritance than anyone realized before molecular cytogenetic techniques were developed. For example, 60% of the cases of X-linked Duchenne and Becker types of muscular dystrophy are due to deletions, not nucleotide substitutions (Den Dunnen et al., 1989). The breakpoints are in nonhomologous regions of the noncoding introns of the massive 2.4-Mb *DMD* gene and are thus thought to arise by unequal (ectopic) sister chromatid exchange. Isolated duplications can also arise by unequal sister chromatid exchange. This mechanism is probably involved in the short duplications that cause nearly 8% of all cases of the X-linked Duchenne and Becker type muscular dystrophies. Unequal meiotic recombination is responsible for a large proportion of the 22q11.2 deletions seen in Di George syndrome and the 7q11.23 duplications seen in Williams-Beuren syndrome, a neurodevelopmental disorder (Baumer et al., 1998). Both intra- and interchromosomal rearrangements lead to the 15q11–q13 deletions seen in Prader-Willi syndrome (Carrozzo et al., 1997).

Cri du Chat (Cat Cry) and Wolf–Hirschhorn Syndromes

The *cri du chat syndrome* (Fig. 15.1) is caused by a partial deletion of 5p. The incidence of this condition in infants is estimated at 1 in 45,000 and its frequency among the mentally retarded at 1.5 in 1000. Niebuhr (1978) reviewed 331 patients with this syndrome. One of the most consistent signs is the mewing, cat-like cry during early infancy. At least 50 phenotypic aberrations have been seen in this syndrome. In addition to the characteristic cry, the most common

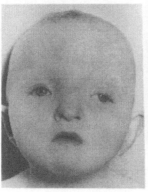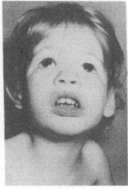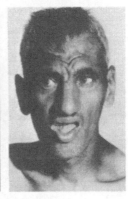

Figure 15.1. Three cri du chat patients (infant, 4 years, and 40 years), showing moon face, hypertelorism, down-slanting palpebral fissures and mouth, and low-set ears (courtesy of R Laxova).

abnormalities are severe mental retardation, hypotonia, microcephaly, hypertelorism, epicanthic folds, moon face, downward-slanting palpebral fissures, strabismus, low-set or poorly developed ears, and cardiac abnormalities.

The size of the deletion of 5p seen in the cri du chat syndrome varies from very small to about 60% of the length of the chromosome arm. The length of the deletion shows little correlation with the severity of the syndrome. The critical segment whose absence causes most features of the syndrome is 5p15.2, but the cat-like cry occurs when the deletion includes proximal 5p15.3 (Church et al., 1995). About 88% of the cases are caused by a de novo deletion (including ring 5) and about 12% are familial, most of these arising by aberrant meiotic segregation in a parent with a translocation (Fig. 15.2). In some cases, meiotic segregation in a balanced translocation carrier has produced both an individual with the cri du chat syndrome and a different abnormal phenotype, due to partial trisomy for the part of 5p that is deleted in the cri du chat sib. As a rule, the partial trisomies are much less severely affected than the partial monosomies (Malaspina et al., 1992).

Wolf–Hirschhorn syndrome is characterized by microcephaly, profound mental retardation, typical facial abnormalities, midline fusion defects, and heart defects. It is caused by a deletion involving the short arm of chromosome 4 (Fig. 15.3). The shortest region of overlap (SRO) for the deletions is only 165 kb long. The entire region has been sequenced and shown to contain a 90-kb gene, called *WHSC1*. This gene is ubiquitously expressed in early

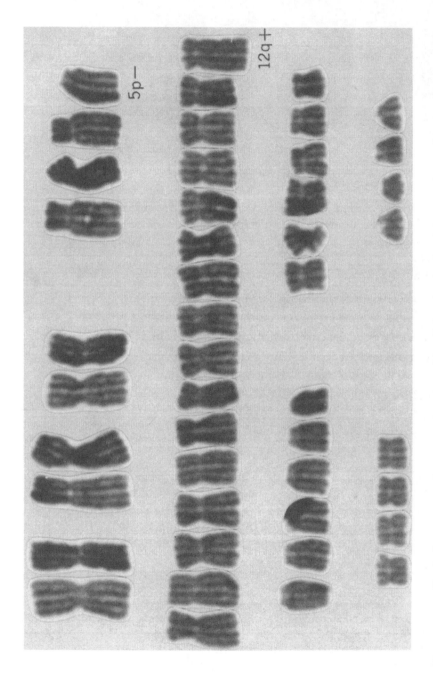

Figure 15.2. Diplochromosome karyotype (confirmed by Q-banding) of a t(5p−;12q+) translocation carrier (courtesy of E Therman).

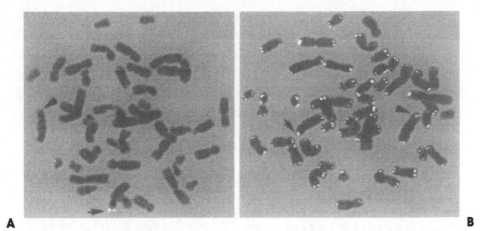

Figure 15.3. FISH studies in a patient with Wolf–Hirschhorn syndrome. (A) A microdissection probe specific for a locus near the end of a normal chromosome 4p (arrow) fails to hybridize to the 4p–chromosome (arrowhead). (B) A telomeric probe shows the presence of TTAGGG repeats at all telomeres, including the 4p– end (arrowhead) (reproduced from Vermeesch et al., 1998, with permission of S. Karger AG, Basel).

development and is homologous to a gene whose mutation produces abnormal development in *Drosophila* (Stec et al., 1998). It is unclear whether deletion of any nearby genes contributes to the phenotype.

Ring Chromosome Phenotypes

Ring chromosomes have been found for all human chromosomes (Wyandt, 1988). Since a ring involves a deletion at both ends of the chromosome, the resulting phenotype overlaps that of the deletion syndrome for each tip of the chromosome. However, the phenotypes associated with ring chromosomes are much more variable, because of the occurrence of sister chromatid exchanges in the ring. These exchanges can lead to double-sized rings and to dicentric rings that undergo bridge-breakage-fusion-bridge cycles and ultimate loss. The duplication of the segment within the ring produces partial trisomy and loss of the ring to monosomy. The severe mental retardation and multiple congenital anomalies seen in a patient whose ring 9 had only a minute deletion were probably due to the double-sized ring chromosome present in many of her cells (Inouye et al., 1979).

Most persons with a ring chromosome have both mental and physical disabilities. Despite this, a few have had children, some of whom have inherited the ring. Rings involving a fairly tiny deletion have been found in a number of normal, even fertile subjects. Such ring chromosomes may predispose to both meiotic nondisjunction and mitotic loss. Of five clinically normal women with 46,XX,r(21), two had children with Down syndrome and an extra r(21), one of them a mosaic. Both a normal 21 and the r(21) came from the mother in the only informative case (Matsubara et al., 1982).

The Critical Regions for Deletion (Segmental Aneusomy) Syndromes

The search for a critical region for the phenotypic effects of a deletion is predicated on the hypothesis that deletion of a single gene or a few closely linked genes accounts for the major phenotypic effects seen in the various deletion syndromes (Fig. 15.4; Fig. 19.2). Of course, deletion of a single gene may have the same effect as a mutation of the gene, and the discovery of individuals whose single-gene disorder is due to a deletion (or disruption by a translocation breakpoint) can be useful in mapping a gene. A recent example involves a rare disorder called Duane syndrome: severe mental retardation, strabismus, and minor limb abnormalities. In two cases, there was a deletion in chromosome 8, with the shortest region of overlap in the two cases less than 3 cM long, a big step towards identifying the gene by positional cloning (Chapter 29). One of the deletions could be demonstrated only by using a chromosome 8 radiation reduction hybrid mapping panel and FISH with a YAC probe (Calabrese et al., 1998).

The nail-patella syndrome is an autosomal dominant disorder with multiple phenotypic effects (nail dysplasia, abnormal patella, and renal dysplasia). It is due to a haploinsufficiency of the LIM homeodomain gene, $LIMX_1B$ (Dreyer et al., 1998). Although this is usually caused by gene mutation, one case was due to a balanced t(9;17)(q34.1;q25) translocation whose 9q breakpoint disrupted the gene (Duba et al., 1998). The autosomal dominant Rubinstein-Taybi syndrome (growth and mental retardation, broad thumbs, and distinctive facies) is usually caused by a mutation in the transcriptional coactivator gene, CBP (cyclic AMP response element binding protein) (Petrij et al., 1995). CBP is also required for proliferating cell nuclear antigen (PCNA) activation/DNA replication (Giles et al., 1998; see also Chapter 3). CBP maps to 16p13.3, and about one in eight

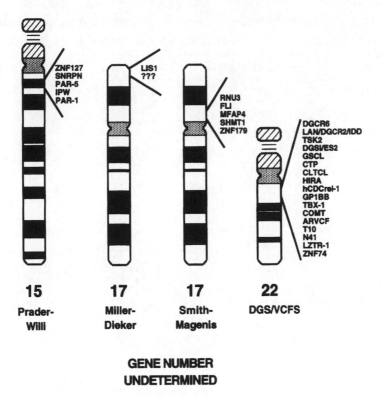

Figure 15.4. Ideograms of the chromosomes involved in four segmental aneusomy syndromes, showing each critical region and genes mapped to each. The question mark for Miller–Dieker indicates that one or more additional genes may account for phenotypic features not yet explained by defects in the LIS1 gene (reproduced from Budarf and Emanuel, Hum Mol Genet 6:1657–1665, 1997, with permission of Oxford University Press).

cases of this syndrome is caused by a submicroscopic interstitial deletion in 16p13.3 (Taine et al., 1998).

Table 15.1 summarizes information on a number of deletion, or segmental aneusomy, syndromes. Some are clearly the result of loss of a single gene, resulting in haploinsufficiency of a gene whose dosage is important or loss of the single functional copy of an imprinted gene. Other syndromes may be due to loss of multiple genes (contiguous-gene syndromes). There are a number of well-established X-linked contiguous-gene syndromes (Ledbetter and Ballabio, 1995; see also Chapter 19) and several candidates involving autosomes. A deletion of the elastin (*ELN*) gene in 7q11.23 probably accounts for the vascular and con-

Table 15.1. Deletion Syndromes Caused by Segmental Aneusomy, Leading to Haploinsufficiency of One Gene or Possibly of Several Contiguous Genes

Syndrome	Critical region	Single gene	Contiguous genes
Nail-patella*	9q34.1	LIMX₁B	No
Rubinstein–Taybi*	16p13.3	CBP	No
Smith–Magenis	17p11.2	?	?
Cri du chat	5p15	?	?
Wolf–Hirschhorn	4p16.3	WHSC₁	?
Prader–Willi	15q11–q13(pat)[†]	SNRPN	No
Angelman	15q11–q13(mat)[†]	UBE3A	No
Beckwith–Wiedemann	11p15.5(pat)[†]	IGF2	No
Miller–Dieker	17p13.3	LIS₁	?
Di George	22q11.2	UFD₁L	?

*Usually due to a point mutation

[†]Imprinted region involved (see Chapter 21)

nective tissue abnormalities in Williams syndrome but not for the mental retardation, facial abnormalities, or hypercalcemia. Deletion of the *LIM₁* kinase and the replication factor subunit 2 (*RCF2*) genes from this region may also contribute to the syndrome (Budarf and Emanuel, 1997).

Smith–Magenis syndrome (SMS), with short stature, minor facial and skeletal anomalies, and mental retardation, is caused by an interstitial microdeletion in 17p11.2, which included the anonymous DNA marker locus D17S258 in all 62 patients studied by Juyal et al. (1996). The smaller deletion in two of these narrowed the critical region considerably. The critical region contains several genes (Fig. 15.4), but their roles are unclear (Budarf and Emanuel, 1997). Cloning the breakpoint enabled Chen et al. (1997) to identify a novel junction fragment in 26 SMS patients and to show that this common segmental aneusomy is usually the result of homologous ectopic recombination involving a flanking repeat cluster. Langer–Giedion syndrome, with sparse hair, microcephaly, mental retardation, multiple cartilaginous exostoses (outgrowths), and short stature, is usually caused by a deletion that includes 8q24.1.

Three fascinating deletion syndromes are discussed in Chapter 21, because they involve an important mechanism called *genomic imprinting*. These are Beckwith–Wiedemann, Prader–Willi, and Angelman syndromes (see also Table 15.1).

The last two are particularly informative because they involve deletions of almost exactly the same critical region. That is, the shortest region of overlap (the deletion common to all cases) is only 4.3 kb and 1.5 kb long in Prader-Willi and Angelman syndromes, respectively, and these are only 5.9 kb apart. The significance of this is explained in Chapter 21.

Miller–Dieker Syndrome

Miller–Dieker syndrome is characterized by lissencephaly, severe mental retardation, seizures, heart defects, growth retardation, and distinction facial appearance. Lissencephaly is a brain malformation caused by abnormal neuronal migration that leaves the cerebral cortex with a smooth surface. There is profound mental retardation and some dysmorphic facial features. About 85% of individuals with Miller–Dieker syndrome and 15% of those with isolated lissencephaly have been shown by FISH to have a deletion in 17p13.3 that includes the D17S379 marker locus (Ledbetter and Ballabio, 1995). Positional cloning in the 350-kb critical region has identified the gene responsible, called *LIS1*. Molecular analysis showed a normal *LIS1* gene in the normal parents of a lissencephalic daughter who had a de novo t(8;17) (p11.2;p13.3) translocation. The translocation breakpoint was in intron 1, disrupting the *LIS1* gene. This confirms that lissencephaly is caused by a haploinsufficiency of the *LIS1* gene (Kurahashi et al., 1998). However, some features of the phenotype may not be explicable in terms of a defect in the *LIS1* gene (Budarf and Emanuel, 1997).

Di George/Velocardiofacial/CATCH22 Syndromes

The Di George/velocardiofacial/CATCH22 syndromes involve defective development of neural crest derivatives from the third and fourth pharyngeal pouches, with cardiac defects, abnormal facial appearance, thymic hypoplasia, cleft palate, and hypocalcemia. The cause is a deletion in 22q11. It is very common, occurring in about one in 3500 live births, making it one of the most common genetic causes of cardiac and craniofacial anomalies. Three-fourths of the deletions are maternal in origin, for unknown reasons that may be relevant to the high frequency of this deletion (Demczuk et al., 1995). The deletion of multiple genes may be required for this complex disease phenotype. The deleted region is usually over 1.5 Mb long and contains more than 20 genes (Lindsay and Baldini, 1998). Several are candidates for playing a role in causing the syndrome. The strongest candidate gene is the *UFD1L* (ubiquitination fusion degradation-like) gene, whose product is involved in degrading proteins that have been bound to ubiquitin. *UFD1L* was deleted in all 182 Di George syndrome patients studied, including one in whom the deletion was only 20 kb long but removed exons 1 to 3 of the *UFD1L* gene (Yamagishi et al., 1999). A totally different deletion, del(4)(q34.2), also produces the velocardiofacial syndrome (Tsai et al., 1999). Presumably, the two regions contain one or more genes that act in the same developmental pathway.

Critical Region for Charcot–Marie–Tooth Type 1A Duplication Syndrome

Charcot–Marie–Tooth (CMT) disease is marked by progressive weakness of the muscles of the extremities. It is the most common form of heritable peripheral neuropathy, with an incidence of one in 2500. Most familial and sporadic cases of type 1A (CMT1A), which is marked by slowed nerve conduction velocity, is usually caused by a duplication of a 1.5-Mb region of chromosome 17p11.2. This includes the *PMP22* gene, which encodes the 22-kilodalton (kDa) peripheral myelin protein. Several related patients without the duplication have a point mutation of the *PMP22* gene, supporting the view that this gene plays an important role in the disease (Roa et al., 1993). However, the *heme A: farnesyltransferase* (*COX10*) gene is disrupted during the ectopic homologous recombination between the 24-kb proximal and distal repeats that flank the region duplicated in CMT1A (Reiter et al., 1997), and this may contribute to the phenotype. It is interesting that on the many occasions when ectopic recombination between these hotspots for recombination leads to a deletion of the intervening region, a similar phenotype (HNPP) is produced: hereditary neuropathy with a tendency to pressure palsies (Lorenzetti et al., 1995).

Critical Region for Down Syndrome

Some individuals with Down syndrome (DS) do not have complete trisomy 21 but only a partial trisomy 21, or duplication, of all or part of 21q22. This has led to the search for a critical region (Fig. 29.2) containing one gene or a small number of genes whose presence in triple dose produces the major features of DS. Identification of such a gene or genes would provide clues to the pathogenesis of the disorder and methods of treatment. There have been several promising leads using this approach.

The superoxide dismutase (*SOD*) gene is located in the critical region and is overexpressed in DS. Neurons from DS brains start to differentiate normally in culture but then develop increased intracellular levels of reactive oxygen and lipid peroxidase, and they die, suggesting a defect in their ability to detoxify reactive oxygen (Busciglio and Yankner, 1995). This is due to the overdose of SOD. Transgenic mice that overexpress SOD show increased lipid peroxidation and cell death in response to lipopolysac-charides and develop bone marrow and thymus defects like those in children with DS (Peled-Kamar et al., 1995). The human *minibrain* (*MNB*) gene also maps to the DS critical region and its murine homologue is expressed in the neuronal regions of the mouse that correspond to the regions that are abnormal in DS (Guimerá et al., 1996). Mutation of the *Drosophila* homologue produces a minibrain, hence the name given to the gene. Functional screening of a 2-Mb portion of the DS critical region in human 21q22.2 was carried out by preparing transgenic mice containing low copy numbers of one of four YACs encompassing the region. Transgenic mice with either of two of the YACs had learning and memory disabilities, which were most severe with a 570-kb YAC containing the human *MNB* gene (Smith et al., 1997).

A third gene on chromosome 21 that is overexpressed in DS is *ETS2*, a transcription factor that is also a protooncogene (Chapter 27). Transgenic mice that overexpress the homologous *Ets2* gene develop skeletal abnormalities resembling those seen in human DS and in a mouse model of DS, trisomy 16 (Sumarsono et al., 1996). Thus, three different genes in the DS critical region appear to be responsible for quite different components of the complex DS phenotype. In addition, a transporter gene named *SLC5A3* has been mapped very precisely to a region of a few kilobases near the distal end of 21q22.1 (Berry et al., 1996). Clearly, a triple dose of any single gene is unlikely to produce DS, or, by extension, any other trisomy syndrome. Triplications of genes that are rather widely scattered on 21q may contribute to the DS phenotype (Korenburg et al., 1994). Nevertheless, the concept of a single critical region continues to be an extremely useful guide to research on the phenotypic effects of trisomies or partial trisomies, as well as deletions.

References

Baumer A, Dutley F, Balmer D, et al. (1998) High level of unequal meiotic crossovers at the origin of the 22q11.2 and 7q11.23 deletions. Hum Mol Genet 7:887–894

Berry GT, Mallee JJ, Blouin J-L, et al. (1996) The 21q22.1 STS marker, VNO2 (EST 00541 cDNA), is part of the 3′ sequence of the human Na+/myo-inositol co-transporter (SLC5A3) gene. Cytogenet Cell Genet 73:77–78

Brewer C, Holloway S, Zawalynski P, et al. (1998) A chromosomal deletion map of human malformations. Am J Hum Genet 63:1153–1159

Brown SA, Warburton D, Brown LY, et al. (1998) Holoprosencephaly due to mutations in ZIC2, a homologue of Drosophila odd-paired. Nat Genet 20:180–183

Budarf ML, Emanuel BS (1997) Progress in the autosomal segmental aneuploidy syndromes (SASs): single or multi-locus disorders? Hum Mol Genet 6:1657–1665

Busciglio J, Yankner BA (1995) Apoptosis and increased generation of reactive oxygen species in Down's syndrome neurons in vitro. Nature 378: 776–779

Calabrese G, Stuppia L, Morizio E, et al. (1998) Detection of an insertion deletion of region 8q13–q21.2 in a patient with Duane syndrome: implications for mapping and cloning a Duane gene. Eur J Hum Genet 6: 187–193

Carrozzo R, Ross E, Christian SL, et al. (1997) Inter- and intrachromosomal rearrangements are both involved in the origin of 15q11–q13 deletions in Prader–Willi syndrome. Am J Hum Genet 61:228–231

Chen KS, Manian P, Koeth T, et al. (1997) Homologous recombination of a flanking repeat cluster is a mechanism for a common contiguous gene syndrome. Nat Genet 17:154–163

Church DM, Bengtsson U, Nielsen KV, et al. (1995) Molecular definition of deletions of different segments of distal 5p that result in distinct clinical features. Am J Hum Genet 56:1162–1172

Demczuk S, Lévy A, Aubry M, et al. (1995) Excess of deletions of maternal origin in the Di George/velo-cardio-facial syndromes. A study of 22 new patients and a review of the literature. Hum Genet 96:9–13

Den Dunnen JT, Grootscholten PM, Bakker E, et al. (1989) Topography of the Duchenne muscular dystrophy (DMD) gene: FIGE and cDNA analysis of 194 cases reveals 115 deletions and 13 duplications. Am J Hum Genet 45:848–854

Dreyer SD, Zhou G, Baldini A, et al. (1998) Mutations in LIMX1B cause abnormal skeletal patterning and renal dysplasia in nail-patella syndrome. Nat Genet 19:47–50

Duba HC, Erdel M, Löffler J, et al. (1998) Nail patella syndrome in a cytogenetically balanced t(9;17)(q34.1;q25). Eur J Hum Genet 6:75–79

Epstein CJ (1995) Down syndrome (trisomy 21). In: Scriver CR, Beaudet AL, Sly WS, Valle D (eds), The metabolic and molecular bases of inherited disease, McGraw-Hill, New York, 7th edn, pp 749–794

Giles RH, Peters DJM, Breuning MH (1998) Conjunction dysfunction: CBP/p300 in human disease. Trends Genet 14:178–183

Guimerá J, Casas J, Pucharcòs C, et al. (1996) A human homologue of Drosophila minibrain (MNB) is expressed in the neuronal regions affected in Down syndrome and maps to the critical region. Hum Mol Genet 5: 1305–1310

Hecht F, Hecht BK (1987) Aneuploidy in humans: dimensions, demography and dangers of abnormal numbes of chromosomes. In: Vig BK, Sandberg AA (eds) Aneuploidy, Part A: incidence and etiology. Liss, New York, pp 9–49

Inouye T, Matsuda H, Shimura K (1979) A ring chromosome 9 in an infant with malformations. Hum Genet 50:231–235

Juyal RC, Figueroa LE, Hauge X, et al. (1996) Molecular analyses of 17p11.2 deletions in 62 Smith-Magenis syndrome patients. Am J Hum Genet 58:998–1007

Korenberg JR, Chen X-N, Schipper R, et al. (1994) Down syndrome phenotypes: the consequences of chromosomal imbalance. Proc Natl Acad Sci USA 91:4997–5001

Kurahashi H, Sakamoto M, Ono J, et al. (1998) Molecular cloning of the chromosomal breakpoint in the *LIS1* gene of a patient with isolated lissencephaly and balanced t(8;17). Hum Genet 103:189–192

Ledbetter DH, Ballabio A (1995) Molecular cytogenetics of contiguous gene syndromes: mechanisms and consequences of gene dosage imbalance. In: Scriver C, Beaudet AL, Sly WS, Valle D (eds) The metabolic and molecular basis of inherited disease, 7th ed, McGraw-Hill, New York, pp 811–839

Lindsay EA, Baldini A (1998) Congenital heart defects and 22q11 deletions: which genes count? Mol Med Today 4:350–358

Lorenzetti D, Pareyson D, Sghirlanzoni A, et al. (1995) A 1.5 Mb deletion in 17p11.2–p12 is frequently observed in Italian families with hereditary neuropathy with liability to pressure palsies. Am J Hum Genet 56:91–98

Malaspina D, Warburton D, Amador X, et al. (1992) Association of schizophrenia and partial trisomy of chromosome 5p. Schizophrenia Res 7:191–199

Matsubara T, Nakagome Y, Ogasawara N, et al. (1982) Maternally transmitted extra ring(21) chromosome in a boy with Down's Syndrome. Hum Genet 60:78–79

Niebuhr E (1978) The cri du chat syndrome. Hum Genet 44:227–275

Peled-Kamar M, Latem J, Okon E, et al. (1995) Thymic abnormalities and enhanced apoptosis of thymocytes and bone marrow cells in transgenic mice overexpressing Cu/Zn-superoxide dismutase: implications for Down syndrome. EMBO J 14:4985–4993

Petrij F, Giles RH, Dauwerse HG, et al. (1995) Rubinstein-Taybi syndrome caused by mutations in the transcriptional co-activator CBP. Nature 376:348–351

Reiter LT, Murakami T, Koeuth T, et al. (1997) The human *COX10* gene is disrupted during homologous recombination between the proximal and distal CMT1A-REPs. Hum Mol Genet 6:1595–1603

Roa BB, Garcia CA, Suter U, et al. (1993) Charcot–Marie–Tooth disease type 1A. Association with a spontaneous point mutation of the *PMP22* gene. N Engl J Med 329:96–101

Smith DJ, Stevens ME, Sudanagunta SP, et al. (1997) Functional screening of 2 Mb of human chromosome 21q22.2 in transgenic mice implicates *minibrain* in learning defects associated with Down syndrome. Nat Genet 16:28–36

Stec I, Wright TJ, Van Ommen GJB, et al. (1998) *WHSC1*, a 90 kb SET domain–containing gene, expressed in early development and homologous to a *Drosophila* dysmorphy gene maps in the Wolf–Hirschhorn syndrome critical region and is fused to *IgH* in t(4;14) multiple myeloma. Hum Mol Genet 7:1071–1082

Sumarsono SH, Wilson TJ, Tymms MJ, et al. (1996) Down's syndrome–like skeletal abnormalities in *Ets2* transgenic mice. Nature 379:534–537

Taine L, Goizet C, Wen ZQ, et al. (1998) Submicroscopic deletion of chromosome 16p13.3 in patients with Rubinstein–Taybi syndrome. Am J Med Genet 78:267–270

Tsai C-H, Van Dyke DL, Feldman GL (1999) Child with velocardiofacial syndrome and del(4)(q34.2): another critical region associated with a velocardiofacial syndrome–like phenotype. Am J Med Genet 82:336–339

Wyandt HE (1988) Ring autosomes: identification, familial transmission, causes of phenotypic effects and in vitro mosaicism. In: Daniel A (ed) The cytogenetics of mammalian autosomal rearrangements. Liss, New York, pp 667–695

Yamagishi H, Garg V, Matsuoka R, et al. (1999) A molecular pathway revealing a genetic basis for human cardiac and craniofacial defects. Science 283:1158–1161

Clinical Importance of Translocations, Inversions, and Insertions

---

Phenotypes of Balanced Translocation Heterozygotes (Carriers)

Robertsonian translocations (RTs) involving two acrocentric chromosomes do not affect the phenotype, apart from occasional male sterility. RTs are about 10 times more frequent in male infertility clinics than in the newborn population (Zuffardi and Tiepolo, 1982). RTs do attach to the sex vesicle (Guichaoua et al., 1990), and this may lead to spreading of inactivation from the XY bivalent into the adjacent autosomal segment. The incidence of RTs is not increased in mentally retarded patient populations. The loss of two acrocentric short arms has no phenotypic effect, indicating that they contain no essential single-copy genes and that loss of some of the multicopy rRNA genes is well tolerated. In a family in which the parents were first cousins and heterozygous carriers of a t(13ql4q), three children were phenotypically normal despite their homozygosity for the translocation chromosome and consequent loss of four of the 10 acrocentric short arms (Martinez-Castro et al., 1984).

Most persons with balanced reciprocal translocations are also phenotypically normal. However, surveys of mentally retarded populations indicate that they have five times the incidence of balanced reciprocal translocations, mainly de novo, as a consecutive series of newborns (Funderburk et al., 1977). Similarly, their incidence in infertile males is nearly four times that seen in controls (Zuffardi and Tiepolo, 1982). What accounts for these sporadic effects? A position effect may be responsible for inactivating a gene or genes near one of the breakpoints, with a newly acquired heterochromatic environment silencing the gene. Alternatively, one of the reciprocal translocation breakpoints may disrupt a gene, or an exonuclease may destroy DNA at the break site before rejoining takes place. More extensive exonuclease activity could disrupt or delete a gene a short distance away from the breakpoint (there are no genes, only satellite DNA, near the breakpoints of RTs). The presence of such subtle deletions has been demonstrated in several translocation and inversion carriers with abnormal phenotypes (Kumar et al., 1998).

Aberrant Meiotic Segregation in Reciprocal Translocation Carriers

Homologous chromosome segments tend to pair in prophase of meiosis I, even in translocation heterozygotes (carriers). The possible outcomes, disregarding crossing over and secondary nondisjunction (1:3 and 0:4 segregations), are illustrated in Figure 16.1. The two translocation chromosomes and their two normal homologues form a quadrivalent that can be seen as a cross-like figure at pachytene. Very small translocated segments may remain unpaired or fail to form a chiasma. This results in the formation of either two bivalents or a trivalent and a univalent. However, if one terminal chiasma is present in each arm of the pachytene cross, the metaphase configuration will be a ring of four; if a chiasma fails to form in one arm, the result will be a chain.

A ring or chain of four may orient itself in different ways on the spindle (Fig. 16.1). One orientation (called *alternate*) on the MI metaphase spindle gives rise to alternate segregation, producing one daughter cell with the two normal chromosomes and the other with the two reciprocal translocation chromosomes. Generally, these will segregate from each other at the second meiotic division

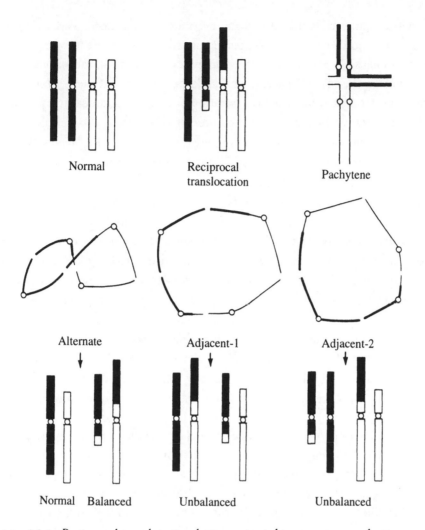

Figure 16.1. Reciprocal translocation between two chromosomes, pachytene configuration of the four chromosomes, and modes of orientation of a ring of four on the spindle in meiotic metaphase 1. Alternate segregation leads to normal and balanced gametes, adjacent-1 segregation to two types of unbalanced gametes. In adjacent-2 segregation, homologous centromeres go to the same pole, producing two generally more extreme types of unbalanced gametes (1 : 3 and 0 : 4 segregation, not shown).

and may lead to progeny with reciprocally unbalanced karyotypes. Adjacent-1 orientation on the spindle leads to the formation of two unbalanced daughter cells, each with one translocation chromosome. The rare adjacent-2 orientation, in which homologous centromeres do not segregate from each other, gives rise to two daughter cells that are usually even more unbalanced. Various types of 1 : 3 segregation are also possible. Sometimes two bivalents are formed, and their independent segregation produces equal numbers of normal, carrier, and each of two types of unbalanced gametes. It should be stressed that there is no theoretical expectation for the segregation ratios of human reciprocal translocations, since the relative frequencies of the different orientations of a specific ring or chain are unknown, so we have to rely on empirical risk figures. Usually, however, types representing complementary meiotic products from the same orientation are equally frequent.

The risk of producing different types of unbalanced karyotypes depends on the pachytene configuration of the translocation quadrivalent. This is determined by the shape of the chromosomes and the distance of each breakpoint from both the centromere (*interstitial segment*) and the telomere (*exchange segment*); these distances affect the probability of crossing over in the exchanged and the interstitial segments. The orientation can be inferred from the observed frequencies of the various outcomes in a large number of translocation families. Cases of karyotypically abnormal children almost always arise by adjacent-1 segregation when the translocated segments are very short or when at least one interstitial segment is long enough for crossing over to be expected. When no crossing over is expected, adjacent-2 segregation becomes almost as frequent as adjacent-1.

When the pachytene configuration is highly asymmetrical, 1 : 3 segregation is favored over 2 : 2 segregation. This is seen if one of the chromosomes is acrocentric, if at least one breakpoint is near the centromere, or if the participating chromosomes are very unequal in length. Translocations showing 1 : 3 segregation have been seen for very few chromosomes. Thus, interchange trisomy has been seen for chromosomes 18, 21, and 22, whereas duplication of 9p is preferred in tertiary trisomy. About 90% of children born with unbalanced karyotypes produced by 1 : 3 disjunction have carrier mothers and about 10% have carrier fathers. Carrier females are considerably more likely to produce unbalanced progeny than are carrier males. The risk of producing further unbalanced liveborn offspring is about the same whether the family has been ascertained through a 1 : 3 segregant or through an unbalanced 2 : 2 segregant: 10–20% for a carrier mother and 0 for a carrier father (Lindenbaum and Bobrow, 1975).

How frequent are reciprocal translocations? In 56,952 newborn infants, 51 balanced reciprocal and insertional translocations were found, and the same number of balanced Robertsonian translocations (Hook and Hamerton, 1977). This is almost one balanced translocation per 1000 neonates. (Unbalanced translocations or insertions were much less common, 0.1/1000, in the same newborn population.) Many of these translocations occurred de novo, both parents having a normal karyotype. Olson and Magenis (1988) have determined that 84% of de novo translocations are paternal in origin, and the high frequency of structural rearrangements in sperm provides a basis for this (Guttenbach et al., 1997). The risk of having a subsequent chromosomally abnormal offspring is very different for balanced translocation carriers ascertained by chance compared with those who come to attention through an unbalanced offspring. Whatever the mode of ascertainment of the reciprocal translocation, the ratio of normal to balanced carrier offspring is about 1:1, since these are complementary meiotic products that usually lead to a normal phenotype.

Aberrant Meiotic Segregation in Robertsonian Translocation Carriers

Robertsonian translocations can involve either both copies of a homologous pair or one copy each of nonhomologous chromosomes. The outcomes of meiotic segregation are quite different in the two cases. Those involving homologues are expected to yield equal numbers of disomic and nullisomic gametes and thus of trisomic and monosomic zygotes. The offspring of rob(13;13)(q10;q10) and rob(21;21)(q10;q10) carriers are all 13-trisomic or 21-trisomic, respectively. The monosomic embryos presumably never implant or spontaneously abort very early. One family in which the mother was a rob(21;21)(q10;q10) carrier had eight children with Down syndrome (Furbetta et al., 1973). Practically all carriers of rob(14;14)(q10;q10), rob(15;15)(q10;q10), and rob(22;22)(q10;q10) have been identified because they have had only spontaneous abortions, indicating the lethality of trisomies as well as monosomies 14, 15, and 22.

An unusual outcome is sometimes observed. For example, a phenotypically normal woman with 45,XX,rob(22;22)(q10;q10) gave birth to a daughter with the same karyotype (Kirkels et al., 1980). In this extremely rare case, either the sperm was monosomic for chromosome 22 or the extra 22 in the daughter was lost at a very early stage. A carrier of a translocation between two *homologous* chromosomes can also have only abnormal offspring or spontaneous abortions. Thus, a woman who was ascertained because of four spontaneous abortions had a whole-arm interchange between the two chromosomes 7, forming t(7p;7p) and t(7q;7q) chromosomes (Niikawa and Ishikawa, 1983). The karyotype, by ISCN (1995) nomenclature, is 46,XX,t(7;7)(p10;q10).

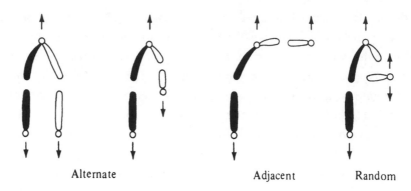

Alternate   Adjacent   Random

Figure 16.2. Different orientations of chromosomes on the meiotic spindle and the resultant segregation of Robertsonian translocations and their homologues. Alternate segregation of t(DqDq) and t(DqGq) leads to normal and carrier gametes; adjacent segregation of t(DqGq) leads to trisomy and monosomy G; while random drifting of a univalent G leads to all four types of gametes.

Whole-arm transfers between nonacrocentric chromosomes are extremely rare in unselected populations or in parents of abnormal offspring. Such translocations, which involve deficiency (monosomy) for one chromosome arm and duplication (trisomy) for another, appear to be lethal during embryogenesis (Niikawa and Ishikawa, 1983).

The meiotic segregation of Robertsonian translocations between nonhomologous chromosomes is depicted in Figure 16.2. Dq21q carriers have a considerably higher of risk of producing unbalanced offspring than do 13qDq carriers. This may be due to an asymmetrical alternate MI configuration for the former and a symmetrical one for the latter. Segregation in the configuration formed by a DqDq translocation would, as a rule, result in one cell with the translocation chromosome and another with the two free chromosomes. In the meiotic configuration formed by Dq21q, the free 21 would be near the same pole as the translocation and would therefore sometimes undergo adjacent segregation. Since chromosome 21 is smaller than a D chromosome, it would more often fail to form a chiasma and thus drift at random in the first meiotic anaphase. This may explain the 21q22q carrier's much higher risk (than the DqDq carrier's) for unbalanced offspring despite the symmetrical alternate meiotic configuration formed by 21q22q. Another factor that may influence the observed (empirical) risk figures is that more 13-trisomic than 21-trisomic zygotes are spontaneously aborted.

The empirical risk of producing unbalanced offspring depends on the mode of ascertainment. In families ascertained in unselected populations, such as newborn infants, the risk for unbalanced offspring is low, whereas in families ascertained through an affected member, the risk is considerably higher. Whatever the mode of ascertainment, the probability of producing a healthy carrier is 50% for 13ql4q, 14q21q, and 21q22q translocation carriers. Surprisingly, in the European Collective Study of Parental Diagnosis, a surplus of paternally derived balanced t(13ql4q) carriers was found (Boue and Gallano, 1984). Most information on segregation of Robertsonian translocations comes from families ascertained through an individual with Down syndrome. The risk figure for a female Dq21q carrier is about 10%, while that for a male carrier is about 2–3%.

Phenotypes Associated with Unbalanced Duplication/Deficiency Karyotypes

An unbalanced translocation involves partial trisomy (duplication) for one chromosome and partial monosomy (deficiency, or deletion) for another, although the deleted segment may be very small or almost nonexistent if one of the breakpoints is near a telomere. Meiotic recombination involving a balanced *complex* translocation can also lead to duplication/deficiencies (Madan et al., 1997).

The phenotypic effects of hundreds of unbalanced duplication/deficiencies have been described. They practically always include mental retardation and multiple congenital anomalies. The potential number of different duplication/ deficiencies is enormous. However, those that allow the birth of a live child are much more restricted; many are apparently lethal, like many deletions.

The lethality of various duplication/deficiencies is reflected in the increased rate of spontaneous abortions and stillbirths in translocation carriers. If one or more of the unbalanced products of aberrant segregation has very severe effects, there may be no recognized pregnancies at all, with sterility or infertility the outcome. Indeed, infertility clinic populations show a slight increase in the frequency of translocations. Reciprocal translocations are seen in about 3% of couples with repeated spontaneous abortions and Robertsonian translocations in nearly 2% (Fryns et al., 1984). This strong selection against many duplication/deficiencies probably reflects the length and gene-richness of the segments involved: the more genes in a segment, the more severe the effects. It may also reflect the presence or absence of a gene that is lethal if expressed in one or three copies (dosage effect). The much greater importance of a relatively few genes in the phenotypic effects of a duplication or deletion is discussed in detail in Chapter 15.

A 1:3 segregation (one of the four chromosomes goes to one pole while three go to the other) in a translocation carrier may give rise to unbalanced offspring with 45 or 47 chromosomes. Such secondary disjunction increases the risk of abnormal offspring in translocation carriers. Eight kinds of offspring are possible from such a segregation. Their karyotypes include monosomy or trisomy for a normal chromosome. An extra translocation chromosome (either one) combined with an otherwise normal complement (47 chromosomes) has been termed *tertiary trisomy*, and a 45-chromosome complement that includes an abnormal chromosome (either one) is *tertiary monosomy*. In *interchange trisomy*, the reciprocal translocation chromosomes are present, plus an extra normal chromosome (either one; 47 chromosomes).

A rare situation in which two different types of abnormal offspring resulted from 1:3 segregation is represented by a woman with t(9q−;21p+) who gave birth to one daughter with an extra 9p chromosome and to another who was 21-trisomic, in addition to a chromosomally normal daughter (Habedank and Faust, 1978). Her translocation had features that are known to promote this type of disjunction, namely, the chromosomes were of different size, one was acrocentric, and the breaks were near the centromere. Another reciprocal translocation that agrees with these rules and practically always segregates 1:3 is t(11;22)(q23;q11). In virtually all families the mother is the carrier (Zackai and Emanuel, 1980).

Aberrant segregation in some translocation carriers has led to multiple spontaneous abortions. In one family with five spontaneous abortions, meiotic segregation in the t(13q−;18q+) father led to three karyotypically abnormal abortuses: 47,+13 (tertiary trisomy), 46,13q− (unbalanced translocation), and 47,t(13q−;18q+)+18 (interchange trisomy). The abortus with 46 chromosomes resulted from an adjacent-1, 2:2 disjunction, the other two from 1:3 segregation (Kajii et al., 1974). Rarely, more than one type of unbalanced offspring is born alive in the same translocation family, for example, the cri du chat deletion syndrome and its countertype with a duplication of the same region of 5p (Malaspina et al., 1992). As expected, the deletion produced more severe effects than the duplication.

Phenotypes of Inversion Heterozygotes (Carriers)

When one or both inversion breakpoints disrupt a gene, either directly or because of exonuclease digestion from the breakpoint(s), the loss of function may affect the phenotype if the gene is X-linked or acts as an autosomal dominant. Almost half the cases of classical hemophilia are caused by a submicroscopic inversion within the factor VIII gene at Xq28 that prevents the gene from producing a functional product (Rossiter et al., 1994). Heterozygous carriers of larger inversions are usually phenotypically normal. The exceptions are cases in which one of the breakpoints disrupts a gene or exonuclease activity disrupts a gene not far from one of the breakpoints.

Aberrant Meiotic Segregation in Inversion and Insertion Carriers

The marked tendency of homologous chromosome regions to pair during prophase of meiosis I indicates that pairing of the inverted segments in an inversion heterozygote leads to the formation of an inversion loop. A crossover within the inversion loop leads to reciprocally unbalanced products. A pericentric inversion carrier may generate reciprocal duplication/deficiencies (Figs. 16.3 [see color insert] and 16.4), while a paracentric inversion carrier may generate a dicentric chromosome and an acentric fragment. One might expect these unbalanced products to be as frequent as the two balanced products (normal and carrier). However, chromosomally unbalanced progeny are rarely seen, in part because heterologous pairing takes place in the inverted region, leaving a much shorter region of homology in which crossing over can take place. This is based on electron microscopic analyses of a pericentric inversion and sperm genotyping for microsatellite markers extending to both sides of a paracentric inversion (Brown et al., 1998). Another powerful approach that has provided direct evidence of crossover suppression in pericentric inversions is two-color FISH analysis of sperm using YAC probes that map proximal and distal to the inverted segment (Fig. 16.3).

The reproductive risk for carriers of complex chromosome rearrangements is high: a 50% risk of spontaneous abortion and a 20% risk of a karyotypically abnormal child, based on some 60 families (Madan et al., 1997). Meiotic

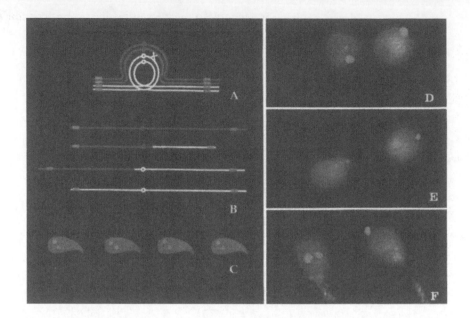

Figure 16.3. Recombination within a pericentric inversion, as visualized by two-color FISH. (A–C) Diagrams of a crossover within the inverted segment, the recombinant and nonrecombinant chromosomes, and nonrecombinant (green-red) and recombinant (two red or two green) sperm. (D–F) Images from two-color FISH showing recombinant and nonrecombinant sperm (reproduced from Jaarola et al., Am J Hum Genet 63:218–224, copyright 1998, American Society of Human Genetics, with permission of the University of Chicago Press) (See color insert).

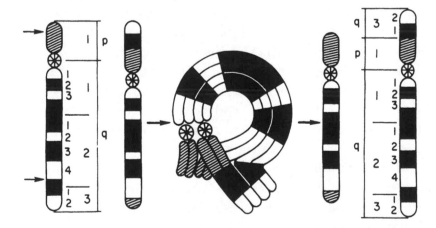

Figure 16.4. Breakpoints in a pericentric inversion of chromosome 14, and two types of abnormal chromosomes (one with a duplication, the other with a deletion) produced by crossing over in the inversion loop (Trunca and Opitz, 1977).

crossing over in a heterozygous carrier of an intrachromosomal insertion leads to chromosomally unbalanced gametes, with a 15% empirical risk in each pregnancy of a child with an unbalanced karyotype. Madan and Menko (1992) reviewed 27 cases of intrachromosomal insertion. Their review has helpful diagrams of the mechanism of origin of unbalanced karyotypes, showing how a carrier can produce abnormal offspring with different recombination products (see Fig. 16.5). Two sibs with facial anomalies and developmental delay were found to have a duplication (partial trisomy) of part of chromosome 2q. This was the result of aberrant meiotic segregation in the father, who had an interstitial segment of chromosome 2q inserted into the long arm of chromosome 5 (Barnicoat et al., 1997). His karyotype was either 46,XY,inv ins(5;2)(q22.3;q32.1q24.3) or 46,XY,dir ins(5;2)(q22.1;q24.3q32.1).

Inverted insertions are especially dangerous, in terms of the risk of karyotypically abnormal progeny. In one family, meiotic segregation of an inverted insertion of a segment from chromosome 13 into 3q could be followed in three generations; individuals with both the partial monosomy and the partial trisomy types of recombinations, as well as carriers and normals, were observed (Toomey et al., 1978).

Sperm Chromosomes in Meiotic Segregation Analysis

The analysis of sperm chromosome content has greatly increased the accuracy of meiotic segregation analysis in normal individuals and balanced carriers of reciprocal and Robertsonian translocations, as well as inversions. Sperm chromosomes are made visible by fertilizing hamster eggs with human sperm (Martin, 1988a). This technique allows a direct analysis of the results of the meiotic process, whereas fetal or newborn chromosomes have undergone severe selection before they come under study. A disadvantage of the sperm technique is that only a limited number of sperm from a limited number of carriers can be analyzed.

In general, the segregations that give rise to unbalanced gametes are found much more frequently in sperm karyotypes than in fetuses or newborns, based on studies using in vitro sperm penetration of hamster eggs. In 13 translocation carriers, all segregation types (alternate, adjacent-1, adjacent-2, 3:1, and 4:0) have been found, whereas the 4:0 segregation has never been seen in a fetus or live-born child (Martin, 1988b). In five RT carriers, the ratio of normal to

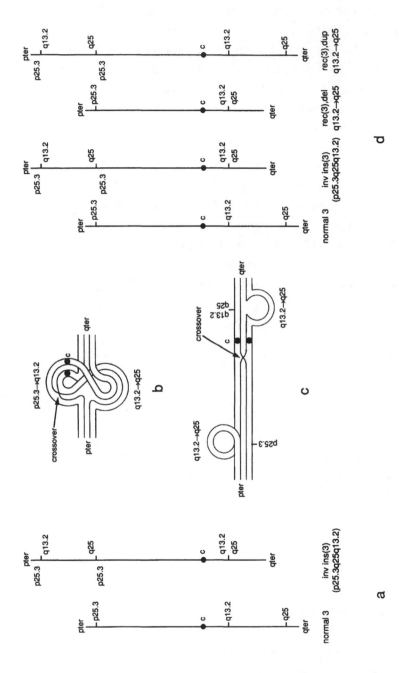

Figure 16.5. How abnormal recombinant chromosomes arise from an inverted insertion. (a) Normal 3 and inv ins(3)(p25.3q25q13.2) in the father; (b, c) alternative pachytene configurations with pairing along the entire chromosome or failing in the inverted segment; (d) meiotic products following a single crossover in the noninserted segment (p25.3–q13.2): the parental types and two recombinants, rec(3)del and rec(3)dup (reproduced from Madan and Menko, Intrachromosomal insertions: a case report and a review, Hum Genet 89:1–9, copyright Springer-Verlag 1992).

balanced chromosome constitutions does not differ from 1 : 1 except in the t(14q;21q) carrier. This may depend on the small number of sperm analyzed or on the fact that this translocation is asymmetrical, in contrast to the others, which are symmetrical. The unbalanced chromosome constitutions range from 7.7% to 27.0%, with a mean of 15.8% (Pellestor, 1990). This is considerably higher than the frequency (5.1%), of unbalanced embryos found in prenatal diagnoses performed when one parent is a known RT carrier (Boue and Gallano, 1984). Thus, there is a strong selection against unbalanced carriers between sperm and embryo, as well as between embryo and liveborn child.

Direct two- or three-color FISH analysis of sperm chromosomes permits much more rapid and accurate analysis of virtually unlimited numbers of sperm from any number of men (Guttenbach et al., 1997). *Interchromosomal effects* are presumptive influences one chromosome abnormality has on the behavior of non-homologous chromosomes. Evidence for such effects is extremely limited and largely anecdotal. Analyses of sperm chromosomes indicate that Robertsonian and most reciprocal translocations do not exert interchromosomal effects (Pellestor, 1990).

References

Barnicoat AJ, Abusaad I, Mackie CM, et al. (1997) Two sibs with partial trisomy 2q. Am J Med Genet 70:166–170

Boue A, Gallano P (1984) A collaborative study of the segregation of inherited chromosome structural rearrangements in 1356 prenatal diagnoses. Prenat Diagn 4:45–67

Brown GM, Leversha M, Hulten M, et al. (1998) Genetic analysis of meiotic recombination in humans by use of sperm typing: reduced recombination within a heterozygous paracentric inversion of chromosome 9q32–q34.3. Am J Hum Genet 62:1484–1492

Fryns JP, Kleczkowska A, Kubien E, et al. (1984) Cytogenetic survey in couples with recurrent fetal wastage. Hum Genet 65:336–354

Funderburk SJ, Spence MA, Sparkes RS (1977) Mental retardation associated with "balanced" chromosome rearrangements. Am J Hum Genet 29:136–141

Furbetta M, Falorni A, Antignani P, et al. (1973) Sibship (21q21q) translocation Down's syndrome with maternal transmission. J Med Genet 10:371–375

Guichaoua MR, Quack B, Speed RM, et al. (1990) Infertility in human males with autosomal translocations: meiotic study of a 14;22 Robertsonian translocation. Hum Genet 86:162–166

Guttenbach M, Engel W, Schmid M (1997) Analysis of structural and numerical chromosome abnormalities in sperm of normal men and carriers of constitutional chromosome aberrations. A review. Hum Genet 100: 1–21

Habedank M, Faust J (1978) Trisomy 9p and unusual translocation mongolism in siblings due to different 3:1 segregations of maternal translocation rcp(9;21) (p11;q11). Hum Genet 42:251–256

Hook EB, Hamerton JL (1977) The frequency of chromosome abnormalities detected in consecutive newborn studies—differences between studies—results by sex and by severity of phenotypic involvement. In: Hook EB, Porter IH (eds) Population cytogenetics. Academic, New York, pp 63–79

ISCN; Mitelman F (ed) (1995) An international system for human cytogenetic nomenclature (1995). Karger, Basel, 1995

Jaarola M, Martin RH, Ashley T (1998) Direct evidence for suppression of recombination within two pericentric inversions in humans: a new sperm-FISH technique. Am J Hum Genet 63:218–224

Kajii T, Meylan J, Mikamo K (1974) Chromosome anomalies in three successive abortuses due to paternal translocation, t(13q−;18q+). Cytogenet Cell Genet 13:426–436

Kirkels VGHJ, Hustinx TWJ, Scheres JMJC (1980) Habitual abortion and translocation (22q;22q): unexpected transmission from a mother to her phenotypically normal daughter. Clin Genet 18:456–461

Kumar A, Becker LA, Depinet TW, et al. (1998) Molecular characterization and delineation of subtle deletions in de novo "balanced" chromosomal rearrangements. Hum Genet 103:173–178

Lindenbaum RH, Bobrow M (1975) Reciprocal translocations in man. 3:1 meiotic disjunction resulting in 47− or 45−chromosome offspring. J Med Genet 12:29–43

Madan K, Menko FH (1992) Intrachromosomal insertions: a case report and a review. Hum Genet 89:1–9

Madan K, Nieuwint AWM, Bever Y van (1997) Recombination in a balanced complex translocation of a mother leading to a balanced reciprocal translocation in the child. Review of 60 cases of balanced complex translocations. Hum Genet 99:806–815

Malaspina D, Warburton D, Amador X, et al. (1992) Association of schizophrenia and partial trisomy of chromosome 5p. Schizophrenia Res 7: 191–199

Martin RH (1988a) Human sperm karyotyping: a tool for the study of aneuploidy. In: Vig BK, Sandberg AA (eds) Aneuploidy, Part B: Induction and test systems. Liss, New York, pp 297–316

Martin RH (1988b) Abnormal spermatozoa in human translocation and inversion carriers. In: Daniel A (ed) The cytogenetics of mammalian autosomal rearrangements. Liss, New York, pp 319–417

Martinez-Castro P, Ramos MC, Rey JA, et al. (1984) Homozygosity for a Robertsonian translocation (13ql4q) in three offspring of heterozygous parents. Cytogenet Cell Genet 38:310–312

Niikawa N, Ishikawa M (1983) Whole-arm translocation between homologous chromosomes 7 in a woman with successive spontaneous abortions. Hum Genet 63:85–86

Olson SB, Magenis RE (1988) Preferential paternal origin of de novo structural chromosome rearrangements. In: Daniel A (ed) The cytogenetics of mammalian autosomal rearrangements. Liss, New York, 583–599

Pellestor F (1990) Analysis of meiotic segregation in a man heterozygous for a 13;15 Robertsonian translocation and a review of the literature. Hum Genet 85:49–54

Rossiter JP, Young M, Kimberland ML, et al. (1994) Factor VIII gene inversions causing severe hemophilia A originate almost exclusively in male germ cells. Hum Mol Genet 3:1035–1039

Toomey KE, Mohandas T, Sparkes RS, et al. (1978) Segregation of an insertional chromosome rearrangement in 3 generations. J Med Genet 15:382–387

Trunca C, Opitz JM (1997) Pericentric inversion of chromosome 14 and the risk of partial duplication of 14q (14q31–14qter), Am J Med Genet 1:217–228

Zackai EH, Emanuel BS (1980) Site-specific reciprocal translocation, t(11;22) (q23;q11), in several unrelated families with 3 : 1 meiotic disjunction. Am J Med Genet 7:507–521

Zuffardi O, Tiepolo L (1982) Frequencies and types of chromosome abnormalities associated with human male infertility. In: Crosignani PG, Rubin BL (eds) Genetic control of gamete production and function. Academic, London, pp 261–273

17

Sex Determination and the Y Chromosome

The Y Chromosome and Y Heterochromatin: The Y Body

The Y chromosome is a small acrocentric chromosome whose chromatids tend to adhere in metaphase spreads because of the presence of a large block of heterochromatin. This makes up the distal half or so of the long arm and is visible by most C-banding procedures. It is especially bright by Q-banding, due to the AT-richness of its satellite DNA (Fig. 17.1). The Y chromosome contains some DNA sequences that are found only on the Y. It is therefore possible to look for Y-DNA in maternal blood during and after pregnancy. Y-DNA is detectable as early as 5–7 weeks of gestation and disappears by 2 months post partum (Thomas et al., 1995). There is, on average, only about one fetal cell per ml of blood in pregnancies with a normal fetus. However, the

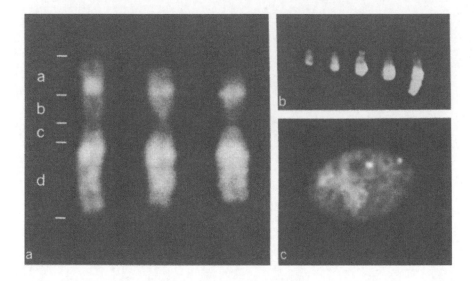

Figure 17.1. (a) Q-banded X chromosomes (Therman et al., 1974). (b) Variation in the size of the Y chromosome (Dutrillaux, 1977). (c) Two Y bodies in a buccal cell from a 47,XYY male (courtesy of E. Therman).

number is more than five times as high, on average, in pregnancies with a 21-trisomic fetus (Bianchi et al., 1997). Fetal DNA appears to enter the maternal circulation in the form of viable cells. A number of groups have used powerful fluorescence-activated flow sorting to isolate the fetal cells, but the cells' rarity has limited the development of this approach for early prenatal diagnosis. However, it has been used to diagnose trisomy 21 (Elias et al., 1992).

After quinacrine staining, the length of the brightly fluorescent distal segment of the Y corresponds in size to the bright Y body that can be seen in interphase nuclei of the same individual. A Y-bearing sperm is distinguishable from an X-bearing sperm by its bright Y body. In cells with two Y chromosomes, two Y bodies can be seen (Fig. 17.1b). The heterochromatic segment of the Y varies in length from minute (rare) to two or three times as long as the average Y and is unchanged from father to son. The average length of the Y differs in different populations or ethnic groups (Fig. 17.1b). For example, a long Y chromosome is present in a much higher fraction of Oriental than of Caucasian males (Ibraimov and Mirrakhimov, 1985). In addition to size variants, other heteromorphisms are common, including an inversion, which moves the centromere close to the heterochromatin. About one in 200 males has such an inversion, but in a Muslim Indian community in South Africa, some 30% of the males do

(Bernstein et al., 1986). A much rarer heteromorphism is the presence of a nucle-olus organizer region and satellite at the end of Yq. In a French Canadian family, a satellited Y chromosome has been inherited by every male for 12 generations, with no phenotypic effect (Genest, 1973).

The Two Pseudoautosomal Regions

Most of the Y chromosome fails to pair with the X, and vice versa. The meiotic XY sex bivalent at diplotene-metaphase I shows only end-to-end attachment of the short arms (Fig. 9.4), a typical terminal chiasma configuration, as first noted almost 90 years ago. That is, both X and Y have a long differential segment not found on the other chromosome and a very short (2.6-mb) homologous pairing region, which forms a short synaptonemal complex (Fig. 17.2) with recombina-tion nodules and bars (Holm and Rasmussen, 1983). A single, virtually obliga-tory crossover occurs in the terminal region on the short arm in every male meiosis, with the recombination rate in this region 10-fold greater in males than in females (Rouyer et al., 1986). Thus, genes in this region (Fig. 17.3) do not

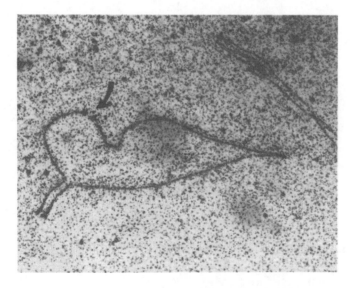

Figure 17.2. Early pachytene spermatocyte showing the XY pairing segment con-sistently seen at the tip of the short arms of the X and Y chromosomes and a sec-ondary association (arrow) at the other end (Chandley et al., 1984, reproduced with permission of S. Karger AG, Basel).

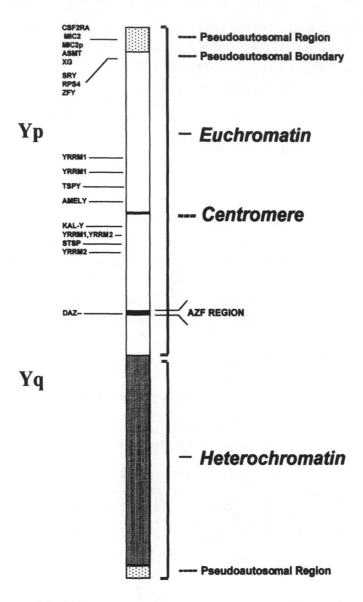

Figure 17.3. Diagram of the Y chromosome showing various genes in the pseudoautosomal region on the short arm (PAR1) and other regions of the Y (reproduced from Mulhall et al., Hum Reprod 12:503–508, 1997, with permission of Oxford University Press).

show the typical X- or Y-linked patterns of inheritance. Such a region is called a *pseudoautosomal region;* this one is PAR1 (Figs. 17.3 and 18.2) (Burgoyne, 1998). Genes in PAR1, like those in other early-replicating regions, are expressed whether they are on either the active or inactive X chromosome. This has been demonstrated by analysis of transcription in rodent-human hybrid cells containing an inactive X or an active X as the only human chromosome (Brown et al., 1997). Diminished recombination in the pseudoautosomal region on Xp and Yp is associated with nondisjunction of X and Y (Hassold et al., 1991).

A second pseudoautosomal region (PAR2) is present near the telomere of the long arms of both X and Y. It is very short and does not appear to be important for normal segregation of X and Y, though it may be for their initial pairing. The telomeres of Xq and Yq sometimes associate in meiosis (Fig. 17.2). Although a short synaptonemal complex can only rarely be seen there, Freije et al. (1992) demonstrated a recombination frequency of 2% in male meiosis between markers only 100 kb apart in the 320-kb PAR2 at the tips of Xq and Yq. This is a much higher frequency than observed between these markers on the two X chromosomes in female meiosis. The sequence organization of PAR2 has been worked out (Kralog et al., 1994). Genes in the pseudoautosomal regions on both long and short arms of the sex chromosomes stay active on the inactive X; that is, they escape inactivation, as described below. The pairing regions on the X and Y replicate early and synchronously (Schempp, 1985).

Sex Determination: Rearrangements Localize the Male-Determining Gene to Yp

The most important function of the Y chromosome is sex determination. In addition, there are genes on the Y chromosome that are essential for male fertility. Even microdeletions in the Y may cause infertility (Chandley, 1998). The male-determining role of the Y chromosome became clear when XO individuals were shown to be female and XXY individuals male. The female phenotype of individuals with an isochromosome Yq or deletion of Yp shows that the male or testis determining factor (*TDF*) is on the short arm of Y. This has been confirmed by studies of XX males with Klinefelter syndrome. An abnormal meiotic interchange between X and Y chromosomes in which a part of Yp is translocated to Xp accounts for most of them, but not all (Petit et al., 1987). Two hotspots for ectopic Xp-Yp recombination are responsible for the X-Y interchange in nearly half the cases of XX males. The underlying cause appears to be the persistence of X-Y homologous sequences proximal to the distal pseudoautosomal region (Wang et al., 1995). Corresponding studies in XY females show that *TDF* is missing from the Y (Disteche et al., 1986). The most probable mechanism for producing such abnormal X and Y chromosomes is an exceptional unequal exchange, in which the Y breakpoint is proximal to *TDF.* Abnormal interchange between the protein kinase genes *PRKX* and *PRKY*, which have remained highly homologous although they are far from the pseudoautosomal region, account for almost one-third of all XX males and XY females (Schiebel et al., 1997). A number of XO males have also been described; they have a part of Yp translocated to an autosome (Abbas et al., 1990). It is unclear whether ectopic recombination between similar sequences is involved in their origin.

SRY, the Only Male-Determining Gene on the Y Chromosome

For nearly 10 years, the H-Y antigen was considered a strong candidate for *TDF.* Later, *ZFY* (zinc finger Y), a gene cloned from the *TDF* region, was for a time considered a likely candidate. However, overwhelming evidence now indicates that the *TDF* gene is *SRY* (sex-determining region on the Y) (for example, see Hawkins et al., 1992). This gene is located in region 1A1A, extremely close to the pseudoautosomal region on Yp (Lahn and Page, 1997). This gene encodes a testis-specific transcript that triggers the differentiation of Sertoli cells from somatic cells in the genital ridge, thus establishing a program of gonadal differentiation that produces a testis (Sinclair et al., 1990). It is unlikely that this latest *TDF* candidate will be relegated to the same "has-been" status as its predecessors, because Koopman et al. (1991) demonstrated in transgenic mice that transfer of the homologous mouse gene, *Sry*, alone is able to cause male differentiation in XX mice. Furthermore, a small fraction of human XY females have a de novo point mutation in their *SRY* gene not present in the father, and no Y deletion. Nearly 30 such mutations in *SRY* have been reported (Hawkins et al., 1992). A 46,XY individual with true hermaphroditism (both ovarian and testicular tissue present in the gonads) was found to be a genetic mosaic produced by a postzygotic (somatic) point mutation in *SRY*, providing still further confirmation of the sex-determining role of *SRY* (Braun et al., 1993).

The only conserved region of the *SRY* gene product among various mammalian species is its 60-amino-acid-long HMG box, a protein domain first detected in HMG1 (high-mobility-group protein 1; see Chapter 5). The HMG box binds to specific DNA sequences and produces a sharp bend in the DNA, strongly influencing the local chromatin structure. The HMG box of the SRY protein specifically recognizes a short sequence within the promoters of two sex-specific genes. It down-regulates production of P450 aromatase, the enzyme that converts the male hormone testosterone to the female hormone estradiol, and thus enhances the level of testosterone. SRY also up-regulates the production of Mullerian inhibitory substance and thus blocks the differentiation of the uterus and upper vagina (Haqq et al., 1993).

XY individuals with an intact *SRY* gene but a duplication of a small region of Xp21 show male-to-female sex reversal. The minimal duplication producing this effect is no more than 160 kb long. One of the genes in this region is *DAX1*, which encodes an unusual member of a nuclear hormone receptor superfamily. Evidence from transgenic mice indicates that the *DAX1* gene product antagonizes *Sry* action in mammalian sex determination (Swain et al., 1998). Understanding this interaction may clarify the critical dosage compensation mechanism for X-linked genes.

Autosomal Genes Involved in Male Sex Determination or Differentiation

SRY is a member of a large family of genes that contain an SRY-like HMG box, called *SOX* genes. Most of them are activators of transcription. However, *SRY* shows so little sequence conservation with them, except for the DNA-binding HMG box itself, that it may function by inhibiting the action of another *SOX* gene—perhaps the X-linked *SOX3* gene, from which it apparently arose during the evolution of mammalian sex determination (Graves, 1998). That point remains unresolved, but it is clear that genes other than *SRY* are involved in sex determination and that at least one of them, the autosomal *SOX9* gene at 17q24.3–q25.1, also belongs to the *SOX* family. Homozygosity for a mutation in *SOX9* is seen in XY campomelic dwarfs who develop as females (Kwok et al., 1995). *SOX9* is expressed in the developing cartilage of all mice (hence the dwarfism in homozygous mutants) and in the embryonic gonads of XY (but not XX) mice at the time *SRY* expression is turned on. Further studies are needed to show whether *SOX9* induces (or potentiates) *SRY* expression or is dependent upon it.

Male-to-female sex reversal in rare XY individuals with a normal *SRY* gene indicates that other genes are involved in sex determination. Monosomy for the short arm of chromosome 9 has this effect, and the critical region has been narrowed down to a 3–5 cM region between marker D9S143 and 9pter (Guioli et al., 1998). The *DMRT1* gene maps to this region (Raymond et al., 1998). This gene is of particular interest because it shows significant homology with genes in the roundworm, *Caenorhabditis elegans* (*mab-3*), and in the fruit fly, *Drosophila melanogaster* (*dsx*), that determine male sexual development. Even more interesting is that the homologous gene in the chicken maps to the Z chromosome and is absent from the W chromosome; thus, ZZ birds, with two doses of *DMRT1*, develop as males, while ZW birds, with a single dose, develop as females (Nanda et al., 1999). It is therefore possible that *DMRT1* is of fundamental importance in male sex determination in all metazoans.

There are other causes of XY sex reversal. One of the best known is a mutation of the X-linked androgen receptor gene at Xq11–q12 (Brown et al., 1989). This causes XY individuals to develop into females with the *testicular feminization* syndrome. Affected individuals usually have a female external appearance and often come to the attention of a physician because of primary amenorrhea or sterility. They have androgen-producing testes in the abdominal cavity or the inguinal canal, but the mutant gene renders the target organs insensitive to androgens. XY females of greater than average height and with *gonadal dysgenesis* are said to have Swyer syndrome. They have streak gonads whose lack of oogonia precludes the development of hormone-producing ovarian follicles and secondary sex characteristics (Damiani et al., 1990). Females who have a Y chromosome or the portion of the Y that contains the *TDF* are especially susceptible to malignant gonadal disease. Patients with an androgen insensitivity syndrome have a 10% probability of developing testicular tumors, and in XY gonadal dysgenesis the streak gonads have been reported to undergo malignant transformation in 30% of cases. An early cytological or molecular diagnosis of the presence of Y chromosome material would allow the prophylactic removal of the susceptible tissues.

Other Genes on the Y Chromosome

The Y chromosome was long considered to have almost no genes. This was based on its very short euchromatic part and the absence of male-to-male transmission of any genetic trait (with the possible exception of hairy ears). In addition, the absence of genetic recombination except in the short pseudoautosomal region means that mutations will accumulate throughout the rest of the Y over time and inactivate any but essential genes. The expectation was that the Y chromosome would therefore contain only the male-determining gene and a few genes important for male fertility. However, intensive molecular analysis has shown there are considerably more genes on the Y than expected: now more than 30 (Lahn and Page, 1997). Some of these are in the PAR1 pseudoautosomal region, as expected. Those in the nonrecombining region are of two kinds. One class is specifically expressed in the testis, and many of these are present in multiple copies on the Y. An example is the *DAZ* gene family, which is in a region of Yq that is frequently deleted in azoospermia, hence the name. High-resolution fiber FISH of Y-chromatin has shown that a linear array of at least seven DAZ genes is present, not all in the same orientation (Fig. 8.3). Another example is provided by the *CDY1* and *CDY2* Y-linked genes that show testis-specific expression. These genes contain no introns and are highly homologous to the autosomal, intron-containing *CDYL* (CDY-like) gene that was mapped to chromosome 6 using radiation hybrids. The intronless *CDY1* and *CDY2* genes clearly arose from *CDYL* by retrotransposition of its mRNA, which naturally lacks introns (Lahn and Page, 1999). The other genes are expressed in many tissues and are homologous to genes on the X chromosome despite the lack of recombination between X and Y.

References

Abbas N, Novelli G, Stella NC, et al. (1990) A 45,X male with molecular evidence of a translocation of Y euchromatin onto chromosome 1. Hum Genet 86:94–98

Bernstein R, Wadee A, Rosendorff J, et al. (1986) Inverted Y chromosome polymorphism in the Gujerati Muslim Indian population of South Africa. Hum Genet 74:223–229

Bianchi DW, Williams JM, Sullivan LM, et al. (1997) PCR quantitation of fetal cells in maternal blood in normal and aneuploid pregnancies. Am J Hum Genet 61:822–829

Braun A, Kammerer S, Cleve H, et al. (1993) True hermaphroditism in a 46,XY individual, caused by a postzygotic somatic point mutation in the male gonadal sex-determining locus (*SRY*): molecular genetics and histological findings in a sporadic case. Am J Hum Genet 52:578–585

Brown CJ, Carrel L, Willard HF (1997) Expression of genes from the human active and inactive X chromosomes. Am J Hum Genet 60:1333–1343

Brown CJ, Goss SJ, Lubahn DB, et al. (1989) Androgen receptor locus on the human X chromosome: regional localization to Xq11–12 and description of a DNA polymorphism. Am J Hum Genet 44:264–269

Burgoyne PS (1998) The mammalian Y chromosome: a new perspective. BioEssays 20:363–366

Chandley AC (1998) Chromosome anomalies and Y chromosome microdeletions as causal factors in male infertility. Hum Reprod 13 Supp 1:45–50

Chandley AC, Goetz P, Hargreave TB, et al. (1984) On the nature and extent of XY pairing at meiotic prophase in man. Cytogenet Cell Genet 38:241–247

Damiani D, Billerbeck AEC, Goldberg ACK, et al. (1990) Investigation of the ZFY gene in XX true hermaphroditism and Swyer syndrome. Hum Genet 85:85–88

Disteche CM, Casanova M, Saal H, et al. (1986) Small deletions of the short arm of the Y chromosome in 46,XY females. Proc Natl Acad Sci USA 83:7841–7844

Dutrillaux B (1977) New chromosome techniques. In: Yunis JJ (ed) Molecular structure of human chromosomes. Academic, New York, pp 233–265

Elias S, Price J, Dockter M, et al. (1992) First trimester diagnosis of trisomy 21 in fetal cells from maternal blood. Lancet 340:1033

Freije D, Helms C, Watson MS, et al. (1992) Identification of a second pseudoautosomal region near the Xq and Yq telomeres. Science 258:1784–1787

Genest P (1973) Transmission héréditaire, depuis 300 ans, d'un chromosome Y à satellites dans une lignée familiale. Ann Génét 16:35–38

Graves JAM (1998) Interactions between SRY and SOX genes in mammalian sex determination. BioEssays 20:264–269

Guioli S, Schmitt K, Critcher R, et al. (1998) Molecular analysis of 9p deletions associated with XY sex reversal: refining the localization of a sex-determining gene to the tip of the chromosome. Am J Hum Genet 63: 905–908

Haqq CM, King C-Y, Donald PH, et al. (1993) SRY recognizes conserved DNA sites in sex-specific promoters. Proc Natl Acad Sci USA 90:1097–1101

Hassold TJ, Sherman SL, Pettay D, et al. (1991) XY chromosome nondisjunction in man is associated with diminished recombination in the pseudoautoaomal region. Am J Hum Genet 49:253–260

Hawkins JR, Taylor A, Berta P, et al. (1992) Mutational analysis of SRY: nonsense and missense mutations in XY sex reversal. Hum Genet 88:471–474

Holm PB, Rasmussen SW (1983) Human meiosis V. Substages of pachytene in human spermatogenesis. Carlsberg Res Commun 48:351–383

Ibraimov AI, Mirrakhimov MM (1985) Q-band polymorphism in the autosomes and the Y chromosome in human populations. In: Sandberg AA (ed) The Y chromosome, Part A: Basic characteristics of the Y chromosome. Liss, New York, pp 213–287

Koopman P, Gubbay J, Vivian N, et al. (1991) Male development of chromosomally female mice transgenic for Sry. Nature 351:117–121

Kralog K, Galvagni F, Brown WRA (1994) The sequence organization of the long arm pseudoautosomal region of the human sex chromosomes. Hum Mol Genet 3:771–778

Kwok C, Weller PA, Guioli S, et al. (1995) Mutations in SOX9, the gene responsible for campomelic dysplasia and autosomal sex reversal. Am J Hum Genet 57:1028–1036

Lahn BT, Page DC (1997) Functional coherence of the human Y chromosome. Science 278:675–680

Lahn BT, Page DC (1999) Retroposition of autosomal mRNA yielded testis-specific gene family on human Y chromosome. Nat Genet 21:429–433

Mulhall JP, Reijo R, Alagappan R, et al. (1997) Azoospermic men with deletion of the DAZ gene cluster are capable of completing spermatogenesis: fertilization, normal embryonic development and pregnancy occur when retrieved spermatozoa are used for intracytoplasmic sperm injection. Hum Reprod 12:503–508

Nanda I, Shan Z, Schartl M, et al. (1999) 300 million years of conserved synteny between chicken Z and human chromosome 9. Nat Genet 21:258–259

Petit C, Chappelle A de la, Levilliers J, et al. (1987) An abnormal terminal X-Y interchange accounts for most but not all cases of human XX maleness. Cell 49:595–602

Raymond CS, Shamu CE, Shen MM, et al. (1998) Evidence for evolutionary conservation of sex-determining genes. Nature 391:691–695

Rouyer F, Simmler M-C, Johnsson C, et al. (1986) A gradient of sex linkage in the pseudoautosomal region of the human sex chromosomes. Nature 319:291–295

Schempp W (1985) High-resolution replication of the human Y chromosome. In: Sandberg AA (ed) The Y chromosome, Part A: Basic characteristics of the Y chromosome. Liss, New York, pp 357–371

Schiebel K, Winkelmann M, Mertz A, et al. (1997) Abnormal XY interchange between a novel isolated protein kinase gene, PRKY, and its homologue, PRKX, accounts for one third of all (Y+) XX males and (Y−) XY females. Hum Mol Genet 6:1985–1989

Sinclair AH, Berta P, Palmer MS, et al. (1990) A gene from the human sex determining region encodes a protein with homology to a conserved DNA binding motif. Nature 346:240–244

Swain A, Narvaez V, Burgoyne P, et al. (1998) Dax1 antagonizes Sry action in mammalian sex determination. Nature 391:761–767

Therman E, Sarto GE, Patan K (1974) Center for Barr body condensation on the proximal part of the human Xq: a hypothesis. Chromosoma 44: 361–366

Thomas MR, Tutschek B, Frost A, et al. (1995) The time of appearance and disappearance of fetal DNA from the maternal circulation. Prenatal Diagn 15:641–646

Wang I, Weil D, Levilliers J, et al. (1995) Prevalence and molecular analysis of two hot spots for ectopic recombination leading to XX maleness. Genomics 28:52–58

18

The X Chromosome, Dosage Compensation, and X Inactivation

The X chromosome is of medium size, making up 5.3% of the haploid karyotype. It is submetacentric, with a centromere index of 0.38 and a distinctive banding pattern (Fig. 17.1). In females, one X chromosome is condensed throughout interphase and is frequently visible in epithelial cells as a Barr body, or X heterochromatin. It is visible as a drumstick-shaped extrusion in 1–5% of polymorphonuclear white blood cells (Fig. 18.1a). The Barr body consists of a loop-shaped X chromosome in which the two telomeres lie close together at the nuclear membrane (Walker et al., 1991). Barr bodies can be scored in cells scraped from the mouth (buccal smears; Fig. 18.1b,c) or vagina, in cultured fibroblasts (Fig. 18.1d,e,f), or in follicle cells attached to a plucked hair. In normal females, a Barr body is visible in only 20–50% of buccal cells, in 30–80% of fibroblasts, and in over 90% of cells in amniotic membranes. In every individual (male or female) with two or more X chromosomes, the maximum

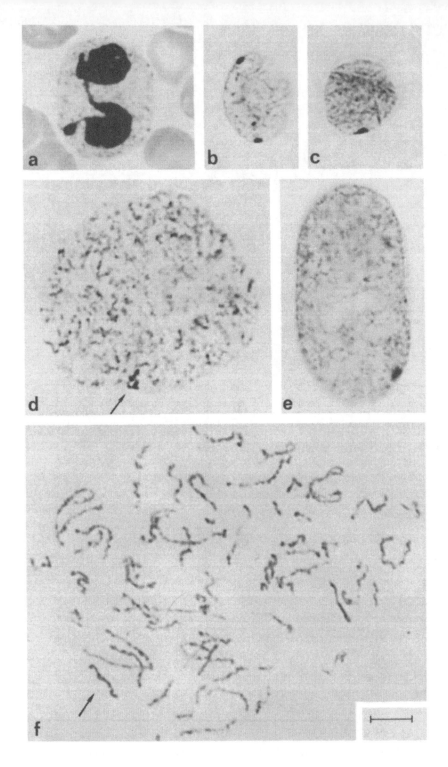

Figure 18.1. Inactive X chromosomes in interphase nuclei. (a) Drumstick in a neutrophil. (b) Two Barr bodies in a buccal nucleus from a 47,XXX woman. (c) Barr body **Figure 18.1.** (*cont.*) in a buccal nucleus from a 46,XX woman. (d) Heteropycnotic chromosome (arrow) in early prophase. (e) Barr body in a fibroblast nucleus. (f) Heteropycnotic chromosome (arrow) in late prophase (Feulgen staining; bar = 5 μm).

number of Barr bodies is one less than the number of X chromosomes. That is, one X remains euchromatic and the additional ones are heterochromatic. The heterochromatic X chromosomes replicate later in S than the euchromatic X, as demonstrated over 35 years ago by autoradiography (Chapter 3). The allocyclic, or out-of-step, behavior of the inactive X chromosome expresses itself in other ways, some described in Chapter 3. In both prophase and metaphase, the inactive X is often more condensed, and thus shorter, than the active X.

The Single Active X (Lyon) Hypothesis

Gene dosage effects can have a profound effect on the phenotype, as shown by autosomal deletions, duplications, and trisomies. However, the presence of a single X chromosome in males, in contrast to the two X chromosomes in females, has no apparent deleterious effect. Lyon (1961) proposed the single active X hypothesis, which provides a mechanism for preventing gene dosage differences between males and females (Fig. 18.2). The hypothesis, based on observations of X-linked genes in mice and Barr bodies in humans, has inspired an enormous amount of research. It has several key features. Each somatic cell has one active X chromosome, and any additional X chromosomes are inactivated early in embryonic development, at an estimated 1000- to 2000-cell stage, if not earlier (Lyon, 1974). The inactivation of the paternal or maternal X chromosome occurs at random. The same X remains active in all the descendants of any cell that underwent X inactivation, thus constituting a clone of cells with the same X active. Most genes on an inactive X are not transcribed; the chromosome is facultatively heterochromatic and may or may not be visible as a Barr body. The low percentage of cells with Barr bodies is not due to reactivation of the inactive X in the other cells, because only one allele at various X-linked loci is expressed in clonal lines despite the abundance of cells without a Barr body.

The single active X hypothesis implies that a female heterozygous for a gene on the X chromosome is a *genetic mosaic* of two different cell populations, one with the maternal X active, the other with the paternal X active. Like XY males, each of the two populations is functionally hemizygous for all the genes subject to X inactivation. The patchy appearance of the skin or eye defect in females heterozygous for such X-linked disorders as anhydrotic ectodermal dysplasia or ocular albinism is easily explained by the Lyon hypothesis. This hypothesis has

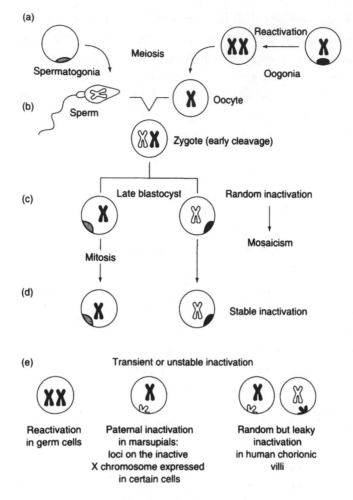

Figure 18.2. Diagram illustrating the single active X hypothesis. Illustration of the single active X hypothesis. (a) The clumped mass at periphery indicates the inactive X in spermatogonia and early oogonia. Note the reactivation of the inactive X in late oogonia. (b) Fusion of sperm and oocyte produces a zygote in which both the paternal (light) and maternal (dark) X chromosomes are active. (c and d) Random inactivation of either the paternal or maternal X occurs, and the same X remains inactive through most subsequent mitotic division differentration. (e) Cell types in which X inactivation is transient or unstable (reprinted from Trends Genet, v10, Migeon, X-chromosome inactivation: molecular mechanisms and genetic consequences, pp 230–235, copyright 1994, with permission of Elsevier Science).

withstood so many critical tests at the cellular level that the analysis of clonal populations is now the standard test of whether an X-linked gene is subject to or escapes X inactivation (facultative heterochromatinization). Davidson et al. (1963) cloned fibroblasts from a woman who was heterozygous at the glucose-6-phosphate dehydrogenase (*G6PD*) locus on the X chromosome. On starch-gel electrophoresis, the initial mass culture contained the enzyme products of both alleles, whereas each clone of cells contained the product of only one or the other allele. Cancer cells are usually clonal, as shown by the presence of the product of only one allele at one or several X-linked gene loci (for example, see Nagel et al., 1995).

Since X inactivation is as likely to involve the maternal X as the paternal X, it is possible to calculate the number of stem cells in a particular cell lineage at the time X inactivation takes place by analyzing the range in the proportion of cells with paternal X inactivation. The greater the range, or skewing from the 50% expected, the fewer the stem cells. Thus, if there were only one stem cell, 100% of cells in that lineage would have the same X inactivated. If there were only two stem cells, the paternal X would be inactive in 100% of cells in one-fourth of females, on average; the maternal X would be inactive in 100% of cells in another one-fourth; and so on. In this way, Puck et al. (1992) estimated that mature T lymphocytes are derived from a pool of only about 10 bone marrow stem cells.

Skewed X Inactivation

Skewed X inactivation is clinically important because it can lead to the expression of X-linked recessive disorders in females who are heterozygous for the mutant allele. Skewed X inactivation is common in female fetuses or newborns who have uniparental disomy of chromosome 15 and whose mothers' pregnancies were marked by confined placental mosaicism, with meiotic origin of the trisomic cell line (Chapter 21). The cause is presumed to be a reduction in the size of the pool of early embryonic cells (Lau et al., 1997).

Migeon et al. (1989) observed skewed X inactivation in females heterozygous for the incontinentia pigmenti mutation, which is lethal in male embryos. They suggested that cells in which the mutant allele is the active one tend to die, leaving only cells in which the normal allele is active in XX embryos; XY embryos, having no normal allele, die. Cell lethality of a mutant allele may also account for the skewed X inactivation that Pegoraro et al. (1997) observed in

16 females in one family. In each of the 16 females, the paternal X was the active X chromosome in more than 95% of the cells. Linkage analysis localized the mutation to Xq28, and molecular studies showed it to be an 800-kb deletion just distal to the *G6PD* locus. The females carrying the deletion were normal except for a doubling of the incidence of spontaneous abortion (to 32%). These presumably represented XY fetuses with the deletion, and the skewed X inactivation again arose because of a cell-lethal karyotype, or genotype. Exclusive (highly skewed) inactivation of the same X chromosome is characteristic of most X;autosome translocations and *XIST* mutations, as discussed below; skewing also occurs in monozygotic twins.

The Critical Region for X Inactivation: X Inactivation Center and the *XIST* Gene

In female carriers of an unbalanced X;autosome translocation (in which only one of the two reciprocal products is present), the Xcen–q13 region is always included in the translocation chromosome. This chromosome, not the normal X, undergoes inactivation in these cells. The shortest possible segment of Xq remaining after deletion or translocation that is still capable of inactivation is the distal end of Xq13, suggesting that this region contains an essential center of X inactivation (Therman et al., 1979). This has been verified by molecular methods (Brown et al., 1991), and positional cloning has been used to identify the gene responsible, called *XIST*. The *XIST* gene does not have a protein product. Instead, the large *XIST* RNA transcript remains in the nucleus, coats the X chromosome from which it is transcribed, and inactivates it (Brown et al., 1992). This involves a major alteration in the chromatin structure, mediated by DNA methylation, which leads to the binding of a methylation-specific DNA-binding protein that recruits a histone deacetylase. Histone H4 on the inactive X thus shows a lack of acetylation (Jeppesen and Turner, 1993). A different histone, called macroH2A1, is preferentially concentrated in the nucleosomes of the inactive X, as shown by combining FISH (using an X-specific probe) with immunofluorescence (using antibodies to macroH2A1). The immunofluorescence identified a *macrochromatin body* (MCB) in a very large percentage of XX cells but in less than 1% of XY cells. Three MCBs were seen in XXXXY cells. MCBs are visible even in cell types in which Barr bodies are not distinguishable, which could make them useful in searching for X chromosome mosaicism (Costanzi and Pehrson, 1998).

In studies of cleavage-stage human embryos derived from an in vitro fertilization program, RNA transcripts of the *XIST* gene are barely detectable as early as the one-cell stage, and more readily detectable by the eight-cell stage, in both XX and XY embryos. Thus, expression is not limited to XX cells or to a single X (Daniels et al., 1997). However, a marked increase in the *Xist* RNA level occurs in mice just before X inactivation. Sheardown et al. (1997) presented evidence that this is mediated by an increase in *Xist* RNA stability in XX embryos. An alternative explanation is that an antisense RNA is transcribed from the opposite strand of the *Xist* gene, starting from a point 15 kb downstream, called the *Tsix* gene. *Tsix* is expressed from both alleles until just before the onset of X inactivation, when its expression from the future inactive X ceases (Lee et al., 1999). *XIST* RNA is abundant in human testes, though not in leukocytes, suggesting that the inactivation of the single X in spermatocytes is mediated by the same mechanism as that for additional X chromosomes (Richler et al., 1992).

Given the importance of the *XIST* gene for X inactivation, its deletion would be expected to result in failure of inactivation of the remainder of the deleted X. Further, a translocation should lead to failure of inactivation of any segment no longer contiguous with *XIST*. The presence of the region containing the center of X inactivation in virtually every X chromosome fragment in unbalanced karyotypes and the failure to find cases of 46,X,iso(Xp) or 46,X,tel(Xp) suggest that dosage effects of normally inactivated genes on X are lethal (Therman and Sussman, 1990). Tiny ring X chromosomes are informative in this regard. They have a profound phenotypic effect, with severe mental retardation, congenital anomalies, and growth retardation. The expression of the *XIST* gene is absent or greatly reduced in these individuals, supporting the idea that failure of inactivation of one or more genes carried by the ring X is responsible for the phenotype (Migeon et al., 1994).

Once the *XIST* gene has initiated X inactivation, its continued activity is not required. Thus, an isodicentric X chromosome that had lost the *XIST* locus in a leukemic cell line has remained late replicating (Rack et al., 1994). X inactivation appears to be locked in by DNA methylation (Kaslow and Migeon, 1987).

The *XIST* gene is the only gene known to be active on the inactive X and inactive on the active X. Is it also unique in its replication pattern? Boggs and Chinault (1994) used the interphase FISH technique (Chapter 8) to show that the two *XIST* alleles in XX cells replicate very asynchronously, just like the alleles of genes that undergo normal X inactivation, and unlike the synchronous replication of genes that escape X inactivation, such as *ZFX* and *RPS4X*. Torchia et al. (1994) obtained comparable results using interphase FISH and probes for *XIST*, *FMR*1, and *factor* 8C (hemophilia A) loci and concluded that the active *XIST* allele was early replicating. However, Hansen et al. (1995) obtained different results, using a different method. They grew cells in the presence of bromodeoxyuridme (BrdU) for 1 hour and used flow sorting to separate the cells into G1, S1, S2, S3, S4, and G2/M fractions. Antibodies to BrdU were then used to isolate the DNA in each fraction that had replicated during each interval and gene probes used to determine when a specific gene was replicated. Their results indicated that the active *XIST* gene replicates late and its silent allele replicates early (Fig. 8.2). They confirmed this by using a modification of the FISH technique (Gartler et al., 1999).

Reactivation of the X Chromosome

The inactive X chromosome is reactivated in oocytes at some time before meiosis. Both X chromosomes are transcribed, and neither shows heteropycnotic behavior. In some tissues, a few normally inactive X-linked genes are active in some cells: X inactivation is incomplete, or "leaky." Moreover, the whole inactive X in trophoblastic cells may become reactivated in human-mouse hybrid cells (Migeon et al., 1986). Rarely, a few genes are reactivated spontaneously. However, treatment with the demethylating agent 5-azacytidine can reactivate X-linked genes more consistently (Mohandas et al., 1981).

Regions That Escape X Inactivation: Functional Map of the X Chromosome

Observations on structural abnormalities of the X chromosome have contributed greatly to understanding of the functional map of the X (Fig. 18.3). A few examples will serve to illustrate this. A tiny deletion of the tip of Xp causes short stature in both females and males; males also have other abnormalities. This suggests the presence of an expressed (noninactivated) gene in this region. A candidate is the *SHOX* gene, which maps to the pseudoautosomal region in humans but is autosomal in mice and might therefore account for impaired growth in XO humans but not in XO mice (Rao et al., 1997). Xp and Xq deletions generally produce similar Turner syndrome phenotypes, apart from short stature

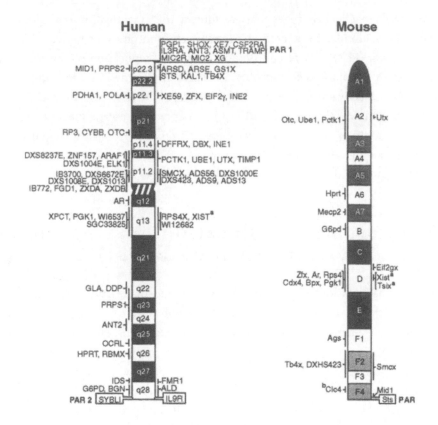

Figure 18.3. Functional maps of the human and mouse X chromosomes. Genes that show X-inactivation are listed to the left of each ideogram, while genes that escape X inactivation are listed to the right of each ideogram (courtesy of CM Disteche).

(Table 19.1), and no specific abnormalities are induced by Xq deletions. Therman and Susman (1990) suggested that a deleted X chromosome has a surplus of the factor needed for inactivation and thus inactivates regions that usually remain active. Against this notion is the fact that the *XIST* RNA normally traverses interstitial regions that remain active to reach more distant regions it does inactivate, so its amount is not critical. Furthermore, if these regions were fully inactivated, the phenotypes of 45,X, 46,X,Xp−, and 46,X,Xq− females should always be indistinguishable, and that is not the case (Table 19.1). The spreading of inactivation into autosome segments can also leapfrog over regions that escape inactivation, for reasons that remain to be discovered. Further study and new approaches are needed to resolve this problem.

Homologous loci of genes that are subject to X inactivation replicate asynchronously, just like the active and inactive X chromosomes themselves (Schmidt and Migeon, 1990). Therefore, synchronous early replication of any region on the two X chromosomes probably indicates that genes in the region have escaped X inactivation. This is certainly true for the genes in the pseudoautosomal region near the tip of Xp: All of them appear to escape inactivation (Fig. 18.3). In addition, several other early-replicating regions on X escape inactivation, including the nonpseudoautosomal region of Xp22.3 and regions Xp11.2–p11.4 and Xq13 (reviewed by Disteche, 1995).

X;Autosome Translocations and Spreading of Inactivation (Position Effect)

X;autosome translocations involving every autosome have been observed, although the autosome is either 21 or 22 in one-third of the translocations. A reciprocal t(X;autosome) usually produces a balanced chromosome complement in which there is no loss or gain of either an X or an autosome segment. Meiotic segregation in a balanced translocation carrier (heterozygote) can lead to unbalanced karyotypes (Chapter 16). In adult female carriers of balanced X;autosome translocations, the normal X chromosome is inactivated in almost all cases, at least in lymphocytes. The reason for this may be that inactivation of the autosome segment attached to the portion of the X chromosome with the X-inactivation center would result in functional monosomy for the autosome segment, while failure of inactivation of the other part of the X chromosome would result in functional trisomy for that segment of the X. Thus, while the normal X and the t(X;A) chromosome containing the inactivation center may initially undergo random inactivation, the chromosomally unbalanced cells may die or grow slowly, so that the better-balanced cell line replaces the other line.

In carriers of unbalanced X;autosome translocations, who often are offspring of balanced carriers, the abnormal chromosome is almost always inactivated, presumably because spreading of inactivation to the autosome segment of the single translocation chromosome would prevent functional trisomy of the autosome segment. Whether this functional trisomy is really that detrimental at a cellular level is questionable, and the reason for skewing of X inactivation in these cases is not clear. Nevertheless, this explanation for the observed replication patterns is consistent with the results of more extensive studies which indicate many balanced X;autosome translocation carriers have a minority cell line in which one of the translocation chromosomes is inactivated. In one case, the normal X was late replicating in all lymphocytes of a woman with a 46,X,t(X;3)(q28;q21) karyotype. However, the normal X was late replicating in only two-thirds of cultured fibroblasts. In the other one-third, the X-chromosome portion but not the autosomal portion of the Xq+ chromosome was late replicating. This indicates that inactivation had not spread to the autosomal segment, thus preventing functional partial monosomy for that segment (Hellkuhl et al., 1982).

Spreading of inactivation into an autosome segment translocated to the X chromosome is a well-known but still poorly understood phenomenon. Allderdice et al. (1978) showed by autoradiographic analysis that inactivation could spread throughout virtually the entire long arm of a chromosome 14 to which the long arm of the X was attached, thus accounting for the relatively normal phenotype of a 46,der(X)t(X;14),der(X)t(X;14),Y male despite the presence of almost three complete copies of chromosome 14. The same method showed that spreading of inactivation was responsible for the absence of Down syndrome in a t(X;21)(q27;q11) translocation carrier (Couturier et al., 1979). Replication studies suggest that spreading of inactivation can leapfrog over autosomal regions that remain early replicating. This is not surprising, since the same is true of the X chromosome, in which early-replicating regions containing genes that escape X inactivation are interspersed with late-replicating, inactive regions throughout its length (Fig. 18.3).

White et al. (1998) used molecular probes to extend this kind of analysis a step further. They studied a woman whose karyotype was 46,X,der(X) t(X;4)(q22;q24). She was phenotypically normal despite duplication (partial trisomy) of the distal half of chromosome 4, which is usually associated with severe retardation of growth and mental development. Spreading of inactivation was demonstrated by showing that three genes and 11 ESTs (expressed sequence tags for anonymous genes) scattered along the 100-Mb segment of 4q were not expressed in a somatic cell hybrid containing only the der(X) translocation chromosome. However, three genes and three ESTs interspersed among the inactive genes were expressed. That is, 14/20 autosomal genes had been inactivated, while 6/20 escaped inactivation, a much higher fraction than that found for X-linked genes. White and associates suggested this was evidence for a difference between autosomal and X chromosomal DNA, but it also indicates a great deal of similarity in the DNA, its organization, and the mechanisms of inactivation of genes on the two types of chromosomes.

Keohane et al. (1999) used FISH with both RNA and DNA probes to show that sites of *XIST* RNA, histone H4 deacetylation, and late replication coincided and that they marked the boundaries of inactivation in an X;autosome translocation and an autosomal insertion into an X. Using RNA FISH, Duthie et al. (1999) showed that *XIST* RNA is localized mainly to R-bands, not to constitutive heterochromatin, and that the spread of *XIST* RNA into the autosomal segment of an X;autosome translocation matched the spread of inactivation. The DNA demethylating agent 5-azacytidine hampers mitotic condensation of the inactive X (Haaf et al., 1993). It would be interesting if it did the same for inactive autosomal regions in X;autosome translocation chromosomes.

References

Allderdice PW, Miller OJ, Miller DA, et al. (1978) Spreading of inactivation in an (X;14) translocation. Am J Med Genet 2:233–240

Boggs BA, Chinault AC (1994) Analysis of replication timing properties of human X-chromosomal loci by fluorescence *in situ* hybridization. Proc Natl Acad Sci USA 91:6083–6087

Brown CJ, Hendrick BD, Rupert JL, et al. (1992) The human *XIST* gene: analysis of a 17 kb inactive X-specific RNA that contains conserved repeats and is highly localized within the nucleus. Cell 71:527–542

Brown CJ, Lafreniere RG, Powers VE, et al. (1991) Localization of the inactivation centre on the human X chromosome in Xq13. Nature 349:82–84

Costanzi C, Pehrson JR (1998) Histone macroH2A1 is concentrated in the inactive X chromosome of female mammals. Nature 393:599–601

Couturier J, Dutrillaux B, Garber P, et al. (1979) Evidence for a correlation between late replication and autosomal gene inactivation in a familial translocation t(X;21). Hum Genet 49:319–326

Daniels R, Zuccotti M, Kinis T, et al. (1997) *XIST* expression in human oocytes and preimplantation embryos. Am J Hum Genet 61:33–39

Davidson RG, Nitowsky HM, Childs B (1963) Demonstration of two populations of cells in the human female heterozygous for glucose-6-phosphate dehydrogenase variants. Proc Natl Acad Sci USA 50:481–485

Disteche CM (1995) Escape from X-inactivation in human and mouse. Trends Genet 11:17–22

Duthie SM, Nesterova TB, Formstone EJ, et al. (1999) *Xist* RNA exhibits a banded localization on the inactive X chromosome and is excluded from autosomal material in *cis*. Hum Mol Genet 8:195–204

Gartler SM, Goldstein L, Tyler-Freer SE, et al. (1999) The timing of *XIST* replication: dominance of the domain. Hum Mol Genet 8:1085–1089

Haaf T, Werner P, Schmid M (1993) 5-Azacytidine distinguishes between active and inactive X chromosome condensation. Cytogenet Cell Genet 63: 160–168

Hansen RS, Canfield TK, Gartler SM (1995) Reverse replication timing for the XIST gene in human fibroblasts. Hum Mol Genet 4:813–820

Hellkuhl B, Chapelle A de la, Grzeschik K-H (1982) Different patterns of X chromosome inactivity in lymphocytes and fibroblasts of a human balanced X;autosome translocation. Hum Genet 60:126–129

Jeppesen P, Turner BM (1993) The inactive X chromosome in female mammals is distinguished by a lack of histone H4 acetylation, a cytogenetic marker for gene expression. Cell 74:281–289

Kaslow DC, Migeon BR (1987) DNA methylation stabilizes X chromosome inactivation in eutherians but not in marsupials: evidence for multistep maintenance of mammalian X dosage compensation. Proc Natl Acad Sci USA 84:6210–6214

Keohane AM, Barlow AL, Waters J, et al. (1999) H4 acetylation, *XIST* RNA and replication timing are coincident and define X;autosome boundaries in two abnormal X chromosomes. Hum Mol Genet 8:377–383

Lau AW, Brown CJ, Peñaherrera M, et al. (1997) Skewed X-chromosome inactivation is common in fetuses and newborns with confined placental mosaicism. Am J Hum Genet 61:1353–1361

Lee JT, Davidow LS, Warshawsky D (1999) *Tsix*, a gene antisense to *Xist* at the X-inactivation centre. Nat Genet 21:400–404

Lyon MF (1961) Gene action in the X-chromosome of the mouse (*Mus musculus* L.). Nature 190:372–373

Lyon MF (1974) Mechanisms and evolutionary origins of variable X-chromosome activity in mammals. Proc R Soc Lond B 187:243–268

Migeon BR (1994) X-chromosome inactivation: molecular mechanisms and genetic consequences. Trends Genet 10:230–235

Migeon BR, Schmidt M, Axelman J, et al. (1986) Complete reactivation of X chromosomes from human chorionic villi with a switch to early DNA replication. Proc Natl Acad Sci USA 83:2182–2186

Migeon BR, Axelman J, Beur SJ, et al. (1989) Selection against lethal alleles in females heterozygous for incontinentia pigmenti. Am J Hum Genet 44:100–106

Migeon BR, Luo S, Jani M, et al. (1994) The severe phenotype of females with tiny ring X chromosomes is associated with inability of these chromosomes to undergo X inactivation. Am J Hum Genet 55:497–504

Mohandas T, Sparkes RS, Shapiro LJ (1981) Reactivation of an inactive human X-chromosome: evidence for X inactivation by DNA methylation. Science 211:393–396

Nagel S, Borisch B, Thein SL, et al. (1995) Somatic mutation detected by mini- and microsatellite DNA markers reveal clonal intratumor heterogeneity in gastrointestinal cancers. Cancer Res 55:2866–2870

Pegoraro E, Whitaker J, Mowery-Rushton P, et al. (1997) Familial skewed X-inactivation: a molecular trait associated with high spontaneous abortion rate maps to Xq28. Am J Hum Genet 61:160–170

Puck JM, Stewart CC, Nussbaum RL (1992) Maximum likelihood analysis of human T-cell X chromosome inactivation patterns: normal women versus carriers of X-linked severe combined immunodeficiency. Am J Hum Genet 50:742–748

Rack KA, Chelly J, Gibbons RJ, et al. (1994) Absence of the XIST gene from late-replicating isodicentric X chromosomes in leukemia. Hum Mol Genet 3:1053–1059

Rao E, Weiss B, Fukami M, et al. (1997) Pseudoautosomal deletions encompassing a novel homeobox gene cause growth failure in idiopathic short stature and Turner syndrome. Nat Genet 16:54–63

Richler C, Soreq H, Wahrman J (1992) X inactivation in mammalian testis is correlated with inactive X-specific transcription. Nat Genet 2:192–195

Schmidt M, Migeon BR (1990) Asynchronous replication of homologous loci on human active and inactive X chromosomes. Proc Natl Acad Sci USA 87:3685–3689

Sheardown SA, Duthie SM, Johnston CM, et al. (1997) Stabilization of *Xist* RNA mediates initiation of X chromosome inactivation. Cell 91:99–107

Therman E, Sarto GE, Palmer CG, et al. (1979) Position of the human X inactivation center on Xq. Hum Genet 50:59–64

Therman E, Susman B (1990) The similarity of phenotypic effects caused by Xp and Xq deletions in the human female: a hypothesis. Hum Genet 85: 175–183

Torchia BS, Call LM, Migeon BR (1994) DNA replication analysis of FMR1, XIST, and factor 8C loci by FISH shows nontranscribed X-linked genes replicate late. Am J Hum Genet 55:96–104

Walker CL, Cargile CB, Floy KM, et al. (1991) The Barr body is a looped X chromosome formed by telomere association. Proc Natl Acad Sci USA 88:6191–6195

White WM, Willard HF, Van Dyke DL, et al. (1998) The spreading of X inactivation into autosomal material of an X;autosome translocation: evidence for a difference between autosomal and X chromosomal DNA. Am J Hum Genet 63:20–28

19

Phenotypic Effects of Sex Chromosome Imbalance

The sex chromosomes show a much wider range of viable aneuploidy than do the autosomes, for several reasons. Each diploid somatic cell has only one active X, and most of the genes on any additional X chromosomes are inactivated. The Y chromosome contains very few genes. Mosaicism, with a normal cell line present, is much more common for sex chromosomes than for autosomes. Figure 19.1 summarizes the known nonmosaic numerical sex chromosome abnormalities. In addition to the examples in this figure, the chromosome constitution XYYYY has been found in a few highly abnormal individuals (Noël et al., 1988).

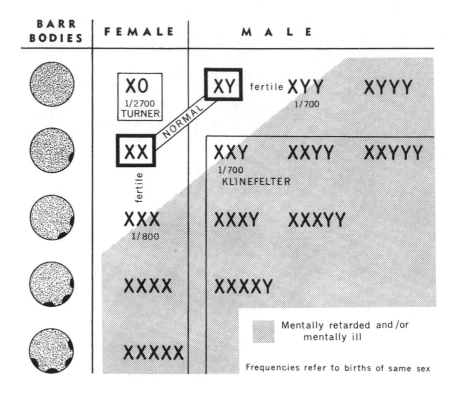

Figure 19.1. The variety of human sex chromosome aberrations and the correlation between the number of X chromosomes and the maximum number of Barr bodies per cell. The rare XYYYY karyotype is not included.

Turner Syndrome

Individuals with short stature (under 5 feet, or 150 cm), webbed neck, low hairline, increased carrying angle at the elbow (cubitus valgus), elevated pituitary gonadotrophins, and failure to undergo secondary sexual development in adolescence are said to have Turner syndrome. The sexual infantilism is due to the absence or near-absence of oocytes, leaving only a streak gonad rich in connective tissue (*gonadal dysgenesis*). In a study of 211 patients clinically diagnosed with Turner syndrome, 97 were XO, 15 were XO/XX or XO/XXX mosaics, 86 had a structurally abnormal X (mainly isochromosome, deletion, or ring), and 13 had a structurally abnormal Y (Jacobs et al., 1997). The parent of origin of the single X chromosome had no effect on birth weight, height, neck webbing, cardiovascular or renal anomalies, or thyroid autoimmunity, ruling out imprinting effects of the sort described in Chapter 21 (Mathur et al., 1991). However, a maternally imprinted (paternally expressed) gene between Xp11.2 and Xqter that escapes inactivation may be involved in cognitive development, affecting verbal and social skills in both XO and XY individuals (Skuse et al., 1997).

One of the prerequisites for production of normal gametes is complete pairing of chromosomes in meiotic prophase. Even in XX fetuses, some 30% of oocytes show pairing abnormalities. This may be the chief cause of oocyte degeneration. In early embryos, the oocytes number some 7 million; at birth the number has decreased to 2 million (Speed, 1988). In most 45,X subjects there are extremely few if any oocytes, and this deficiency is apparent even before the pachytene stage of meiosis (Speed, 1986). A gene dosage effect, rather than a pairing anomaly, may be the cause of this early destruction of oocytes by apoptosis (programmed cell death). This effect is mediated by caspases and preventable by the *BCL2* gene product, an inhibitor of the p53– mediated pathway to apoptosis (Chapter 28). Lacking oocytes, the gonads are connective tissue streaks that can produce no ovarian follicles or corpora lutea, so there is primary amenorrhea or, if a few oocytes remain, secondary amenorrhea. Additional phenotypic abnormalities in Turner syndrome include shield chest, abundant pigmented nevi, and defects of the urinary and cardiovascular systems, such as coarctation of the aorta (Table 19.1). Mental retardation is not a feature of Turner syndrome. Over 99% of 45,X zygotes are lost by spontaneous abortion or fetal death, for unexplained reasons (Chapter 11). Prenatal nuchal edema (fluid accumulation at the base of the neck) is common, and over half of 45,X newborns have lymphedema of the hands and feet.

Is There a Critical Region for Turner Syndrome?

Genes that are subject to X inactivation are unlikely to be responsible for the somatic features of Turner syndrome, because their dosage is the same in 45,X and 46,XX individuals. They could, of course, play a role in gonadal dysgenesis, because both X chromosomes are active in oocytes. Consequently, genes that escape X inactivation and any others that have functional homologues on Y are the best candidate genes for the somatic Turner phenotype. About 20 of these genes are known, including the cluster in the 2.6-Mb pseudoautosomal region on Xp. In fact, short stature is present when this pseudoautosomal region is deleted (Ogata and Matsuo, 1995). The *SHOX* (short stature-homeobox containing) gene

Table 19.1. Turner Syndrome Features of Nonmosaic Adults with 45,X, 46,X,Xp-, and 46,X,Xq- Karyotypes, All Ascertainment Types Pooled

Feature	45,X (%) $n = 332$	46,X,Xp- (%) $n = 52$	46,X,Xq- (%) $n = 67$
Gonadal dysgenesis	91	65	93
Short stature	100	88	43
Short neck	77	38	21
Low hairline	72	19	9
Shield chest	74	35	13
Cubitus valgus	77	25	16
Short metacarpals	55	29	12
Pigmented nevi	64	27	19
Webbed neck	42	2	1
Autoimmune thyroid disease	18	6	3
Color blindness	11		3

Source: Revised from Therman and Sussman, 1990

maps to the pseudoautosomal region of the X and Y chromosomes. Patients with the fairly mild condition, Leri-Weill dyschondrosteosis, are heterozygous for a deletion or, more rarely, a point mutation of this gene, while patients with the more severe one, Langer mesomelic dwarfism, are homozygous (Shears et al., 1998). The reduced copy number of this gene in 45,X individuals makes it a candidate gene for their short stature. Rao et al. (1997) found that 36 individuals with short stature and various rearrangements of Xp22 or Yp11.3 all had a deletion of a 170-kb region in PAR1, and they isolated the *SHOX* gene from this region. Furthermore, one of 91 individuals with short stature of unknown origin had a mutation of *SHOX*, confirming its importance for normal development. Of course, stature is a multifactorial trait influenced by other factors and multiple genes. One of these, the *RPS4X* gene, encodes ribosomal protein S4, a protein essential for growth. *RPS4X* escapes X inactivation and has an expressed homologue on Y, consistent with a role in Turner syndrome. However, four X,i(Xq) individuals had the full Turner phenotype despite having three copies of RPS4X and an increased level of its mRNA, indicating that haploinsufficiency of this gene (or probably any other gene on Xq) is not responsible for the short stature or other features of Turner syndrome (Geerkens et al., 1996).

Other features of Turner syndrome are not associated with deletion of the pseudoautosomal region, but most of them are associated with a deletion in Xp. Zinn et al. (1998) determined the size of the deletion in 28 nonmosaic subjects with partial deletions of Xp, by using FISH with a battery of Xp probes. Deletion of a region in the much more proximal Xp11.2–p22.1 was critical for ovarian failure (gonadal dysgenesis), short stature, high arched palate, and possibly autoimmune thyroid disease but not for lymphedema, webbed neck, or coarctation of the aorta. Some of these features are more commonly associated with deletions of Xq, which can also cause gonadal dysgenesis. Turner syndrome clearly requires the loss of genes from more than one region of the X chromosome.

At least one other gene on X affects ovarian function. Women who are heterozygous for an X;autosome translocation with a breakpoint in the short arm or the proximal long arm of the X usually have normal ovarian function, but those with a breakpoint in Xq13–q26, the critical region for ovarian function, almost invariably have gonadal insufficiency with reduced fertility. The exceptions are individuals with the breakpoint in Xq22, suggesting that there are two critical regions separated by a small intercalary segment in Xq22 (Madan, 1983). However, some women with a deletion of all or part of the critical region have had normal fertility, casting some doubt on this critical-region hypothesis. One explanation may be that genes for female fertility have accumulated on the X chromosome. Alternatively, normal ovarian function may depend upon the maintenance of continuity between genes from each part of the critical region (Madan, 1983). Future studies are needed to clarify this intriguing situation.

Polysomy X

The genes on the X that escape X inactivation probably play an important role in the causation of abnormal phenotypes in some of the individuals with abnormal X chromosome numbers. Subjects with three X chromosomes do not have a well-defined phenotype, although they are sometimes mildly or moderately mentally retarded. They have normal fertility, and almost all their children have normal karyotypes, instead of half their daughters having 47,XXX and half their sons 47,XXY karyotypes, as expected. A likely explanation is that the extra chromosome is preferentially segregated to the polar body at the first meiotic division in the oocyte, by an unknown mechanism. Subjects with more than three X chromosomes have severe mental retardation and various somatic anomalies but normal sexual development. Nielsen et al. (1977) reviewed the 26 known cases of 48,XXXX. The 25th case of 49,XXXXX was reported only in 1997. It was diagnosed prenatally after ultrasound had revealed hydrocephaly, with dilatation of the third and fourth ventricles of the brain. FISH analysis with an X-specific probe revealed five hybridization signals (dots) in amniotic fluid cell nuclei, and standard karyotyping confirmed the diagnosis (Myles et al., 1997).

Klinefelter Syndrome

Klinefelter syndrome is characterized by tiny testes devoid of sperm cells, sterility, fairly normal production of male hormones, a tendency to breast development (gynecomastia), and sometimes a eunuchoid habitus. The 47,XXY karyotype is the major cause of Klinefelter syndrome. The extra sex chromosome seems to lower the IQ to some extent, although only a minority of affected individuals have even mild mental retardation. About one in 700 males is XXY, and many of them would like to have children. Male infertility and sterility are usually due to a low sperm count in the semen (oligospermia) or a total absence of sperm (azoospermia), and that is the case for XXY men. Nevertheless, a few XXY males have fathered children, as confirmed by genetic markers, including DNA fingerprinting (Terzoli et al., 1992). Some XXY males do produce a small number of sperm, and while their low sperm counts make fertility very unlikely, the in vitro fertilization technique called *intracytoplasmic sperm injection* (ICSI) has enabled these men to become biological fathers (Hinney et al., 1997). Three-color FISH analysis of decondensed sperm nuclei in one XXY male revealed a deficiency of X-bearing sperm and a small but significant increase in XX sperm, XY sperm, and sperm with no sex chromosomes (Guttenbach et al., 1997).

Males with one Y chromosome and more than two X chromosomes also develop Klinefelter syndrome, but they are more severely retarded and display other somatic abnormalities. For example, 49,XXXXY males have an IQ in the range of 20 to 50, extensive skeletal anomalies, severe hypogenitalism, strabismus, wide-set eyes, and other anomalies. Well over 100 such males have been reported. XXXY males are not as frequent, for unknown reasons.

Phenotypes Associated with Multiple Y Karyotypes

The 47,XYY condition was, unfortunately, sensationalized by early investigators, and consequently by the news media. XYY individuals were incorrectly depicted as having an increased likelihood of engaging in violent behavior. XYY, like XXY and XXYY, is sometimes associated with mild mental retardation, and it is these groups who have a somewhat higher probability of coming to the attention of the police than do normal males, although generally for nonviolent offenses (Welch, 1985). Most XYY males lead normal lives, and their only distinguishable feature may be their height: They usually are considerably taller than their male relatives, often being over 6 feet tall (180 cm). With a few exceptions, their fertility is not impaired. Early studies on testis biopsies suggested that the extra Y chromosome is lost from the male germline at an early stage. However, some XYY cells do enter meiosis, showing an unpaired X and a YY bivalent (Speed and Chandley, 1990). Three-color FISH analysis with centromeric probes specific for chromosomes X, Y, and 18 showed 1% YY disomic sperm and 0.3% XY disomic sperm, both considerably higher frequencies than seen in controls (Blanco et al., 1997).

Individuals with aneuploidies like XXYY, XXXYY, XYYY, and XYYYY are rare. They are usually retarded and also display somatic anomalies. The ones with more than one X chromosome have Klinefelter syndrome as well.

Deletions and Duplications of the X Chromosome: Risks Associated with Hemizygosity

XY males have a single X chromosome. Consequently, a deletion or duplication of a segment of the X usually has a far more serious phenotypic effect in males than in females, who have a second X and the potential for preferential (skewed) X inactivation to reduce the proportion of cells in which the abnormal X is active. Deletions in the X in an XY individual produce nullisomy for the deleted segment. This can lead to embryonic lethality or to multiple X-linked disorders in a single individual, the so-called contiguous gene syndromes (Ledbetter and Ballabio, 1995). Figure 19.2 illustrates contiguous gene syndromes for regions Xp22, Xp21, and Xq21. Each involves at least five known genes, in contrast to the sometimes many fewer seen in the segmental aneusomies of autosomes (Chapter 15). Deletions involving either the long or the short arm of the X can produce some features of Turner syndrome (Table 19.1).

Duplications of a segment of X can also have serious effects in males. An inherited duplication of the gene-rich Xq27–qter region produced severely affected sons in three unrelated families. They had intrauterine growth retardation, marked hypotonia, marked developmental delay, and microcephaly. The duplication was characterized using FISH with multiple X-specific probes (Fig. 19.3;

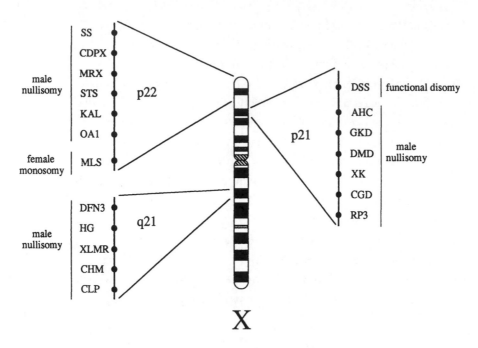

Figure 19.2. Contiguous gene syndromes produced by segmental aneusomy of the X chromosome (Ledbetter and Ballabio, Molecular cytogenetics of contiguous gene syndromes: mechanisms and consequences of gene dosage imbalance, copyright 1995 McGraw-Hill, with permission of the McGraw-Hill Companies).

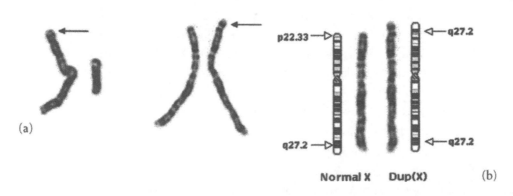

Figure 19.3. Duplication of Xq27.2–qter, transmitted from mother to son. (a) The sex chromosomes of both. (b) The mother's sex chromosomes and corresponding ideograms. Arrows indicate the breakpoints (Goodman et al., Am J Med Genet 80:379, copyright 1998. Reprinted by permission of Wiley-Liss, Inc., a subsidiary of John Wiley & Sons, Inc.).

Goodman et al., 1998). Inversions in X affect the phenotype only when they disrupt a gene (Chapter 16). However, aberrant segregation in carriers of a pericentric inversion of the X have produced at least nine individuals with duplication/deficiencies in the X. They usually have features of Turner syndrome, including ovarian (gonadal) dysgenesis (Madariaga and Rivera, 1997).

X;Autosome and Y;Autosome Translocations

Male carriers of an X;autosome translocation, like their female counterparts, usually have a normal somatic phenotype, but they are even more likely than female carriers to have severe infertility. In fact, most t(X;A) males have azoospermia, regardless of the location of the breakpoint in the X (Madan, 1983). The same is true of men with a Y;autosome translocation. Meiosis in these males usually shows considerable *asynapsis* (failure of pairing) around the breakpoint regions and fairly large segments of the autosome included in the heterochromatic sex vesicle. This facultative heterochromatinization of an autosome segment is presumably accompanied by inactivation of the genes in this segment. These changes are followed by degeneration of the spermatocytes soon after the pachytene stage (Delobel et al., 1998). There is a clear relationship between male sterility and the autosome–sex vesicle association in both X;autosome and Y;autosome translocations (Gabriel-Robez and Rumpler, 1996).

Male infertility and sterility are associated with any meiotic abnormality that interrupts the development of spermatocytes at some stage, usually during the first meiotic division. Many infertile men can now become biological fathers through ICSI. In a study of 1792 men, carried out prior to ICSI, 72 (4%) had a constitutional chromosome abnormality, including 31 translocations; 10 inversions; and 7 XXY, 6 XYY, and 5 XO/XY mosaics (Tuerlings et al., 1998). In view of the slightly increased risk of chromosomally abnormal sperm in infertile men, prenatal cytogenetic studies are usually recommended.

References

Blanco J, Rubio C, Simon C, et al. (1997) Increased incidence of disomic sperm nuclei in a 47,XYY male assessed by fluorescent in situ hybridization (FISH). Hum Genet 99:413–416

Delobel B, Djlelati R, Gabriel-Robez O, et al. (1998) Y-autosome translocation and infertility: usefulness of molecular, cytogenetic and meiotic studies. Hum Genet 102:98–102

Gabriel-Robez O, Rumpler Y (1996) The meiotic pairing behavior in human spermatocyte carriers of chromosome anomalies and their repercussions on reproductive fitness. II. Robertsonian and reciprocal translocations. A European study. Ann Génét 39:17–25

Geerkens C, Just W, Held KR, et al. (1996) Ullrich-Turner syndrome is not caused by haploinsufficiency of RPS4X. Hum Genet 97:39–44

Goodman BK, Shaffer LG, Rutberg J, et al. (1998) Inherited duplication Xq27–qter at Xp22.3 in severely affected males: molecular cytogenetic evaluation and clinical description in three unrelated families. Am J Med Genet 80:377–384

Guttenbach M, Michelman HW, Hinney B, et al. (1997) Segregation of sex chromosomes into sperm nuclei in a man with 47,XXY Klinefelter's karyotype: a FISH analysis. Hum Genet 99:474–477

Hinney B, Engel W, Guttenbach M, et al. (1997) Pregnancy after intracytoplasmic sperm injection with sperm from a man with a 47,XXY Klinefelter's karyotype. Fertil Steril 68:718–720

Jacobs P, Dalton P, James R, et al. (1997) Turner syndrome: a cytogenetic and molecular study. Am J Hum Genet 61:471–483

Ledbetter DH, Ballabio A (1995) Molecular cytogenetics of contiguous gene syndromes: mechanisms and consequences of gene dosage imbalance. In: Scriver C, Beaudet AL, Sly WS, Valle, D (eds) The metabolic and molecular bases of inherited disease, 7th edn, McGraw-Hill, New York, pp 811–839

Madan K (1983) Balanced structural changes involving the human X: effect on sexual phenotype. Hum Genet 63:216–221

Madariaga ML, Rivera H (1997) Familial inv(X)(p22q22): ovarian dysgenesis in two sisters with del Xq and fertility in one male carrier. Clin Genet 52: 180–183

Mathur A, Stekol L, Schatz D, et al. (1991) The parental origin of the single X chromosome in Turner syndrome: lack of correlation with parental age or clinical phenotype. Am J Hum Genet 48:682–686

Myles TD, Burd L, McCorquodale MM, et al. (1997) Dandy-Walker malformation in a fetus with pentasomy X (49,XXXXX) prenatally diagnosed by fluorescence in situ hybridization technique. Fetal Diagn Ther 10:333–336

Nielsen J, Homma A, Christiansen F, et al. (1977) Women with tetra-X (48,XXXX). Hereditas 85:151–156

Noël B, Bénézech M, Bouzon MT, et al. (1988) Un garçon de sept ans 49,XYYYY. Ann Génét 31:111–116

Ogata T, Matsuo N (1995) Turner syndrome and female sex chromosome aberrations: deduction of the principal factors involved in the development of clinical features. Hum Genet 95:607–629

Rao E, Weiss B, Fukami M (1997) Pseudoautosomal deletions encompassing a novel homeobox gene cause growth failure in idiopathic short stature and Turner syndrome. Nat Genet 16:54–63

Shears DJ, Vassal HJ, Goodman FR, et al. (1998) Mutation and deletion of the pseudoautosomal gene SHOX cause Leri-Weill dyschondrosteosis. Nat Genet 19:70–73

Skuse DH, James RS, Bishop DVM, et al. (1997) Evidence from Turner's syndrome of an imprinted X-linked locus affecting cognitive function. Nature 387:705–708

Speed RM (1986) Oocyte development in XO foetuses of man and mouse: the possible role of heterologous X-chromosome pairing in germ cell survival. Chromosoma 94:115–124

Speed RM (1988) The possible role of meiotic pairing anomalies in the atresia of human fetal oocytes. Hum Genet 78:260–266

Speed RM, Chandley AC (1990) Prophase of meiosis in human spermatocytes analysed by EM microspreading in infertile men and their controls and comparisons with human oocytes. Hum Genet 84:547–554

Terzoli G, Lalatta F, Lobbiani A, et al. (1992) Fertility in a 47,XXY patient: assessment of biological paternity by deoxyribonucleic acid fingerprinting. Fertil Steril 58:821–822

Therman E, Susman B (1990) The similarity of phenotypic effects caused by Xp and Xq deletions in the human female: a hypothesis. Hum Genet 85:175–183

Tuerlings JHAM, France HF de, Hamers A, et al. (1998) Chromosome studies in 1792 males prior to intra-cytoplasmic sperm injection: the Dutch experience. Eur J Hum Genet 6:194–200

Welch JP (1985) Clinical aspects of the XYY syndrome. In: Sandberg AA (ed) The Y chromosome, Part B. Clinical aspects of Y chromosome abnormalities. Liss, New York, pp 323–343

Zinn AR, Tonk VS, Chen Z, et al. (1998) Evidence for a Turner syndrome locus or loci at Xp11.2–p22.1. Am J Hum Genet 63:1757–1766

20

Fragile Sites, Trinucleotide Repeat Expansion, and the Fragile X Syndrome

F ragile sites are chromosomal regions that show breaks when cells are exposed to certain drugs or grown in media with a deficiency of folate. The more than 80 *common fragile sites* can be induced in anyone, while *rare fragile sites* are seen in only a small proportion of individuals and are inherited in a Mendelian fashion. Fragile sites appear as unstained or stretched regions in the chromosomes (Figs. 20.1 and 22.2). Aphidicolin, an inhibitor of DNA polymerase, induces common fragile sites in a small percentage of cells, while camptothecin, an inhibitor of topoisomerase I, enhances the percentage of cells showing them. Lack of folic acid or thymidine in the culture medium induces folate-sensitive rare fragile sites, such as FRAXA and FRAXE on the X chromosome, while BrdU induces other rare fragile sites, such as FRA10B and FRA16B on chromosomes 10 and 16, respectively. Distamycin A, a peptide that binds in the minor groove of AT-rich DNA, induces some fragile sites, including FRA16B. Most rare fragile

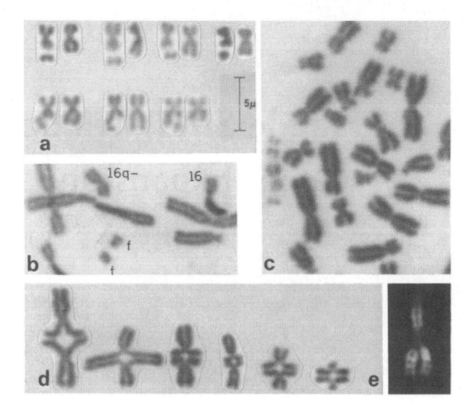

Figure 20.1. (a) Chromosome 16 with a fragile site, and its normal homologue, from several cells. (b) Break at the fragile site and the replicated fragment. (c) Allocyclic C-group chromosome with banding. (d) Mitotic chiasmata of different chromosome pairs. (e) Triradial chromosome 1 resulting from partial endoreduplication.

sites, even in the homozygous condition, have no phenotypic effects. The exceptions are FRAXA, the fragile site at Xq27.3, which is responsible for the fragile X mental retardation syndrome, and the nearby FRAXE at Xq28, which is sometimes associated with mild mental retardation.

Common Fragile Sites: Methods of Induction

The common fragile sites are induced by drugs that inhibit DNA replication or drugs that attack DNAase I hypersensitive sites, which are associated with transcriptionally active genes. The common fragile sites are important because they are hotspots for chromosome breakage and perhaps for integration of foreign DNA. Chromosomes have a tendency to break at fragile sites, and this can lead to the formation of deletions and translocations (Glover and Stein, 1988). For example, breakage at the fragile site at 11q23.3, FRA11B, can lead to del(11)(q23.3–qter) and the Jacobsen 11q– deletion syndrome (Jones et al., 1995). FRA8E at 8q24.11 is near deletion breakpoints in the Langer-Giedion syndrome (Hill et al., 1997). The common fragile sites appear to correspond to chromosomal breakpoints associated with cancer (Yunis, 1984). Sister chromatid exchanges may be increased at these sites (Glover and Stein, 1987), which may act as branchpoints in triradial chromosomes (Kuhn and Therman, 1982).

The most common fragile site is FRA3B, in 3p14.2. Breaks at this site occur at the highest frequency when DNA replication is blocked by folate deficiency or by aphidicolin, which inhibits two DNA polymerases. The site contains no trinucleotide or other repeat motifs, but FISH analysis shows it is late replicating (LeBeau et al., 1998). Molecular analysis of the replication timing of each allele of 21 marker loci in 3p14.2 has confirmed this and shown that the allele on the homologue that shows more breaks in 3p14.2 is later replicating than the allele on the other homologue (Fig. 20.2). Exposure to aphidicolin delays replication of FRA3B still more, suggesting that the inducible fragility of this site is due to failure to replicate all the DNA in the region (LeBeau et al., 1998; Wang et al., 1999). FRA3B is often a site of spontaneous breakage, with loss of alleles distal to the breakpoint. This can be detected for loci that are heterozygous in an individual by looking for loss of heterozygosity (LOH) in selected cell populations. This is seen in over 60% of renal and other carcinomas and presumably contributes to the malignancy (Shridhar et al., 1997). Most of the very common rearrangements and mutations involving 3p14.2 are probably triggered by carcinogens in cigarette smoke that act on the FRA3B fragile site (Sozzi et al., 1997). FRA3B extends over a considerable distance in 3p14.2 and includes a spontaneous integration site for herpes virus type 16 (HPV16). This site is commonly deleted in uterine cervical cancers, which are usually HPV16-associated. It may be that HPV16 integration leads to specific breakage at this site, with loss of distal alleles and tumor formation (Wilke et al., 1997).

The insertion of telomeric TTAGGG repeats into an interstitial location can produce a novel fragile site (Chapter 14). The rare insertion of a different kind of tandem repeat, for example, an rRNA gene cluster (Fig. 20.3), can also lead

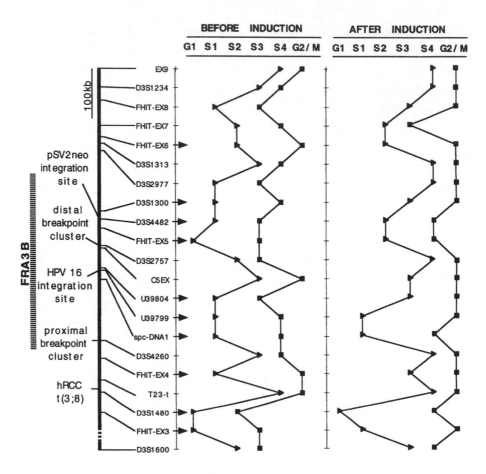

Figure 20.2. The time of replication of the two alleles at each of 21 marker loci encompassing the FRA3B fragile site and most of the *FHIT* gene (exons 3–9), before and after induction with aphidicolin. Arrows indicate loci with maximum allelic differences in replication timing. Breaks and gaps occurred preferentially on the chromosome with the later-replicating allele (reproduced from Wang et al., Hum Mol Genet 8:431–437, 1999, with permission of Oxford University Press).

to chromosome instability. An interesting feature of this is the formation of micronuclei containing these genes in individuals who have an insertion of an rRNA gene cluster into the middle of the long arm of chromosome 7 (Guttenbach et al., 1998).

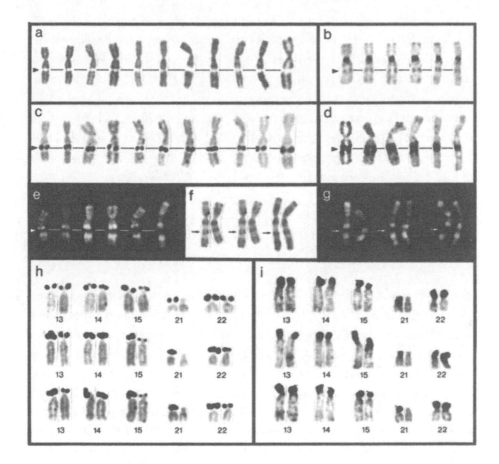

Figure 20.3. Insertion of an NOR into an interstitial site (indicated by arrowheads) on 7q. The altered chromosome after (a) Giemsa staining, (b) C-banding, (c) silver staining, (d) in situ hybridization with (GACA)n, and (e) Q-banding. Both chromosomes 7 from three cells after (f) R-banding and (g) Q-banding. The acrocentric chromosomes from three cells after (h) silver staining and (i) in situ hybridization with (GACA)n (Guttenbach et al., 1998, reproduced with permission of S. Karger AG, Basel).

Heritable (Rare) Fragile Sites

Rarely, more than one heritable fragile site is seen in a family (Romain et al., 1986) Many of the heritable fragile sites, including FRAXA and FRAXE, are made visible by inducing a folate deficiency. All these folate-sensitive sites are due to the presence of expanded repeats of the CCG trinucleotide (reviewed in Yu et al., 1997). Heritable fragile sites may undergo spontaneous breakage. One breakpoint in a ring X chromosome that arose in the sister of a male with the fragile X syndrome was close to the CCG repeat in the gene (Mornet et al., 1993). FRA16B is a rare fragile site at 16q22.1. It is an expansion, with up to 75 repeats, of a 33-bp AT-rich minisatellite repeat (Yu et al., 1997). This may explain its sensitivity to distamycin A, which binds preferentially to AT-rich DNA. It is also inducible by daunomycin A or bromodeoxyuridine (BrdU). FRA10B contains repeats of a 42-bp unit; it is not inducible by daunomycin A (Hewett et al., 1998).

FRAXA, the Fragile X Syndrome, and the FMR1 Gene

The presence of a fragile site at Xq27.3 is associated with particular, though somewhat variable, clinical findings, called the *fragile X syndrome*. Minor features include slightly increased head size, prominent ears and jaw, hyperextensible joints, and cardiac mitral valve prolapse. A useful diagnostic feature is macroorchidism (large testes), present in most of the adult males. The most important features are neurological. Mental retardation is present in about 90% of the males, with IQs sometimes as low as the 20–60 range. The incidence of the fragile X syndrome is about 1 in 3000 males, based on molecular analysis of the size of the CCG repeat (Morton et al., 1997). Many heterozygous (carrier) females are also affected, even though the disorder is much less severe in females than in males (Nussbaum and Ledbetter, 1995). The frequency of the FRAXA fragile site in females, who have two X chromosomes, is twice that in males, who have only one, and about half of women with FRAXA are affected (Rousseau et al., 1994).

Individuals with the fragile X syndrome do not express the *FMR1* gene that is adjacent to the FRAXA site (Pieretti et al., 1991). Similarly, individuals with the FRAXE fragile site do not express the adjacent *FMR2* gene (Gecz et al., 1996). The *FMR1* gene in normal individuals is replicated late in S, according to some studies (Hansen et al., 1997), although FISH replication studies suggest much earlier replication (Torchia et al., 1994; Yashaya et al., 1998). Nevertheless, all agree that the mutant allele in patients with fragile X syndrome replicates much later, in what is normally G2. The zone of delayed replication in band Xq27 can extend more than 400 kb 5′ of *FMR1* and, depending on the extent of the CCG expansion, can involve one, two, or three replicon-sized domains (Hansen et al., 1997).

The Fragile X Phenotype Reflects the Number of CCG Repeats

Analysis of families segregating for the fragile X syndrome yielded striking results, called the *Sherman paradox*. The risk of an abnormal phenotype depends on the position in the pedigree; affected individuals receive the mutant allele from the mother, not the father; and the disorder tends to occur earlier in each generation and its severity to increase (*anticipation*). The reason for this became clear only after the discovery that males with the fragile X syndrome have a greatly increased number of repeats of the CCG trinucleotide (CGG in the other strand). Normally, individuals have 6–50 repeats, while affected males have 200–2000. Individuals with an intermediate number, roughly 50–200, are almost always phenotypically normal despite the increase in copy number (Fu et al., 1991). Meiotic expansion of the number of copies is rare in individuals with 6–50 copies but relatively common in individuals with 50–200 copies. The intermediate expansion has therefore been called a *premutation* and is a necessary step for the further expansion of the repeats that produces clinical disease. The CCG repeat array is usually interrupted after every 9 or 10 CCGs by an AGG, and this stabilizes the array. When the length of uninterrupted CCGs (CGGs in the other strand) reaches 34–38, the instability threshold is passed and further expansion occurs very readily (Eichler et al., 1994).

Mechanism of Expansion of Trinucleotide and Other Repeats

One mechanism of expansion of trinucleotide repeats may be strand slippage secondary to mispairing during replication. Other mechanisms have also been proposed (Sinden, 1999). All of there involve the formation of alternative DNA structures within the triplet repeats. The alternative structures found in CAG, CGG, or CTG triplet repeats inhibit processing of Okazaki fragments by flap endonuclease 1 (FEN-1), leading to site-specific expansions (Spiro et al., 1999). Strand slippage has been established for variations in size of other types of microsatellites. Microsatellites are tandem repeats of a DNA region only 1–6 bp long: mono-, di-, tri-, tetra- penta-, and hexanucleotide repeats. They are very abundant throughout the genome and highly variable in length. These features, and their ease of scoring using polymerase chain reaction (PCR) and gel electrophoresis, have made microsatellites the most widely used genetic markers (Chapter 29). Strand slippage usually activates a mismatch repair system, and a deficiency in mismatch repair is associated with high rates of size variation in microsatellites. This size variation occurs only in the perfect repeat (no other sequences interspersed) part of the microsatellite, not in the region of imperfect repeats (Eichler et al., 1994). The same is true of the trinucleotide repeat class of microsatellites, as in the fragile X syndrome; alleles with imperfect repeats are more stable than alleles with perfect repeats.

Why should there be an instability threshold at 34–38 uninterrupted CCGs? One explanation is that the Okazaki fragments that are so central to replication of the lagging strand of DNA (Chapter 3) play an important role in the dynamic mutation of short repeats. Thus, Okazaki fragments that are anchored to a unique DNA sequence flanking short repeats permit, at most, minor expansions, whereas Okazaki fragments that fall entirely within large repeats can undergo major expansions. The threshold for dynamic mutation corresponds to the approximate length of an Okazaki fragment (Sutherland et al., 1998).

Expansion of CCG Premutation Only in Maternal Meiosis

A striking feature in fragile X families is that females with a premutation can have sons and daughters carrying the full mutation, or *expansion*, but males with a premutation never do. That is, the trinucleotide expansion appears to take place in the female germline, not in the male germline. A surprising finding is that four men with the full mutation of the *FMR1* gene in somatic tissues had a much lower premutational level of CCG expansion in sperm. This could mean that the germline has maintained only this lower number of repeats and that the full mutation arises postzygotically, as a developmental somatic mutation (Reyniers et al., 1993). If so, this must occur very early in embryogenesis, because somatic variability in CCG repeat length is established early and is maintained unchanged in clones established from 13- and 21-week-old fetuses (Wohrle et al., 1993). It seems more likely that the male germline is acting like any other specialized tissue in establishing its own level of expansion. This provides a simpler explanation of the pedigree data.

Mechanism of Inactivation of the *FMR1* Gene

The CCG trinucleotide repeat is close to a CpG island near the 5' end of the *FMR1* gene. Like other CpG islands, this one is normally unmethylated despite the large number of potential sites of methylation. This is also true of premutations. However, the situation is different when the repeat is fully expanded: The CpG island becomes hypermethylated, and the *FMR1* gene is silenced. CCG trinucleotide repeat amplification and hypermethylation of an adjacent CpG island are also characteristic of the FRAXE mental retardation syndrome (Knight et al., 1993; Barnicoat et al., 1997).

The *FMR1* protein product, FMRP, associates with polyribosomes and fosters mRNA translation into proteins. In the absence of FMRP, or in the presence of an abnormal FMRP, translation is impaired (Feng et al., 1997). The importance of FMRP for understanding the fragile X syndrome may be that it is produced near synapses in response to neurotransmitter activation and may thus be important for the development of synaptic connections (Weiler et al., 1997).

Other Trinucleotide Expansion Disorders

A growing number of diseases have been associated with trinucleotide expansions (Table 20.1), and at least some of them also show anticipation. This has been most clearly established for myotonic dystrophy, which is associated with expansion of a CTG repeat. Here, the length variations from one generation to the next are most likely to occur in early embryonic mitotic divisions in both somatic and germline cells. As in the fragile X syndrome, there is a bias against very high expansion levels in sperm (Jansen et al., 1994). The CTG expansion occurs in the 3' region of the protein kinase gene and alters the adjacent chromatin structure, with loss of the DNAase-hypersensitive sites so important for transcriptional activity (Otter and Tapscott, 1995). There is slightly preferential segregation of the larger allele into gametes in female meiosis (segregation distortion of 56.5 : 43.5 instead of 50 : 50) but none in male meiosis (Chakraborty et al., 1996). Segregation distortion is even more marked (73 : 23 in favor of the larger allele) in spinocerebellar ataxia 3 and Haw River syndromes, but in both

Table 20.1. Diseases Due to Expansions of Trinucleotide Repeats

Disease	Fragile site or gene	Repeat unit	Number of repeats		
			Normal	Premutation	Disease
Fragile X syndrome	FRAXA/FMR1	CGG	6–52	59–230	230–2000*
FRAXE syndrome	FRAXE	CCG	4–39	31–61?	200–900*
Jacobson syndrome	FRA11B	CGG	11	80	100–1000*
Kennedy syndrome	SMBA	CAG	14–32	?	40–55[†]
Myotonic dystrophy	DM	CTG	5–37	50–80	80–3000[‡]
Huntington disease	HD	CAG	10–34	36–39	40–121[†]
Spinocerebellar ataxia 1	SCA1	CAG	6–39	None	40–81[†]
Spinocerebellar ataxia 2	SCA2	CAG	14–31	None	34–59[†]
Spinocerebellar ataxia 3	SCA3	CAG	13–44	?	60–84[†]
Spinocerebellar ataxia 6	SCA6	CAG	4–18	?	21–28[†]
Spinocerebellar ataxia 7	SCA7	CAG	7–17	?	38–130[†]
Haw River syndrome	HRS/DRPLA	CAG	7–25	?	49–75[†]
Friedreich ataxia	FRDA	GAA	6–29	34–40?	200–900

*Expansion silences the gene by promoter methylation

[†]Expansion produces a polyglutamine expansion in protein product, precipitates in neurons

[‡]Expansion 3′ to coding region, alters gene expression or RNA processing

Source: Adapted from Sinden, 1999, copyright 1999, used by permission, the University of Chicago Press

these disorders it occurs in male meiosis and not in female meiosis (Ikeuchi et al., 1996).

Huntington disease (HD) is another trinucleotide expansion disorder. The HD gene, IT15, maps to 4p16.3 and contains an unstable CAG trinucleotide repeat expansion (Huntington's Disease Collaborative Research Group, 1993). Leeflang et al., (1999) determined the size of the (CAG)n repeat in more than 3500 individual sperm from males in the large Venezuelan HD pedigree, whose number of repeats varied (n = 37–72). The size of the repeat had changed in 82% of the sperm overall and in 98% of the sperm from men with n > 50. This high frequency of dynamic mutation strongly suggests that the change in repeat length occurs in the abundant mitotic divisions in the germline rather than in the final meiotic division. For more details about dynamic mutation and some of the resultant neurological diseases, see Wells and Warren (1997).

References

Barnicoat AJ, Wang Q, Turk J, et al. (1997) Clinical, cytogenetic, and molecular analysis of three families with FRAXE. J Med Genet 34:13–17

Chakraborty R, Stivers DN, Deka R, et al. (1996) Segregation distortion of the CTG repeats at the myotonic dystrophy locus. Am J Hum Genet 59:109–118

Eichler EE, Holden JJA, Popovich BW, et al. (1994) Length of uninterrupted CGG repeats determines instability in the FMR1 gene. Nat Genet 8:88–94

Feng Y, Absher D, Eberhart D, et al. (1997) FMRP associates with polyribosomes as an mRNP, and the 1304N mutation of severe fragile X syndrome abolishes the association. Mol Cell 1:109–118

Fu Y-H, Kuhl DPA, Pizzutti A, et al. (1991) Variation in the CCG repeat at the fragile X site results in genetic instability: resolution of the Sherman paradox. Cell 67:1047–1058

Gecz J, Gedeon AK, Sutherland GR, et al. (1996) Identification of the gene *FMR2*, associated with FRAXE mental retardation. Nat Genet 13:105–108

Glover TW, Stein CK (1987) Induction of sister chromatid exchanges at common fragile sites. Am J Hum Genet 41:882–890

Glover TW, Stein CK (1988) Chromosome breakage and recombination at fragile sites. Am J Hum Genet 43:265–273

Guttenbach M, Nassar N, Feichtinger W, et al. (1998) An interstitial nucleolus organizer region in the long arm of human chromosome 7: cytogenetic characterization and familial segregation. Cytogenet Cell Genet 80:104–112

Hansen RS, Canfield TK, Fjeld AD, et al. (1997) A variable domain of delayed replication of FRAXA fragile X chromosomes: X-inactivation-like spread of late replication. Proc Natl Acad Sci USA 94:4587–4592

Hewett DR, Handt O, Hobson L, et al. (1998) *FRA10B* structure reveals common elements in repeat expansion and chromosome fragile site genesis. Mol Cell 1:773–781

Hill A, Harada Y, Takahashi E, et al. (1997) Assignment of fragile site 8E (FRA8E) to human chromosome band 8q24.11 adjacent to the hereditary multiple exostoses gene and two overlapping Langer-Giedion syndrome deletion endpoints. Cytogenet Cell Genet 78:56–57

Huntington's Disease Collaborative Research Group (1993) A novel gene containing a trinucleotide repeat that is expanded and unstable in Huntington's disease chromosomes. Cell 72:971–983

Ikeuchi T, Igarashi S, Takiyama Y, et al. (1996) Non-Mendelian transmission of dentato-pallidoluysian atrophy and Machado-Joseph disease. The maternal allele is preferentially transmitted in male meiosis. Am J Hum Genet 58:730–733

Jansen G, Willems P, Coerwinkel M, et al. (1994) Gonosomal mosaicism in myotonic dystrophy patients: involvement of mitotic events in (CTG)n repeat variation and selection against extreme expansion in sperm. Am J Hum Genet 54:575–585

Jones C, Penny L, Maltina T, et al. (1995) A fragile site within the proto-oncogene CBL2 is associated with a chromosome deletion syndrome. Nature 376:145–149

Knight SJ, Flannery AV, Hirst MC, et al. (1993) Trinucleotide repeat amplification and hypermethylation of a CpG island in FRAXE mental retardation. Cell 74:127–134

Kuhn EM, Therman E (1982) Origin of symmetrical triradial chromosomes in human cells. Chromosoma 86:673–681

LeBeau MML, Rassool FV, Neilly MG, et al. (1998) Replication of a common fragile site, FRA3B, occurs late in the S phase and is delayed further upon induction: implications for the mechanism of fragile site induction. Hum Mol Genet 7:755–761

Leeflang EP, Tavare S, Marjoram P, et al. (1999) Analysis of germline mutation spectra at the Huntington's disease locus supports a mitotic mutation mechanism. Hum Mol Genet 8:173–183

Mornet E, Boggs A, Deluchat C, et al. (1993) Molecular analysis of a ring chromosome X in a family with fragile X syndrome. Hum Genet 92:373–378

Morton NE, Burdey S, Webb TP, et al. (1997) Fragile X syndrome is less common than previously estimated. J Med Genet 34:1–5

Nussbaum RL, Ledbetter DH (1995) The fragile X syndrome. In: Scriver CR, Beaudet AL, Sly WS, Valle D (eds) The metabolic and molecular bases of inherited disease. McGraw-Hill, New York, pp 795–810

Otter AD, Tapscott SJ (1995) Triplet repeat expansion in myotonic dystrophy alters the adjacent chromatin structure. Proc Natl Acad Sci USA 92: 5465–5469

Pieretti M, Zhang FP, Fu YH, et al. (1991) Absence of expression of the FMR-1 gene in fragile X syndrome. Cell 66:817–822

Reyniers E, Vits L, De Boulle K (1993) The full mutation in the FMR-1 gene of male fragile X patients is absent in their sperm. Nat Genet 4:143–146

Romain DR, Columbano-Green LM, Smythe RH, et al. (1986) Studies on three rare fragile sites: 2ql3, 12ql3, and 17pl2 segregating in one family. Hum Genet 73:164–170

Rousseau F, Heitz D, Tarleton J, et al. (1994) A multicenter study on genotype-phenotype correlations in the fragile X syndrome, using direct diagnosis with probe StB12.3: the first 2253 cases. Am J Hum Genet 55:225–237

Shridhar V, Wang L, Rosati R, et al. (1997) Frequent breakpoint in the region surrounding FRA3B in sporadic renal cell carcinoma. Oncogene 14: 1269–1277

Sinden RR (1999) Human genetics '99. Trinucleotide repeats: biological implications of the DNA structures associated with disease-causing triplet repeats. Am J Hum Genet 64:346–353

Sozzi G, Sard L, DeGregorio L, et al. (1997) Association between cigarette smoking and FHIT gene alterations in lung cancer. Cancer Res 57: 2121–2123

Spiro C, Pelletier R, Rolfsmeier ML, et al. (1999) Inhibition of FEN-1 processing by DNA secondary structure at trinucleotide repeats. Mol Cell 4: 1079–1085

Sutherland GR, Baker E, Richards RL (1998) Fragile sites still breaking. Trends Genet 14:501–505

Torchia BS, Call LM, Migeon BR (1994) DNA replication analysis of FMR1, XIST, and factor 8C loci by FISH shows nontranscribed X-linked genes replicate late. Am J Hum Genet 55:96–104

Wang L, Darling J, Zhang J-S, et al. (1999) Allele-specific late replication and fragility of the most active common fragile site, FRA3B. Hum Mol Genet 8:431–437

Weiler IJ, Irwin SA, Klintsova AY, et al. (1997) Fragile X mental retardation protein is translated near synapses in response to neurotransmitter activation. Proc Natl Acad Sci USA 94:5395–5400

Wells RD, Warren ST (eds) (1997) Genetic instabilities and hereditary neurological diseases. Academic, San Diego

Wilke CM, Hall BK, Hoge A, et al. (1997) FRA3B extends over a broad range and contains a spontaneous HPV16 integration site: direct evidence for the coincidence of viral integration sites and fragile sites. Hum Mol Genet 5:187–195

Wohrle D, Hennig I, Vogel W, et al. (1993) Mitotic stability of fragile X mutations in differentiated cells indicates early post-conceptional trinucleotide repeat expansion. Nat Genet 4:140–142

Yashaya J, Shalgi R, Shohae M, et al. (1998) Replication timing of the various FMR1 alleles detected by FISH: inferences regarding their transcriptional status. Hum Genet 102:6–14

Yu S, Mangelsdorf M, Hewett D, et al. (1997) Human chromosomal fragile site FRA16B is an amplified AT-rich minisatellite repeat. Cell 88:367–374

Yunis JJ, Sorenq AL (1984) Constitutive fragile sites and cancer. Science 226: 1199–1204

21

Euploid Chromosome Aberrations, Uniparental Disomy, and Genomic Imprinting

Uniparental Disomy

In some individuals, both copies of a particular chromosome come from one parent and none from the other; this is called *uniparental disomy* (UPD) (Engel, 1980). It is not yet clear how common each type of uniparental disomy is, but many deviations from Mendelian inheritance could be explained by it, as described below. Most UPD has a meiotic origin that leads to a trisomic zygote, with loss of one copy of the chromosome in a subsequent mitosis. One-third of the time, the single copy from one parent will be lost, producing UPD. Trisomy of meiotic origin is usually the result of maternal meiosis I nondisjunction, so most UPD of meiotic origin is maternal UPD (Robinson et al., 1997). Paternal UPD is more likely to have a postzygotic origin from a normal zygote by mitotic nondisjunction. A much smaller proportion of maternal UPD arises in this way.

Trisomic cells produced by *mitotic* nondisjunction have two identical copies of the chromosome from one parent. In trisomies due to meiotic nondisjunction, restoration of disomy by loss of a chromosome will produce uniparental *isodisomy* one-third of the time. The two identical chromosomes are, naturally, homozygous at all their gene loci, including any that are mutated. UPD of a maternally derived chromosome has been observed for chromosomes 1, 2, 4, 6, 7, 9, 10, 13–16, 21, 22, and X. Paternal UPD has been observed for chromosomes 1, 5–8, 11, 13–16, 20–22, X, and XY (Engel, 1998). For years, the most commonly observed UPDs were those of chromosomes 7, 11, 14, and 15, perhaps because these are associated with characteristic phenotypes that bring them to medical attention and cytogenetic study. However, many more cases have been discovered since the introduction of earlier prenatal diagnostic studies, based on first-trimester placental biopsies (chorionic villus sampling).

Confined Placental Mosaicism and the Origin of UPD

Usually the embryo and the trophoblast have the same chromosome constitution, but sometimes a karyotypically abnormal cell line is limited to the placenta. The widespread use of chorionic villus sampling early in pregnancy has led to the discovery that a surprisingly high proportion (1–2%) of first-trimester conceptuses are mosaics with a trisomic and a disomic cell line. Fortunately, it is rare to find any evidence of residual mosaicism in the fetuses or newborns resulting from these pregnancies, so this condition has been called *confined placental mosaicism* (CPM). However, about one-third of these fetuses or newborns show UPD for the chromosome in question, as expected if restoration of disomy involves random loss of one of the three copies of the chromosome. These findings have important implications for genetic counseling.

The trisomic cell line in CPM can arise during meiosis or in postfertilization mitoses. Most CPM involving chromosomes 9, 16, and 22 is meiotic in origin, while CPM involving chromosomes 2, 7, 8, 10, and 12 is predominantly somatic in origin. A poor pregnancy outcome is restricted to CPM of meiotic origin, which is also associated with a higher proportion of trisomic cells in the placenta, and fetal UPD. The embryo proper arises from only 3–5 of the 64 cells of the early blastocyst and is clearly more likely to receive trisomic cells if these make up a higher percentage of cells in the blastocyst (Robinson et al., 1997). The most important question for the family seeking counseling is the pregnancy outcome expected with CPM and resultant UPD of a particular chromosome. Just how important are these as causes of disease? It is too early to give more than a rough estimate. Consider chromosome 16, which is inordinately over-represented among human trisomies (Chapter 11). Its frequency is 15 per 1000 recognized pregnancies, and 80–95% arise in maternal meiosis I. Most 16-trisomics are aborted by the twelfth week of pregnancy, but about 10% reduce to disomy, and 3% show no trisomic cells in the fetus (CPM). One-third of these (10 cases observed) have UPD16 associated with fetal loss late in pregnancy or severe intrauterine growth retardation, and a high percentage of trisomic cells in the placenta at delivery. There have even been cases of 16-trisomy/disomy mosaicism in liveborns (Wolstenholme, 1995).

UPD Can Lead to Homozygosity of a Recessive Disease Gene

The occurrence of the autosomal recessive disease cystic fibrosis in a child whose mother carried the mutant gene but whose father did not, led to the discovery of uniparental isodisomy. The child received identical copies of a chromosome 7 from the carrier mother and none from the father (Spence et al., 1988). Auto-somal recessive disorders caused by isodisomy of 11 different chromosomes have been reported, including Bloom syndrome in isodisomy 15, spinal muscular atrophy in isodisomy 5, and the skeletal disorder pycnodysostosis in isodisomy 1. The two chromosomes from one parent do not have to be completely identical, or *isodisomic;* only the disease locus needs to be homozygous. In one case, the skin disease epidermolysis bullosa was caused by heterodisomy 1 in which only a 35-cM (35% recombination) region was homozygous on the two mater-nally derived chromosomes 1, as a result of crossing over (reviewed by Engel, 1998). As many as 1–2% of cases of recessive disease may be due to UPD, with only one parent a heterozygote.

UPD Can Lead to Disease Due to a Novel Mechanism: Genomic Imprinting

The increasing availability of genetic markers has made the identification of UPD much easier, and this has led to the discovery that a series of disorders are

Table 21.1. Imprinted Autosomal Genes and Their Locations

Gene	Location	Imprinted allele
IGF_2	11p15.5*	Maternal[†]
H_{19}	11p15.5	Paternal
KIP_2	11p15.5	Paternal
$IMPT_1$	11p15.5	Paternal
$KVLQ_1$	11p15.5	Paternal
SNRNP	15q11–q13[‡]	Maternal
IPW	15q11–q13	Maternal
ZNF_{27}	15q11–q13	Maternal
$ZNF_{27}AS$	15q11–q13	Maternal
NDN	15q11–q13	Maternal
UBE_3A	15q11–q13	Maternal
HIC_{-1}	17p13.3	?
GNAS	20q13.3	Paternal

*Imprinted gene cluster about 1.5 Mb in size

[†] Tissue-specific pattern of imprinting

[‡] Imprinted cluster about 1 Mb in size

due to the absence of a maternal (or paternal) copy of a particular chromosome, chromosome region, or gene. This indicates that in one sex an imprint is placed on the gene or genes during gametogenesis. This process, called *genomic imprinting*, permanently inactivates (or more rarely, activates) the gene. The children described earlier whose cystic fibrosis was due to UPD7mat had an additional trait, short stature, which is not usually seen in cystic fibrosis but has been seen in many other UPD7mat cases without cystic fibrosis, and even in association with UPD7qmat (Eggerding et al., 1994). UPD7mat accounts for about 10% of the cases of the Russell-Silver syndrome, in which short stature is a prominent feature. For extensive reviews of genomic imprinting, see Reik and Surani (1997) or Bartolomei and Tilghman (1997).

At least 12 autosomal genes have been shown to be imprinted (Table 21.1). An imprinting map has been developed as a guide in evaluating the potential danger of confined placental mosaicism (Ledbetter and Engel, 1995). The most striking feature of imprinted genes is their presence in clusters. One could regard the very large regions on the X chromosome that are subject to inactivation as even larger clusters of imprinted genes, since they too show monoallelic expression. The known autosomal clusters occur on chromosome 11 at p15.5 and on chromosome 15 at q11–q12. Clustering may have evolved so that a single regulatory element could mediate the coordinated imprinting of an entire cluster. This is the case for the X chromosome, whose inactivation is mediated by the untranslated RNA product of the *XIST* gene (Chapter 18). It is also true for the cluster at 11p15.5, with the untranslated RNA product of the *H19* gene playing the regulatory role. All these regulatory RNAs bind to chromatin regions containing diverse DNA sequences that are clearly not homologous in sequence to the RNA. The binding must involve as yet unknown proteins and lead to heterochromatinization. Imprinted autosomal alleles therefore replicate asynchronously, in contrast to nonimprinted alleles, which replicate synchronously (Knoll et al., 1994). Small deletions first pinpointed the location of imprinted regions and of the syndromes associated with these segmental aneusomics.

Three Imprinting Disorders: Beckwith–Wiedemann, Prader–Willi, and Angelman Syndromes

The Beckwith–Wiedemann syndrome (BWS) is marked by overgrowth of muscles, tongue, heart, kidney, and liver. Large adrenal cells (cytomegaly) and renal dysplasia are also seen. In about one-fifth of the cases, the cause is paternal UPD11, and specifically the presence of two paternal copies of the 11p15.5 region instead of the usual one paternal and one maternal. One gene in this region that is partially responsible for the phenotypic effects is *IGF2* (insulin-like growth factor 2). This gene is expressed only from the paternal, not the maternal, copy, so the presence of two paternal copies doubles the dose of *IGF2*, stimulating growth (Sun et al., 1997). In many of the cases in which UPD11pat is not present, there may be a breakdown in imprinting, so that the maternal copy of the gene is also active (Weksberg et al., 1993).

The BWS critical region contains not only a gene that is imprinted in male gametogenesis but also one that is imprinted in female gametogenesis. This second imprinted gene is *KIP2*, which is expressed only from the maternal chromosome (Hatada et al., 1996). Therefore, paternal UPD11 leaves no functioning *KIP2* gene. Since the *KIP2* gene product, p57, is an inhibitor of cyclin- dependent kinases (Chapter 2), absence of the gene product can lead to the kind of unregulated cell proliferation and predisposition to childhood tumors seen in Beckwith–Wiedemann syndrome. Loss of *KIP2* function can also be the result of mutation, and such mutations may account for nearly 10% of BWS patients (O'Keefe et al., 1997).

Prader–Willi syndrome (PWS) is marked by hypotonia, obesity, hypogonadism, and sometimes mild mental retardation. Nearly 30% of cases of PWS are due to the absence of the paternal chromosome 15 as a result of UPD15mat. However, almost 70% are due to a deletion of the paternal 15q11–q13 region. This can be the result of either inter- or intrachromosomal rearrangements (Carrozzo et al., 1997). Analysis of many such cases has established a minimal region of overlap of the various deletions, defining a critical region within which the relevant gene or genes must lie. The first expressed gene identified in this region was *SNRPN* (small nuclear ribonucleoprotein polypeptide N). Only the paternal allele of this gene is expressed in fetal brain and heart. Thus, absence of the paternal copy of *SNRPN* may be responsible for PWS (Reed and Leff, 1994; Glenn, 1996). This is strongly supported by the finding of Prader-Willi syndrome in a patient with a de novo balanced translocation, t(4;15)(q27;q11.2)pat with breakpoint between exons 2 and 3 of the *SNRPN* gene (Fig. 21.1; see also Kuslich et al., 1999). Six additional imprinted genes that are expressed exclusively from the paternal allele have been identified in the Angelman/Prader-Willi critical region. The maternal alleles of all six are late replicating, highly methylated, and transcriptionally silent. Five of these apparently have nothing to do with the PWS phenotype. A sixth, *NDN* (*necdin*), is expressed in brain neurons and may also be involved in generating the PWS phenotype (Jay et al., 1997).

Angelman syndrome (AS), marked by severe mental retardation, seizures, absence of speech, and inappropriate laughter, occurs about once in 15,000 live births. Although most of the more than 300 cases reported have been children, almost 5% of institutionalized adults with profound mental retardation have this disorder (Buckley et al., 1998). Angelman syndrome is most often (70% of cases) caused by a deletion of the maternally derived chromosome 15q11–q13, but it is sometimes due instead to paternal UPD15, to defective imprinting, or, in about 20% of cases, to a mutation of the *UBE3A* gene in the Angelman/Prader-Willi critical region. In one study, 17 of 56 subjects with Angelman syndrome had a mutation of this gene (Fang et al., 1999). *UBE3A* is a ligase that covalently links a protein to ubiquitin, targeting the protein for degradation. Initially, *UBE3A* was considered an unlikely candidate gene for either syndrome, because it was not

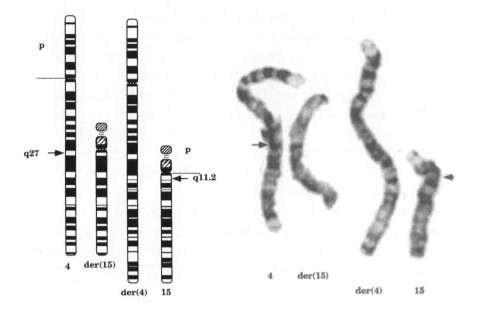

Figure 21.1. A translocation, t(4;15)(q27;q11.2), whose breakpoint disrupts the *SNRPN* gene at 15q11.2 in an individual with Prader–Willi syndrome (reproduced from Kuslich et al., Am J Hum Genet 64:70–76, copyright 1999, American Society of Human Genetics, with permission of the University of Chicago Press).

imprinted in lymphocytes. In fact, imprinting of *UBE3A*, with expression only from the maternal allele, is restricted to the brain. This may explain why the phenotypic effects in Angelman syndrome are similarly restricted (Vu and Hoffman, 1997).

Chimeras, Triploidy, and Tetraploidy

Fusion of two different zygotes into a single embryo produces a *chimera*. When one of the zygotes is XX and the other XY, the chimeric individual may develop as a true hermaphrodite and thus come to medical attention. Most chimeras probably escape detection, although the increasing use of genetic markers should detect more of them. The use of molecular markers has also shed light on the mechanisms leading to the production of chimeras. In one XX/XY true hermaphrodite, microsatellite markers showed that there was a single haploid maternal contribution, a single haploid paternal genome, and both X chromosomes in the XX cells were maternal. The explanation in this case (Fig. 21.2A) involves mitotic division of a haploid oocyte, followed by fertilization of one product by a Y-bearing sperm and diploidization of the other maternal product (Strain et al., 1995). In this case, chimerism was the result of a single fertilization, combined with parthenogenetic diploidization. In a similar case, two different sperm had contributed to the proband's genome, while the presence of a single maternal allele at each of 40 marker loci indicated a single haploid maternal contribution. This could be accounted for if the haploid oocyte divided mitotically and each product were then fertilized by a different sperm, as illustrated in Fig. 21.2B (Giltay et al., 1998).

A triploid zygote may occur as the result of various processes. The egg or the sperm may have an unreduced chromosome number as a result of restitution in either the first or the second meiotic division; the second polar body may reunite with the egg nucleus; or two sperms may penetrate and fertilize the same egg.

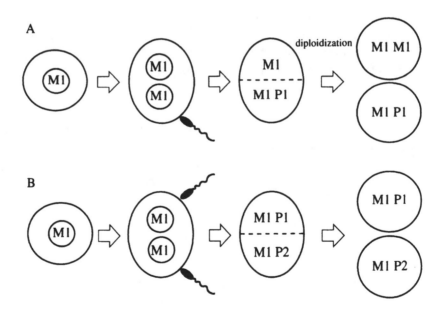

Figure 21.2. How chimeras can arise. In both (A) and (B), an ovum (M1) is parthenogenetically activated. (A) One of the two haploid maternal cells is fertilized by a normal sperm (P1), and the other is diploidized. (B) Each of the two haploid maternal cells is fertilized by a normal sperm (P1 and P2) (reproduced from Giltay et al., Am J Hum Genet 62:937–940, copyright 1998, American Society of Human Genetics, with permission of the University of Chicago Press).

Triploids constitute about 17% of spontaneous abortuses with an abnormal chromosome number (Chapter 11). Only about 1 in 10,000 triploid zygotes results in a liveborn infant, and most of these die within a day. A few triploid infants have survived for a few months. In such rare cases, hidden mosaicism must be considered; indeed, most polyploid infants have been mixoploid, with a diploid cell line too. It is unclear why triploidy and tetraploidy are so deleterious. Altered dosage compensation between the X chromosome and autosomes or between imprinted maternal and paternal autosomal genes may contribute to the phenotypic effects of triploidy. Tripolar mitoses have been described in some dispermic fertilized eggs and could lead to aneuploidy and lethality in this type of triploid, although no trace of it is seen in aborted triploid embryos.

A small number of infants and a 22-month-old girl with apparently nonmosaic tetraploidy have been reported, but the few other liveborn children have been diploid/tetraploid mosaics. Tetraploid embryos are rarer than triploids, probably because there are fewer mechanisms for producing them. The most probable of these is duplication of the diploid complement in a somatic cell at a very early stage of development. Other possible origins, such as the chance fertilization of a rare diploid ovum by an equally rare diploid sperm, are unlikely. The fertilization of one egg by three sperm does occur but leads to the development of a hydatidiform mole, as described below.

Ovarian Teratomas: Both Genomes Maternal in Origin

Ovarian teratomas are very distinctive benign tumors of the ovary. They are quite heterogeneous in their makeup, but all contain disorganized bits and pieces of various tissues, including thyroid, teeth, and hair. In fact, each looks like an ovarian pregnancy that originated by aberrant triggering of embryonic development in a nonovulated, unfertilized egg (*gynogenesis*, or *parthenogenesis*). This is consistent with the 46,XX karyotype of all ovarian teratomas and has been confirmed by analysis of genetic markers: All were maternally derived (Linder et al., 1975). The distance of particular loci from the centromere can be determined by the segregation of markers in teratomas that originated from oocytes that retained the second polar body rather than diploidizing a haploid complement, but this method has been largely supplanted by other mapping methods (Chapter 29).

The absence of any placental tissue in ovarian teratomas suggests that imprinted genes expressed only from the paternal genome are essential for the development of placental tissue. This is consistent with the presence of placental tissue only in *androgenetic* conceptuses, those with paternal genomes only. One imprinted gene, H_{19}, which maps to 11p15, is always hypomethylated on both alleles in ovarian teratomas, whereas in normal fetal and adult organs one allele is hypermethylated and transcriptionally silent (Zhang et al., 1993). This supports the idea that the maternal allele is the expressed one and suggests its overexpression, and perhaps that of other paternally imprinted genes, leads to the abnormal phenotype. Obviously, any imprinted gene that is transcriptionally silent on maternal chromosomes would not be expressed at all in these gynogenetic ovarian teratomas, and that might have an even more deleterious effect.

Hydatidiform Moles: Both Genomes Usually Paternal in Origin

Hydatidiform moles are abnormal trophoblastic growths that are seen mainly in association with abnormal chromosome complements or uniparental disomy and imprinting. In Western countries, 1 in 1500–2000 pregnancies results in the development of a mole, whereas in the Orient, Taiwan for instance, the frequency can be as high as 1 in 200. Moles can resemble malignant tumors because of the presence of giant nuclei that have arisen by endomitosis and endoreduplication (Sarto et al., 1984). In fact, more than 50% of choriocarcinomas arise from moles. Hydatidiform moles are *partial* or *complete*. Partial moles are usually associated with a triploid embryo produced by fertilization of an egg by two sperm (Jacobs et al., 1982). Complete moles have no associated embryo. Although 2.1% of partial and 10% of complete moles become malignant, fortunately, some 80% of these highly malignant tumors respond very favorably to chemotherapy.

Nearly 80% of complete moles arise by fertilization of an "empty" egg (without a nucleus) by an X-bearing sperm and a subsequent doubling of the paternal chromosome set (diploidization) to produce an androgenetic zygote. The origin of the empty egg is not known. The chromosome complement of complete moles is almost always 46,XX, with two identical paternal chromosome sets. Rarely, they are XY, arising by fertilization of an empty egg by two sperm or by a diploid sperm. The androgenetic origin of complete moles was first shown using chromosome polymorphisms as markers (Kajii and Ohama, 1977) and confirmed with genetic markers (Jacobs et al., 1980). These findings indicate that one or more maternally derived chromosomes are essential for normal embryonic development but that the paternal genome contains at least some of the expressed genes important for placental development. A few tetraploid moles have been produced by fertilization of a haploid oocyte by three sperm (Surti et al., 1986). This indicates that the relative dosage of competing maternal and paternal genes is important for normal development.

The association between the absence of a maternal set of chromosomes and the occurrence of a hydatidiform mole suggests that some genes are imprinted (inactivated) on the paternal chromosomes but remain active on the maternal set. If so, then the 20% of moles in which both maternal and paternal sets are present could be explained by a mutant maternal allele for one of these genes, leaving the embryo with no functional copy. A woman who inherited such a mutation from her father might have *only* molar pregnancies, and such families have been described. An alternative explanation is necessary for couples who have recurrent hydatidiform moles but also some normal pregnancies: homozygosity for a mutant, nonimprinted gene. In two families (one consanguineous), a genome-wide scan of molecular markers (Chapter 29) revealed the presence of a region of homozygosity for markers in a 15.2-cM region of 19q13.3–q13.4 in all six women who had repeated molar and nonmolar pregnancies (Moglabey et al., 1999).

References

Bartolomei MS, Tilghman SM (1997) Genomic imprinting in mammals. Annu Rev Genet 30:493–525

Buckley RH, Dinno N, Weber P, et al. (1998) Angelman syndrome: are the estimates too low? Am J Med Genet 80:385–390

Carrozzo R, Ross E, Christian SL, et al. (1997) Inter- and intrachromosomal rearrangements are both involved in the origin of 15q11–q13 deletions in Prader-Willi syndrome. Am J Hum Genet 61:228–231

Eggerding FA, Schonberg SA, Chehab FF, et al. (1994) Uniparental isodisomy for paternal 7p and maternal 7q in a child with growth retardation. Am J Hum Genet 55:253–265

Engel E (1980) A new genetic concept: uniparental (iso?)disomy and its potential effect, isodisomy. Am J Med Genet 6:137–143

Engel E (1998) Uniparental disomies in unselected populations. Am J Hum Genet 63:962–966

Fang P, Lev-Lehman E, Tsai T-F, et al. (1999) The spectrum of mutations in *UBE3A* causing Angelman syndrome. Hum Mol Genet 8:129–135

Giltay JC, Brunt T, Beemer FA, et al. (1998) Polymorphic detection of a parthenogenetic maternal and double paternal contribution to a 46,XX/46,XY hermaphrodite. Am J Hum Genet 62:937–940

Glenn CC, Saitoh S, Jong MTC (1996) Gene structure, DNA methylation, and imprinted expression of the human *SNRPN* gene. Am J Hum Genet 58:335–346

Hatada I, Ohashi H, Fukushima Y, et al. (1996) An imprinted gene p57(*KIP2*) is mutated in Beckwith–Wiedemann syndrome. Nat Genet 14:171–173

Jacobs PA, Szulman AK, Funkhouser J, et al. (1982) Human triploidy: relationship between parental origin of the additional haploid complement and development of partial hydatidiform mole. Ann Hum Genet 46:223–231

Jacobs PA, Wilson CM, Sprenkle JA, et al. (1980) Mechanism of origin of complete hydatidiform moles. Nature 286:714–716

Jay P, Rougelle C, Massacrier A, et al. (1997) The human necdin gene, *NDN*, is maternally imprinted and located in the Prader–Willi syndrome chromosomal region. Nat Genet 17:357–361

Kajii T, Ohama K (1977) Androgenetic origin of hydatidiform mole. Nature 268:633–634

Knoll, JHM, Cheng S-D, Lalande M (1994) Allele specificity of DNA replication timing in the Angelman/Prader–Willi syndrome imprinted chromosomal region. Nat Genet 6:41–46

Kuslich CD, Kobori JA, Mohapatra G, et al. (1999) Prader–Willi syndrome is caused by disruption of the *SNRPN* gene. Am J Hum Genet 64:70–76

Ledbetter DH, Engel E (1995) Uniparental disomy in humans: Development of an imprinting map and its implications for prenatal diagnosis. Hum Mol Genet 4:1757–1764

Linder D, McCaw BK, Hecht F (1975) Parthenogenetic origin of benign ovarian teratomas. N Engl J Med 292:63–66

Moglabey YB, Kircheisen R, Seoud M, et al. (1999) Genetic mapping of a maternal locus responsible for familial hydatidiform moles. Hum Mol Genet 8:667–671

O'Keefe D, Dao D, Zhao L, et al. (1997) Coding mutations in $p57^{KIP2}$ are present in some cases of Beckwith–Wiedemann syndrome but are rare or absent in Wilms tumor. Am J Hum Genet 61:295–303

Reed ML, Leff SE (1994) Maternal imprinting of human *SNRPN*, a gene deleted in Prader–Willi syndrome. Nat Genet 6:163–167

Reik W, Surani A (1997) Genomic imprinting. Oxford, New York

Robinson WP, Barrett IJ, Bernard L, et al. (1997) A meiotic origin of trisomy in confined placental mosaicism is correlated with presence of fetal uniparental disomy, high levels of trisomy in trophoblast and increased risk of fetal intrauterine growth retardation. Am J Hum Genet 60:917–927

Sarto GE, Stubblefield PA, Lurain J, et al. (1984) Mechanisms of growth in hydatidiform moles. Am J Obstet Gynecol 148:1014–1023

Spence JE, Perciaccante RG, Greig GM, et al. (1988) Uniparental disomy as a mechanism for human genetic disease. Am J Hum Genet 42:217–226

Strain L, Warner JP, Johnston T, et al. (1995) A human parthenogenetic chimaera. Nat Genet 11:164–169

Sun F-L, Dean WL, Kelsey G, et al. (1997) Transactivation of *Igf2* in a mouse model of Beckwith–Wiedemann syndrome. Nature 389:809–815

Surti U, Szulman AK, Wagner K, et al. (1986) Tetraploid partial hydatidiform moles: two cases with a triple paternal contribution and a 92,XXXY karyotype. Hum Genet 72:15–21

Vu TH, Hoffman AR (1997) Imprinting of the Angelman syndrome gene, *UBE3A*, is restricted to brain. Nat Genet 17;12–13

Weksberg R, Shen DR, Fei YL, et al. (1993) Disruption of insulin-like growth factor 2 imprinting in Beckwith–Wiedemann syndrome. Nat Genet 5: 143–150

Wolstenholme J (1995) An audit of trisomy 16 in man. Prenatal Diagn 15:109–121

Zhang Y, Shields T, Crenshaw T, et al. (1993) Imprinting of human H19: allele-specific CpG methylation, loss of the active allele in Wilms tumor, and potential for somatic allele switching. Am J Hum Genet 53:113–124

22

Chromosome Changes in Cell Differentiation

T he most striking achievement of embryogenesis is the differentiation of the genetically identical descendants of a fertilized egg into hundreds of different types of cells and eventually into a highly complex human being. Cell differentiation is accompanied by changes in chromatin structure or, less commonly, ploidy level; these mediate the changes in gene expression required for cell differentiation. Thus, the chromatin structure in stem cells of the germline is readily distinguished from that of differentiating oocytes and spermatocytes, and that of stem cells in the bone marrow, intestine, or skin from that of the more differentiated cell types they give rise to. These changes in chromatin conformation and ploidy level are brought about by a number of quite different mechanisms (reviewed by Miller, 1997).

Programmed DNA Loss

Programmed DNA loss is an important mechanism of cell differentiation, although one used by only a few cell lineages. The best known example is erythropoiesis, in which red blood cell precursors undergo heterochromatinization and condensation of their chromatin and finally extrude the entire nucleus. More restricted, and precise, DNA loss occurs in the differentiation of T and B lymphocytes. This involves DNA double-strand breakage, excision of large portions of the T cell receptor gene array and B cell immunoglobulin heavy and light chain gene arrays, and V(D)J rejoining of the remaining DNA to yield a very large number of uniquely modified genes in T and B cells, with any one cell expressing only one allele of a now unique gene. The resulting T cell population has an enormous variety of distinctive receptors, while the B cell population can generate an equally enormous variety of different antibodies (Tonegawa, 1983). It is interesting that the unrearranged kappa light chain gene array is heavily methylated and inactive, but after V(D)J rejoining is completed, the rearranged kappa chain gene is demethylated and becomes capable of transcription (Lichtenstein et al., 1994).

Facultative Heterochromatin: Chromatin Structure and Gene Expression

Facultative heterochromatinization during differentiation is widespread. It is functionally equivalent to programmed loss of the DNA sequences involved but has the advantage of being reversible. The best known examples are X inactivation (Chapter 18) and genomic imprinting (Chapter 21). Both of these involve an epigenetic differentiation process that leads to differential expression of the two parental alleles at a given locus, or at multiple loci, and asynchronous DNA replication of the two alleles, the inactive one becoming late replicating. The olfactory receptor (OR) gene family provides a third example. There are several hundred to a thousand OR genes, scattered in multiple clusters on numerous chromosomes (Chapter 31). Several studies have shown that a single olfactory neuron expresses no more than a very tiny number of these OR genes (Malnic et al., 1999), perhaps only one from each cluster. Highly sensitive reverse transcription–polymerase chain reaction (RT-PCR) analyses indicate that each olfactory neuron expresses only one allele of a given OR gene, and from early embryonic life the two alleles replicate asynchronously, like imprinted or X-linked genes (Chess et al., 1994).

At the most general level, there are two classes of genes: housekeeping and tissue-specific. Housekeeping genes are constitutively expressed in most cell types. Their protein products carry out the various metabolic, structural, and other housekeeping functions required by all cells. In contrast, a restricted array of tissue-specific genes are expressed in a particular cell or tissue and shut down in other cell types. Such gene inactivation is associated with a more condensed chromatin structure, in which the DNA of the gene is not accessible to the transcription factors needed to transcribe RNA from it. The mechanisms involved in such chromatin changes appear to involve methylation of cytosine residues in CpG dinucleotides within the promoter regions just upstream (5′) of genes. When methylated, these sites can bind a specific protein, which in turn binds a histone deacetylase, HDAC1. This enzyme removes acetyl groups from core histones, producing a denser chromatin structure (Chapter 5). Phosphorylation of core histones also leads to a tighter chromatin conformation, but this mechanism is more important for chromosome condensation during mitosis and meiosis than for interphase gene regulation (Chapter 2).

Nature and Mechanism of Genomic Imprinting

Imprinting is thought to occur during gametogenesis, but is this always the case? There is biallelic expression (absence of imprinting) of the SNRPN and UBE3A genes in all tissues except the brain. This means that an early imprint has been removed or that, like X inactivation, imprinting occurs during embryogenesis and may show tissue-specific differences in timing or occurrence. It could also mean that there is a way around the repressive effect of the imprint. The insulin-like growth factor 2 (IGF2) gene is imprinted in fetal liver but shows biallelic expression in adult liver. The reason for this discrepancy is that IGF2 is transcribed from both alleles from promoter P1, used in adult liver, but is transcribed only from one allele from promoters P3 and P4, used in fetal liver (Vu and Hoffman, 1994). It is unclear whether a similar explanation accounts for other examples of apparently tissue-specific imprinting. Rodent-human somatic cell hybrids provide one model system in which to study genomic imprinting, because the imprint on a single human chromosome, 11 or 15, say, is maintained in the hybrid cells (Gabriel et al., 1998).

What triggers imprinting of any gene is not known. However, details of the process of imprinting are beginning to be worked out. For example, the critical region for Angelman and Prader–Willi syndromes appears to contain a specific imprinting center, comparable to that involved in X inactivation. A cluster of genes in the Prader–Willi critical region is inactivated on the maternal chromosome 15, possibly by the untranslated RNA product of one member of the maternal gene cluster (Wevrick et al., 1994). Similarly, the maternal *IGF2* gene is inactivated by the untranslated RNA product of the maternal *H19* gene. In each case, the *cis*-acting RNA may form a complex with RNA-binding and other proteins, such as histone deacetylase. There is evidence that the removal of acetyl groups from core histones may be a general mechanism for imprinting and gene or X-chromosome inactivation (Jeppesen, 1997). DNA methylation also plays a role both in X inactivation and in imprinting, as well as in gene inactivation (Chaillet et al., 1995).

Deletions in the Prader–Willi syndrome have in common the loss of the *SNRPN* gene and result in the inappropriate silencing of the *IPW* and *ZNF127* genes in *cis* (on the same chromosome) that are 150 and 1000 kb away, flanking *SNRPN*. A man with the deletion cannot reset the imprinting program and therefore passes on to his children a maternal imprint. Thus, the region around the *SNRPN* gene produces a signal necessary for resetting the program. Dittrick et al. (1996) found, in a brain cDNA (bd) library, clones that hybridized to a series of transcripts present in the female germ line but absent in the male germ line. These transcripts arose from alternative upstream start sites, called BD exons, of the *SNRPN* gene. Dittrick et al. (1996) proposed a model showing how alternative transcripts of the *SNRPN* gene can lead to imprint switching. Their model, as interpreted by Bartolomei and Tilghman, 1997 (Fig. 22.1) also relates the locations of the close but not overlapping Angelman and Prader–Willi critical regions to the different imprints that normally occur in the male and female germlines.

The silencing of either maternal or paternal alleles is usually associated with methylation of the promoter region of the allele, rendering it inactive and late replicating. Thus, the paternal *H19* gene in the 11p15 region is methylated and silent, while the maternal allele is unmethylated and expressed. However, the *IGF2* gene in the same region, which is expressed only from the paternal allele, is imprinted only indirectly, and not by methylation. The maternal H19 gene produces an untranslated RNA product, which acts in *cis*, like the *XIST* RNA on the inactive X, to inactivate several nearby genes, including *IGF2*. The paternal

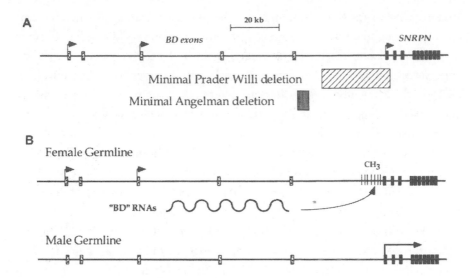

Figure 22.1. Imprinting effects in the Prader–Willi/Angelman region. (A) Exons (black) of the *SNRPN* gene and upstream BD exons (clear) of the *SNRPN* gene are shown, with transcription start sites indicated by arrows. (B) Model showing transcription from BD start sites in the female germline leads to methylation and inactivation of SNRPN, while absence of such BD transcription in the male germline allows *SNRPN* transcription (Bartolomei and Tilghman, 1997, with permission from Annu Rev Genet, v31, copyright 1997, by Annual Reviews, www.annualreviews.org).

H19 is inactive, so the paternal *IGF2* can be expressed (Bartolomei and Tilghman, 1997).

Tissue-Specific Differences in X Inactivation

X inactivation does not occur in early embryonic cells but requires some as yet unknown step in cell differentiation. The process is not yet understood despite intensive study both in humans and in model systems, including induced differentiation of undifferentiated but totipotent embryonic stem cells. The time of X inactivation varies from tissue to tissue; it does not occur in the male germ line until adulthood, and even then not until primary spermatocytes have differentiated from the spermatogonial stem cells. The single X chromosome then remains inactive throughout spermatogenesis. X inactivation is reversed at a certain stage of differentiation of the female germline, again by mechanisms that are poorly understood, and both X chromosomes remain active throughout oogenesis (Fig. 18.2). In the extraembryonic tissues, the paternal X chromosome is preferentially inactivated, perhaps reflecting a paternal imprint that is erased in cells of the embryo itself (Chapter 18).

Erasing an imprint, or reactivating an inactive X chromosome, can be done by demethylation of the relevant DNA, comparable to that described above for immunoglobulin genes after V(D)J rearrangements have been completed. There are other examples of tissue-specific genes that are demethylated only in the cell type in which they are expressed (Kafri et al., 1993). A human enzyme capable of carrying out such demethylations has been identified. The gene for this demethylase is transcribed into mRNA in a wide variety of tissues, although demethylase activity has been demonstrated only in lung cancer cells (Bhattacharya et al., 1999). The high level of this enzyme in some cancers may be responsible for their reduced level of DNA methylation. The resultant alterations in gene expression may contribute to the genomic destabilization so important in the development of cancers (Chapter 26).

Germline-Specific Gene Expression and Sex-Specific Imprinting

In some respects, the germline behaves just like any other tissue whose differentiation requires the activation of specific genes and the inactivation of many others. A growing number of genes have been identified whose expression is limited to the male and/or female germline. One of the first of these was the autosomal phosphoglycerate kinase (*PGK*) gene. It is expressed only in the male germline, and only from the primary spermatocyte stage, when the X-linked *PGK* housekeeping gene is inactivated, along with most of the other genes on the X chromosome. The enzyme PGK is essential for cell survival. Another gene, that for cyclin A1, is expressed in both male and female germlines at the time of meiosis. The cyclin A1 mRNA level rises dramatically in late pachytene spermatocytes and falls to an undetectable level soon after meiosis is over. Somatic cells and germline stem cells, such as spermatogonia, express only cyclin A2 (Sweeney et al., 1996). Cyclins A1 and A2 activate the cyclin-dependent kinases CDK2 and CDK1 to drive the cell through S and G2 (Chapter 2).

One of the most recently identified genes with male germline–specific expression is the *DMRT1* gene. *DMRT1* has been mapped to 9p23.3–p24.1 by a breakpoint that disrupts the gene and caused XY male-to-female sex reversal, indicating its importance for male sex determination. A surprising finding is that *DMRT1* shares homology with a major sex-determining gene in both the roundworm *Caenorhabditis elegans* and the fruit fly *Drosophila melanogaster* (Raymond et al., 1998). Further evidence of the high degree of functional homology between even quite distantly related eukaryotes is provided by another gene with germline-specific expression: *diaphanous* (*DIA*) named for the homologous *Drosophila* gene, whose mutation produces male and female sterility (Bione et al., 1998). The human *DIA* gene maps to Xq21, as shown by disruption of the gene by a t(X;12)(q21;p13) translocation in a woman with secondary amenorrhea due to premature ovarian failure (lack of oocytes) at age 17. *DIA* is expressed only from the active X chromosome (Bione et al., 1998).

The male and female germlines differ remarkably not only in how they conduct meiosis and genetic recombination (Chapter 10) but also in their ability to imprint genes. Imprinting occurs during both spermatogenesis and oogenesis, but different genes are inactivated in the two processes, even when the imprinted genes are in the same region, as seen in the Prader–Willi and Angelman syndrome imprints in 15q11–q13 (Chapter 21). The mechanisms involved in differential imprinting in spermatogenesis and oogenesis, like those of imprinting itself, remain to be discovered.

Embryonic Inactivation of All but One Centromere per Chromosome

When a dicentric chromosome arises in the germline or early embryo, one of the two centromeres is usually inactivated during embryogenesis by what can be called a *differentiation* process. Only one primary constriction is visible (hence the name that is sometimes used, *pseudodicentric*), but a second C-band may mark the position of the inactive centromere (Fig. 22.2b,c). Inactivation of a centromere is usually a fixed event in differentiation, with the same centromere inactive in all somatic cells. Dicentric chromosomes in which the centromeres are some distance apart can be maintained only if one centromere is inactivated. This happens with most isodicentric and dicentric chromosomes, including Robertsonian translocations. The inactivation is not absolute in many of these dicen-

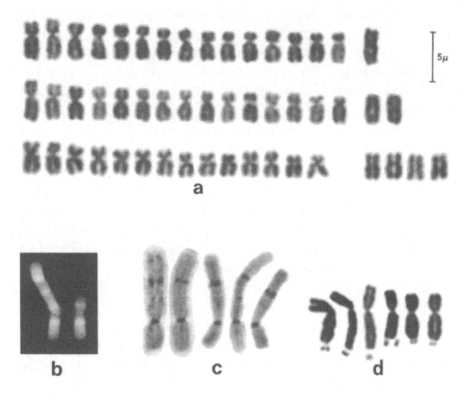

Figure 22.2. (a) The C group, with X chromosomes lacking a functional centromere, from three cells. (b) Q-banded idic(Xp–) joined at Xp telomeres, and normal X. (c) C-banded idic(Xp–) from five cells, showing two C-bands but only one primary constriction (active centromere). (d) X chromosome from six cells, expressing a fragile site (a–c, courtesy E. Therman; D, courtesy S. Roberts).

tric RTs (Chapter 16), and the same is true in some other dicentrics. Sullivan and Willard (1998) studied five stable dicentric X chromosomes made up of two copies of Xq separated by variable lengths of Xp. All cells of the one with 32 Mb of Xp had only one active centromere. The other four, with only 4–12 Mb of Xp, had two active centromeres in 82–87% of cells. In one interesting subject, who had a t(15;20)(pter;pter) chromosome, the centromere from chromosome 15 was always inactivated in lymphocytes, but either centromere could be inactivated in fibroblasts (Rivera et al., 1989). The inactivation process may be irreversible, as in a 6p;19p translocation, in which the chromosomes were attached end to end and the centromere of chromosome 6 was inactive. The dicentric chromosome had a tendency to break at the point of fusion, producing complex mosaicism (Drets and Therman, 1983). The resultant chromosome 6 behaved like an acentric fragment, indicating that the centromere of this chromosome was not reactivated. The mechanism of centromere inactivation is unknown. Is it related to the mechanism that prevents sister chromatids from developing separate functional centromeres at the first meiotic division?

The same process of centromere inactivation may account for the absence of a functional centromere on some X chromosomes, especially in older women. These chromosomes look and act like acentric fragments, drifting at random in anaphase and giving rise to cells in which one X is missing and to cells in which Xs accumulate (Fig. 22.2A). This may account for the increased frequency of 45,X cells in older women (Fitzgerald and McEwan, 1977; Nakagome et al., 1984). Centromere inactivation has rarely been observed in monocentric autosomes (Therman et al., 1986).

Endoreduplication, Polyploidy, and Polyteny

Normal mitosis is characterized by alternating chromosome reproduction and segregation of daughter chromosomes. However, the two processes can be uncoupled in a variety of ways. Here we shall be concerned only with those modifications of mitosis that occur as a part of normal cell differentiation; those important in carcinogenesis are discussed in chapters 25 and 26. Virtually all these modifications are characterized by an absent or defective spindle, and in most cases they result in an increased number of chromosome sets, or *polyploidy*. The most important are the similar processes of endoreduplication and endomitosis.

Polyploidy always occurs in the differentiation of some types of cells, such as trophoblast cells of the placenta and platelet-producing megakaryocytes of the bone marrow. It occurs less often, and to a lesser degree, during the differentiation of a few other types of cells. Polyploidy most commonly arises by *endoreduplication*, in which the chromosomes replicate two or more times between mitoses instead of once. If two rounds of replication have occurred, the diplochromosomes in the next mitosis consist of four chromatids (Fig. 15.2). After three or four rounds of replication without an intervening mitosis, polytene chromosome bundles consisting of 8 or 16 chromatids, respectively, may be produced. Endoreduplication may occur in many tissues, especially in adverse conditions. In some human fibroblast cultures 3–5% of the dividing cells may be tetraploid and a few even octoploid, whereas such divisions are rare in cultured lympho-cytes. Diplochromosomes (with four chromatids) occur only at the first division after endoreduplication, so they are found in only a fraction of the tetraploid cells.

Some bone marrow cells undergo repeated rounds of endoreduplication without cell division to form platelet-producing megakaryocytes with 8C–128C amounts of DNA. Normal megakaryocyte differentiation is induced by the acti-vation of the cytokine receptor c-MPL by the binding of its specific ligand, thrombopoietin (Drachman and Kaushansky, 1997). Differentiation can also be induced in cultured erythroleukemia cells by exposing them to a phorbol ester, such as TPA (12-0-tetradecanoyl phorbol-13-acetate). The failure of mitosis to occur in these cells may be due to a marked fall in the cyclin-dependent kinase CDK1 and a lack of cyclin B1/CDK1 kinase (metaphase promoting factor, MPF) activity, so that the completion of S does not trigger mitosis (Chapter 2; Datta et al., 1996). Without the activation of the anaphase promoting factor, APC, cyclin E is not targeted for destruction by the ubiquitin system, and repeated rounds of replication can occur.

Trophoblast cells of the placenta reach ploidy levels of 64C or higher. The mechanism leading to these polyploid cells is not clear. It may involve *endomito-sis*, a process in which chromosomes duplicate and periodically condense, going through stages called *endoprophase*, *endometaphase*, and *endoanaphase*, but the nuclear membrane never breaks down (Sarto et al., 1982). Molecular analysis might clarify the events actually involved. In rodents, where trophoblast cells reach ploidy levels of 512C, the mechanism involved appears to be endoreduplication, with cells, and nuclei, increasing stepwise in both size and degree of polyploidy. This is accompanied by the development of polytene chromosomes in which all the sister chromatids synthesized from one homologue remain adherent to one another, even though they do not show the density and fine banding seen in *Drosophila* polytene chromosomes. In situ hybridization with four of five gene probes tested showed a single signal from all the amplified copies of tightly bundled sister chromatids in these interphase nuclei, with no pairing of homo-logues (Varmuza et al., 1988). Homologous pairing of mitotic chromosomes is generally absent in humans. The mitotic crossing over seen in Bloom syndrome (Chapter 24) is suggestive of somatic pairing, but the other chromosomes in the metaphase spreads do not show somatic pairing.

Cells with far more than one nucleus arise as a regular developmental change in a variety of cells, most notably placental cytotrophoblast and muscle. Their mechanism of production is cell fusion, which is triggered by still unknown factors. Binucleate and low multinucleate cells can arise by mitosis that is not followed by cell division. This is a common phenomenon in liver (D'Amato, 1989).

Triradials, Multiradials, and ICF Syndrome, a Hypomethylation Disorder

A triradial chromosome, or triradial, is marked by a partial endoreduplication: It has three arms instead of the usual two (Fig. 20.1). Triradials are rare: Stahl-Mauge et al. (1978) found only two triradials in 53,000 cells from normal people. Triradials are much more common after exposure to chromosome-breaking agents and in patients with Fanconi anemia or Bloom syndrome (Chapter 24). The breakpoint in a triradial is often a fragile site. Although several different mechanisms for the origin of triradials have been proposed, most symmetrical triradials probably arise by partial endoreduplication (Kuhn and Therman, 1982), but what triggers this? Treating normal cells with the demethylating agent 5-azacytidine causes partial endoreduplications, sister chromatid exchanges, and decondensation of the pericentromeric heterochromatin of chromosomes 1, 9, and 16 (Hori, 1983). Normally, the DNA in these regions is highly methylated (Miller et al., 1974). Could its demethylation be responsible for all these changes? Studies on a rare genetic disease, ICF syndrome, indicate this is indeed the case and serve to emphasize the importance of regulated DNA methylation in normal differentiation.

The autosomal recessive ICF syndrome of immunodeficiency, centromeric heterochromatin instability, and facial anomalies is characterized by the presence of multiradial chromosomes 1, 9, and 16 and other pericentromeric rearrangements of these chromosomes (Fig. 22.3). The hetetochromatin of these chromosomes is hypomethylated in the ICF syndrome (Miniou et al., 1994), especially satellite 2, a major component of the heterochromatin of chromosome 1 (Hernandez et al., 1997). The ICF syndrome is thus a hypomethylation disorder. The phenotype of normal cells after treatment with 5-azacytidine or 5-azadeoxycytidine is identical to that of ICF cells: There are multiradials with up to seven arms, and 80% of the multiradials involve chromosome 1. The treatment also produces whole-arm deletions, isochromosomes 1, and pericentromeric fusions of chromosomes 1 and 9 or 1 and 16 (Hernandez et al., 1997). The *ICF* gene locus has been mapped to a 9-cM region in 20q11–q13, using homozygosity mapping (Wijmenya et al., 1998). This region contains the *DNA*

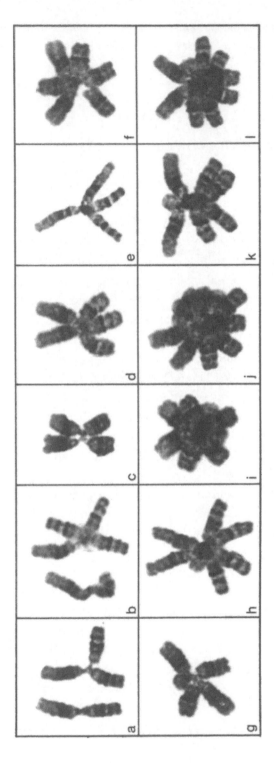

Figure 22.3. Multiradial chromosomes 1 and 16 in cultured lymphocytes from a patient with the ICF syndrome (reproduced from Smeets et al., ICF syndrome: a new case and review of the literature, Hum Genet 94:240–246, copyright 1994, Springer-Verlag).

methyltransferase 3B (*DNMT*3B) gene. Mutations of both alleles of this gene have been identified in each of five unrelated ICF patients (Xu et al., 1999). All five had the typical hypomethylation of satellite 2 on chromosomes 1 and 16 and of satellite 3 on chromosome 9.

References

Bartolomei MS, Tilghman SM (1997) Genomic imprinting in mammals. Annu Rev Genet 31:493–525

Bhattacharya SK, Ramchandani S, Cervoni N, et al. (1999) A mammalian protein with specific demethylase activity for mCpG DNA. Nature 397:579–583

Bione S, Sals C, Manzin C, et al. (1998) A human homologue of the *Drosophila melanogaster diaphanous* gene is disrupted in a patient with premature ovarian failure: evidence for conserved function in oogenesis and implications for human sterility. Am J Hum Genet 62:533–541

Brown CJ, Ballabio A, Rupert JI, et al. (1991) A gene from the region of the X inactivation centre is expressed exclusively from the inactive X chromosome. Nature 349:38–44

Chaillet JR, Bader DS, Leder P (1995) Regulation of genomic imprinting by gametic and embryonic processes. Genes Dev 9:1177–1187

Chess A, Simon I, Cedar H, et al. (1994) Allelic inactivation regulates olfactory receptor gene expression. Cell 78:823–834

D'Amato F (1989) Polyploidy in cell differentiation. Caryologia 42:183–211

Datta NS, Williams JL, Caldwell J, et al. (1996) Novel alterations of CDK1/cyclin B1 kinase complex formation occur during the acquisition of a polyploid DNA content. Mol Biol Cell 7:209–223

Dittrich B, Buiting K, Korn B, et al. (1996) Imprint switching on human chromosome 15 may involve alternate transcripts of the *SNRPN* gene. Nat Genet 14:163–170

Drachman J, Kaushansky K (1997) Dissecting the thrombopoietin receptor: Functional elements of the Mpl cytoplasmic domain. Proc Natl Acad Sci USA 94:2350–2355

Drets ME, Therman E (1983) Human telomeric 6;19 translocation chromosome with a tendency to break at the fusion point. Chromosoma 88: 139–144

Fitzgerald PH, McEwan CM (1977) Total aneuploidy and age-related sex chromosome aneuploidy in cultured lymphocytes of normal men and women. Hum Genet 39:329–337

Gabriel JM, Higgins MJ, Gebuhr TC, et al. (1998) A model system to study genomic imprinting of human genes. Proc Natl Acad Sci USA 95: 14857–14862

Hernandez R, Frady A, Zhang X-Y, et al. (1997) Preferential induction of chromosome 1 multibranched figures and whole-arm deletions in a human pro-B cell line treated with 5-azacytidine or 5-azadeoxycytidine. Cytogenet Cell Genet 76:196–201

Hori T-A (1983) Induction of chromosome decondensation, sister chromatid exchanges and endoreduplications by 5-azacytidine, an inhibitor of DNA methylation. Mutat Res 121:47–52

Jeppesen P (1997) Histone acetylation: a possible mechanism for the inheritance of cell memory at mitosis. BioEssays 19:67–74

Kafri T, Gao X, Razin A (1993) Mechanistic aspects of genome-wide demethylation in the preimplantation monse embryo. Proc Natl Acad Sci USA 90:10558–10562

Kuhn EM, Therman E (1982) Origin of symmetrical triradial chromosomes in human cells. Chromosoma 86:673–681

Lichtenstein M, Keine G, Cedar H, et al. (1994) B cell–specific demethylation: a novel role for the intronic kappa chain enhancer sequence. Cell 76: 913–923

Malnic B, Hirono J, Sato T, et al. (1999) Combinatorial receptor codes for odors. Cell 96:713–723

Miller OJ (1997) Chromosome changes in cell differentiation. Genetics 146: 1–8

Miller OJ, Schnedl W, Allen J, et al. (1974) 5-Methylcytosine localised in mammalian constitutive heterochromatin. Nature 251:636–637

Miniou P, Jeanpierre M, Blanquet V, et al. (1994) Abnormal methylation pattern in constitutive and facultative (X inactive chromosome) heterochromatin of ICF patients. Hum Mol Genet 3:2093–2102

Nakagome Y, Abe T, Misawa S, et al. (1984) The "loss" of centromeres from chromosomes of aged women. Am J Hum Genet 36:398–404

Raymond CS, Shamu CE, Shen MM, et al. (1998) Evidence for evolutionary conservation of sex-determining genes. Nature 391:691–695

Rivera H, Zuffardi O, Maraschio P, et al. (1989) Alternate centromere inactivation in a pseudodicentric (15;20)(pter;pter) associated with a progressive neurological disorder. J Med Genet 26:626–630

Sarto GE, Stubblefield PA, Therman E (1982) Endomitosis in human trophoblast. Hum Genet 62:228–232

Smeets DFCM, Moog U, Weemaes CMR, et al. (1994) ICF syndrome: a new case and review of the literature. Hum Genet 94:240–246

Stahl-Maugé C, Hager HD, Schroeder TM (1978) The problem of partial endoreduplication. Hum Genet 45:51–62

Sullivan BA, Willard HF (1998) Stable dicentric X chromosomes with two functional centromeres. Nat Genet 20:227–228

Sweeney C, Murphy M, Kubelka M, et al. (1996) A distinct cyclin A is expressed in germ cells of the mouse. Development 122:53–64

Therman E, Trunca C, Kuhn EM, et al. (1986) Dicentric chromosomes and the inactivation of the centromere. Hum Genet 72:191–195

Tonegawa S (1983) Somatic generation of antibody diversity. Nature 302: 575–581

Varmuza S, Prideaux V, Kothary R, at al. (1988) Polytene chromosomes in mouse trophoblast giant cells. Development 102:127–134

Vu TH, Hoffman AR (1994) Promoter-specific imprinting of the human insulin-like growth factor-II gene. Nature 371:714–717

Wevrick R, Kerns JA, Francke U (1994) Identification of a novel paternally expressed gene in the Prader–Willi syndrome region. Hum Mol Genet 3:1877–1882

Wijmenya C, Heuvel LPWJ van den, Strengman E, et al. (1998) Localization of the ICF syndrome to chromosome 20 by homozygosity mapping. Am J Hum Genet 63:803–809

Xu G-L, Bestor TH, Bourc'his D, et al. (1999) Chromosome instability and immunodeficiency syndrome caused by mutations in a DNA methyltransferase gene. Nature 402:187–191

23

Somatic Cell Hybridization in Cytogenetic Analysis

For many years, human geneticists dreamed of finding a way to study segregation and recombination in somatic cells, in order to overcome the difficulties of genetic analysis with small families, few genetic markers, limited availability of meiotic material, and no experimental matings. Somatic cell genetics became a powerful way to do this, with the discovery of preferential loss (segregation) of human chromosomes in interspecific human-rodent somatic cell hybrids. The analysis of such hybrids has made it possible to correlate the loss or retention of a specific chromosome or chromosome segment with the loss or retention of a specific gene product or gene, allowing the gene to be mapped to that chromosome or segment. Enormous strides have been made using this approach. It is still one of the most widely used methods for mapping genes and for demonstrating multiple genetic causes of a disease phenotype (*genetic heterogeneity*). Somatic cell hybrids have many other uses in cytogenetic analysis. It is therefore helpful to understand somatic cell hybridization and how it is achieved experimentally.

Cell Fusion

Spontaneous fusion of cells occurs in the normal development of bone, muscle, and placental trophoblast tissues. Spontaneous fusion also occurs between cancer cells or between cancer cells and normal cells (Kovacs, 1985), but it is unclear whether cell fusion plays any role in the origin of the polyploid and aneuploid chromosome constitutions seen in cancer (Chapter 26). Spontaneous fusion of nonmalignant lymphocytes and fibroblasts has so far been observed only in Bloom syndrome (Chapter 24) (Otto and Therman, 1982). The mechanism of spontaneous cell fusion is still poorly understood.

The observation that inactivated parainfluenza (Sendai) virus can induce cell fusion, and the discovery that this can lead to mononucleated hybrid cells, has caused a revolution in cytogenetics and the rapid development of the field of somatic cell genetics (reviewed by Harris, 1995). Cell fusion is now induced mainly with polyethylene glycol (PEG), but electrofusion also works with some cell types (Cervenka and Camargo, 1987). Cell fusion is a widely used tool in complementation analysis (below and Chapter 24) and gene mapping (Chapter 29), and in the creation of hybridomas, lymphocyte hybrids that produce virtually limitless amounts of monospecific (*monoclonal*) antibodies. Their use has advanced many fields, including cytogenetics.

Complementation Analysis in Heterokaryons

When two cells fuse, the hybrid at first has two nuclei in a common cytoplasm. These *heterokaryons* can persist for some time, allowing various types of studies to be carried out. The one most important for cytogenetics is called *complementation analysis*. This is used to determine whether a particular abnormal recessive phenotype that is observed in two unrelated individuals is caused by a mutation in the same gene or by mutations in different genes. Cells cultured from one person are fused with cells from the other person to produce heterokaryons. These are then tested for the abnormal cellular phenotype. If this is still present, the same gene is mutant in the two individuals. However, if the phenotype is now normal, each cell type has supplied the gene product missing in the other cell type; that is, each has complemented the other. This almost always means that different gene loci are mutant in the two individuals. (Very rarely, the mutations may involve the same gene locus but be so far apart that a mitotic recombination between them produces one normal gene copy and one that has both mutations.) This approach has been used to show that mutations in many different genes are responsible for each of several chromosome breakage syndromes (Chapter 24).

Premature Chromosome Condensation and Allocycly

When a cell whose chromosomes are in metaphase is fused with a cell in interphase, the nuclear membrane of the interphase nucleus breaks down (disassembles) and the chromosomes condense, a process called premature chromosome condensation (PCC). The condensation becomes visible some 15–20 min after the fusion and reaches its peak in 1 h (Rao, 1982; Sperling, 1982). The appearance of PCC depends on the stage of the cell cycle at the time of the fusion (Fig. 23.1). PCC chromosomes in G0 or early G1 cells are short, but become longer and thinner as G1 progresses (Fig. 23.1). PCC chromosomes in S-phase cells appear pulverized, or fragmented. However, even at this stage, the chromosomes are continuous; the gaps in them represent the segments that are replicating, as demonstrated by autoradiography after tritiated-thymidine incorporation (Fig. 23.2). PCC chromosomes in G2 cells resemble long, thin prophase chromosomes, which gradually shorten (Fig. 23.3). They can be banded with the usual techniques but sometimes appear spontaneously banded (Fig. 23.3 bottom). The spontaneous banding resembles replication banding. The late-replicating segments have not completed their condensation and remain stretched out.

   PCC induced by cell fusion can be used to study chromosomes during interphase. Since interphase chromosomes are longer than mitotic chromosomes, their DNA and RNA synthesis patterns can be analyzed in greater detail (Sperling, 1982). PCC can be used to study the chromosomes of nondividing differentiated cells, such as neurons. PCC can be useful in studying chromosome breakage. With ordinary cytological techniques, chromosome breakage can be studied, at the earliest, several hours after its induction, whereas with PCC, breaks can be analyzed after only 20–30 min. This yields a better estimate of the

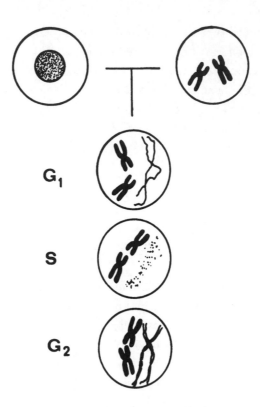

Figure 23.1. Fusion of an interphase and a metaphase cell leads to PCC formation. In G1, the PCC chromosomes are thin and single; in S, they appear "pulverized," in G2, they resemble prophase chromosomes.

original breakage rate, because breaks tend to be repaired. Analysis of PCC at different times after treatment demonstrates a gradual decrease of visible chromosome aberrations (Hittelman et al., 1980).

An otherwise normal metaphase plate may contain one or a few chromosomes that resemble PCC chromosomes in G1, S, or G2 (Figs. 20.1 and 23.2). Such *allocyclic* (out-of-phase) chromosomes are very rare in normal, untreated cells, although ring chromosomes are often allocyclic. The incidence of allocyclic chromosomes is increased by exogenous agents that break chromosomes (Rao, 1982) and in Bloom syndrome (Otto et al., 1981), which results from an endogenous chromosome-breaking genotype (Chapter 24). One hypothesis is that an allocyclic chromosome has undergone a mutation in a hypothetical condensation center that renders it unable to keep up with the rest of the chromosomes during mitosis and therefore forms a micronucleus (Otto et al., 1981). This might

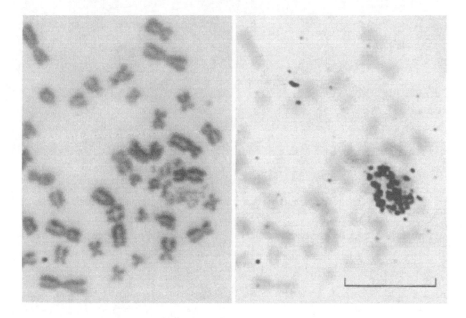

Figure 23.2. Part of a tetraploid lymphocyte metaphase from a patient with Bloom syndrome with an allocyclic chromosome before (left) and after (right) autoradiography. [³H]Thymidine was added to the cell culture 4 hours before fixation. A mitotic chiasma is also present (Otto et al., 1981).

account for the correlated presence of micronuclei and allocyclic chromosomes (Obe and Beek, 1982).

Somatic Cell Hybrids

When both nuclei in a heterokaryon enter mitosis synchronously, they form a single mitotic spindle and distribute a complete set of chromosomes from each nucleus to each pole, producing identical *synkaryons*, or hybrid cells (Fig. 23.4). In order to study hybrid cells, one must be able to propagate selectively only those cells that contain chromosomes from both parents. Generally the human cells used in human-rodent cell mixtures have limited capacity for growth in culture; they will quickly be swamped out by the growth of the parental rodent cells and of the rodent-human hybrid cells. A selection procedure must be used to allow rodent-human hybrid cells to overgrow the rodent cells. The means used include culture media in which one cell type cannot grow or grows pref-

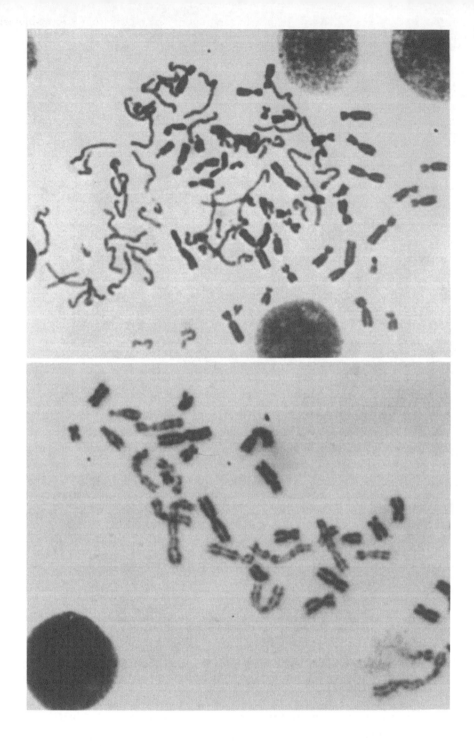

Figure 23.3. Spontaneous fusion of two Bloom syndrome lymphocytes leads to formation of prematurely condensed chromosomes (PCC). (Top) PCC in G1. (Bottom) Naturally banded PCC in G2 (Otto and Therman, 1982).

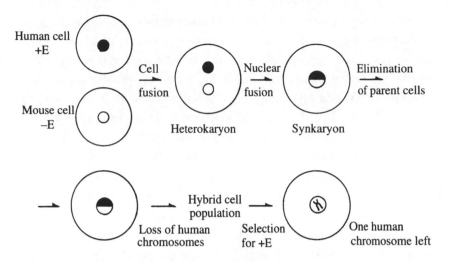

Figure 23.4. Assignment of a gene coding for the enzyme E to a specific human chromosome through fusion of a human cell (nucleus black) with a mouse cell (nucleus white) followed by hybrid cell selection.

erentially and selective killing of one cell type with toxins, antibodies, or viruses to which the other cells are resistant. Human cells can be killed with diphtheria toxin or the cardiac glycoside ouabain, to which the mouse and hybrid cells are resistant. In early studies, the rodent cell line usually had a recessive phenotype, such as a nutritional deficiency, that could be complemented by a wild-type human gene product. By growing the cells in culture medium devoid of the required nutrient, one could select against the parental rodent cells and allow selective growth of hybrid cells containing a specific human chromosome or chromosome segment carrying the gene.

Preferential Human Chromosome Loss from Rodent-Human Hybrids

Weiss and Green first pointed out the preferential loss of human chromosomes from human-mouse cell hybrids more than 30 years ago. Cell hybridization can be regarded as a parasexual process in which meiotic segregation is replaced by random chromosome loss. Neither the reason for nor the exact mechanism of the process of chromosome elimination is understood, but it plays an important role in gene mapping. A large number of human chromosomes are lost fairly quickly and pretty much at random; a variable number of the remaining human chromosomes are lost more slowly. Selective retention of a specific human chromosome can be brought about by the use of a selectable marker. A mouse parental line that is deficient in either hypoxanthine-guanine phosphoribosyl transferase (HPRT) or thymidine kinase (TK) can be fused with normal human cells and the newly hybridized cells grown in hypoxanthine-amethopterin-thymidine (HAT) medium. These enzymes are required for cell survival in HAT medium. Only the hybrid cells that have retained a human chromosome carrying the corresponding functional gene (*HPRT* on the active X or *TK* on a chromosome 17, respectively) will survive. A number of selectable marker systems have been developed that enable one to select for the retention or the elimination of a specific chromosome. New sets of positive/negative selectable markers are continually being developed, some by introducing (*transfecting*) into a random chromosome a gene conferring resistance to neomycin, hygromycin, puromycin, or blasticidin S (Karreman, 1998). This permits selective retention or elimination of this chromosome from hybrid cells.

It would be useful if the human DNA in interspecific hybrid cells could be isolated. A novel way to do this is to clone the human DNA preferentially into yeast artificial chromosomes by a technique called *transformation-associated recombination cloning* (Larionov et al., 1996). This is achieved by cotransformation of yeast with (1) a linearized plasmid containing a human repeat sequence plus a yeast centromere and a yeast telomere and (2) a large fragment of DNA from the hybrid cell. Only fragments containing the same human repeat are able to recombine with the repeat in the linearized plasmid to form a circular yeast artificial chromosome (YAC). This method has been used for selectively cloning the human DNA in hybrids containing a single human chromosome and radiation hybrids containing only 5 Mb of human DNA (Larionov et al., 1996).

Although human chromosomes are preferentially lost from the commonly produced rodent-human hybrids, there are significant exceptions to this generalization. Croce discovered that fusion of cells of the established HT1080 human colon cancer cell line with freshly isolated murine diploid cells led to preferential loss of murine chromosomes. These hybrids, and the corresponding ones that preferentially lose human chromosomes, enabled Miller et al. (1976) to show that transcription of both human and mouse ribosomal RNA genes is dependent upon a species-specific transcription factor not carried on an acrocentric chromosome. Marcus et al. (1976) produced one human-mouse hybrid that accumulated multiple copies of the human Y chromosome despite losing almost all the other human chromosomes. This hybrid was instrumental in the initial cloning of single-copy DNA sequences from the Y chromosome (Bishop et al., 1983).

Induced Chromosome Breakage: Radiation Hybrids

At least one of the parental cell types used in generating most human-rodent hybrids is malignant. The structural chromosome instability of malignant cells, which is associated with defective DNA damage checkpoints (Chapter 26), continues in the hybrid cells and is probably responsible for the spontaneous breakage that has long been observed in these hybrids. Induced chromosome breakage is a much more useful mapping tool. Massive chromosome-shattering doses of ionizing radiation are administered to human cells just prior to fusion with rodent cells (Goss and Harris, 1977) or to hybrid cells containing a single human chromosome (Cox et al., 1990) before fusion with rodent cells to produce hybrids containing tiny human chromosome fragments. The former method is useful for mapping any chromosome, the latter for mapping a single chromosome. For general chromosome mapping, a radiation dose of 10,000 rads (100 Gy) is satisfactory, giving fragment sizes from 2 Mb to entire chromosome arms. For high-resolution mapping or positional cloning, higher doses are preferred; 25,000 rads (250 Gy) yields DNA fragments that are mostly shorter than 3 Mb (Siden et al., 1992).

Microcell Hybrids

Selectable markers are useful in constructing chromosome-specific microcell hybrids. The neomycin resistance gene, *neo*, is a very useful dominant selectable marker. It can be introduced into diploid human fibroblasts using a retroviral vector and will integrate at random into any chromosome. These fibroblasts can then be converted into microcells that are fused to rodent cells to make microcell hybrids. By selecting those that are resistant to neomycin, one obtains microcell hybrids with a specific human chromosome containing the *neo* gene (Lugo et al., 1987). Using a series of seven such hybrids, each containing a different human chromosome, Ning et al. (1993) showed that PCR amplification of human sequences in the hybrids, using a LINE1 oligonucleotide primer, produced such a distinctive pattern of DNA fragment sizes for each chromosome that it constituted a "PCR karyotype." This can be used to check the human chromosomes present in other hybrids.

The hygromycin phosphotransferase gene confers resistance to hygromycin and sensitivity to gancyclovir. Speevak et al. (1995) introduced this gene into random chromosome sites in human diploid fibroblasts, which were then used to generate microcell hybrids containing a single human chromosome. Growth of these hybrids in hygromycin (and confirmation by gancyclovir killing) enabled them to isolate a series of chromosomes containing this dually selectable marker gene.

Chromosome and Gene Transfer: Transgenomes and Transgenes

Cells in culture have the ability to take up whole chromosomes or fragments of chromosomes from the medium and incorporate them into their genomes. More important is their ability to undergo transfection, incorporating much smaller DNA fragments, ranging in size from YACs hundreds of kb in size to single genes (*transfection*). The introduction of single chromosomes or chromosome fragments has been instrumental in the analysis of tumor suppressor genes (Chapter 28). For example, the transfer of a normal chromosome 5 into some colorectal carcinoma cells inhibits the overexpression of the c-*MYC* gene that is characteristic of these tumors and abolishes their tumorigenicity. This indicates that disruption or mutation of a gene on chromosome 5 is important in colon carcinogenesis (Rodriguez-Alfageme et al., 1992). Introduction of a transgenic construct containing the catalytic subunit of the telomerase gene into human retinal epithelial cells or skin fibroblasts prevented their senescence, enabling the cells to continue to divide (Bodnar et al., 1998).

Homologous recombination is the mechanism involved in targeting genes to cells in culture. This leads to the replacement of a defective gene by a normal copy or vice versa. The chicken pre–B cell line DT40 is highly proficient for homologous recombination, whereas most human cells are poor at it. Koi et al. (1997) used microcell-mediated cell fusion to transfer single human chromosomes tagged with *pSVneo* as a selectable marker into DT40 cells. They found that targeted integration of various gene constructs into their homologous loci on the human chromosome then occurred at high frequency. This system could be very useful for mapping and cloning genes and studying their function, and as a source of specific gene products. Gene targeting can be used to study the function of some of the large number of genes important for chromosome structure and function. For example, the p21 protein product of the cell cycle inhibitor gene *CIP1/WAF1* (Chapter 2) was shown to block cell senescence in diploid human fibroblasts, permitting their unlimited growth. To show this, the gene was inactivated by two rounds of targeted homologous recombination (Brown et al., 1997).

Somatic cell genetic approaches to other problems continue to appear. A method has been developed for preparing monosomic cell lines. The *gpt* plasmid is introduced into cultured diploid cells and allowed to integrate at random into various sites. The cells are grown with selection against retention of *gpt* and with partial inhibition of topoisomerase II to promote chromatid nondisjunction during mitosis. This has led to cell lines that are monosomic for chromosome 4, 8, or 21 and to several lines with partial monosomies or deletions (Clarke et al., 1998).

References

Bishop CE, Guellaen G, Geldwerth D, et al. (1983) Single-copy DNA sequences specific for the human Y chromosome. Nature 303:831–833

Bodnar AG, Oullette M, Frolkis M, et al. (1998) Extension of life-span by introduction of telomerase into normal human cells. Science 279:349–352

Brown JP, Wei W, Sedivy JM (1997) Bypass of senescence after disruption of p21(CIP1/WAF1) gene in normal diploid human fibroblasts. Science 277:831–834

Cervenka J, Camargo M (1987) Premature chromosome condensation induced by electrofusion. Cytogenet Cell Genet 45:169–173

Clarke DJ, Giménez-Abián JF, Tönnies H, et al. (1998) Creation of monosomic derivatives of human cultured cell lines. Proc Natl Acad Sci USA 95: 167–171

Cox DR, Burmeister M, Price ER, et al. (1990) Radiation hybrid mapping: A somatic cell genetic method for constructing high-resolution maps of mammalian chromosomes. Science 250:245–250

Goss SJ, Harris H (1977) Gene transfer by means of cell fusion. I. Statistical mapping of the human X-chromosome by analysis of radiation-induced gene segregation. J Cell Sci 25:17–37

Harris, H (1995) The cells of the body. A history of somatic cell genetics. Cold Spring Harbor Laboratory, Plainview

Hittelman WN, Sognier MA, Cole A (1980) Direct measurement of chromosome damage and its repair by premature chromosome condensation. In: Meyn RE, Withers HR (eds) Radiation biology in cancer research. Raven, New York, pp 103–123

Karreman C (1998) A new set of positive/negative selectable markers for mammalian cells. Gene 218:57–61

Koi M, Lamb PW, Filatov L, et al. (1997) Construction of chicken x human microcell hybrids for human gene targeting. Cytogenet Cell Genet 76:72–76

Kovacs G (1985) Premature chromosome condensation: evidence for *in vivo* cell fusion in human malignant tumours. Int J Cancer 36:637–641

Larionov V, Kouprina N, Graves J, et al. (1996) Highly selective isolation of human DNAs from rodent-human hybrid cells as circular yeast artificial chromosomes by transformation-associated recombination cloning. Proc Natl Acad Sci USA 93:13925–13930

Lugo TG, Handelin B, Killary A, et al. (1987) Isolation of microcell hybrid clones containing retroviral vector insertions into specific human chromosomes. Mol Cell Biol 7:2814–2820

Marcus M, Tantravahi R, Dev VG, et al. (1976) Human-mouse cell hybrid with multiple Y chromosomes. Nature 262:63–65

Miller OJ, Miller DA, Dev VG, et al. (1976) Expression of human and suppression of mouse nucleolus organizer activity in mouse-human somatic cell hybrids. Proc Natl Acad Sci USA 73:4531–4535

Ning Y, Lovell M, Cooley LD, et al. (1993) "PCR karyotype" of monochromosomal somatic cell hybrids. Genomics 16:758–760

Obe G, Beek B (1982) Premature chromosome condensation in micronuclei. In: Rao PN, Johnson RT, Sperling K (eds) Premature chromosome condensation. Academic, New York, pp 113–130

Otto PG, Therman E (1982) Spontaneous cell fusion and PCC formation in Bloom's syndrome. Chromosoma 85:143–148

Otto PG, Otto PA, Therman E (1981) The behavior of allocyclic chromosomes in Bloom's syndrome. Chromosoma 84:337–344

Rao PN (1982) The phenomenon of premature chromosome condensation. In: Rao PN, Johnson RT, Sperling K (eds) Premature chromosome condensation. Academic, New York, pp 1–41

Rodriguez-Alfageme C, Stanbridge EJ, Astrin SM (1992) Suppression of deregulated c-*MYC* expression in human colon carcinoma cells by chromosome 5 transfer. Proc Natl Acad Sci USA 89:1482–1486

Siden TS, Kumlien J, Schwartz CE, et al. (1992) Radiation fusion hybrids for human chromosomes 3 and X generated at various radiation doses. Somat Cell Mol Genet 18:33–44

Speevak MD, Bérubé NG, McGowan-Jordan J, et al. (1995) Construction and analysis of microcell hybrids containing dual selectable tagged human chromosomes. Cytogenet Cell Genet 69:63–65

Sperling K (1982) Cell cycle and chromosome cycle: morphological and functional aspects. In: Rao PN, Johnson RT, Sperling K (eds) Premature chromosome condensation. Academic, New York, pp 43–78

24

Chromosome Instability Syndromes

S everal mutant genes are known that greatly increase the incidence of chromosome aberrations and have striking clinical effects. The most extensively studied of these autosomal recessive disorders are *Bloom syndrome* (BS), *Fanconi anemia* (FA), and *ataxia telangiectasia* (AT). The risk of cancer is enhanced in such individuals: one-fourth of BS patients, one-eighth of AT patients, and about one-eighth of FA patients develop cancer, frequently at an early age. The great interest in BS, FA, and AT might seem out of proportion to their rare occurrence. However, their study has shed light on chromosome structure and function, providing insights into the role of disturbed DNA replication and repair in causing chromosome breaks, mutator phenotypes, and cancer.

Bloom Syndrome

This autosomal recessive disease is characterized by low birth weight, stunted growth, sun sensitivity, facial erythema (dilation of blood vessels), immunodeficiency, and a greatly increased risk of cancer (German, 1993). There is a high frequency of chromosome aberrations in BS patients, as first noted in 1965 by German. These are of two types. In the first, random breaks lead to fragments or to reciprocal translocations between nonhomologous chromosomes. The second type is found almost exclusively in BS and consists of a greatly enhanced tendency to have homologous exchanges. All the breakage probably takes place in S–G2 (Therman and Kuhn, 1985). The increased tendency towards exchanges between homologous chromatids expresses itself in a 10- to 20-fold increase in the number of sister chromatid exchanges (SCEs) (Chaganti et al., 1974; Fig. 24.1) and in a 50- to 100-fold increase in the number of quadriradial configurations indicative of mitotic crossing over (Fig. 24.2; Therman and Kuhn, 1981). Spontaneous cell fusion is sometimes seen in BS cells and in malignant cells, but otherwise occurs only in normally differentiating bone, muscle, and trophoblast (Chapter 23).

The high SCE frequency seen in BS fibroblasts in culture is reduced to normal by fusing BS cells with normal cells or by microcell-mediated transfer of a normal chromosome 15 into BS cells (McDaniel and Schultz, 1992). The BS gene, called *BLM*, is genetically very closely linked to *FES*, which has been mapped more precisely to 15q26 (Mathew et al., 1993). Lymphoblastoid BS cells also show a high SCE frequency, but subcultures sometimes revert to a normal frequency. Genetic marker studies showed that the revertant lines all showed loss of heterozygosity (LOH) for 15q markers distal to the *BLM* locus, presumably as a result of somatic crossing over between homologous chromosomes each carrying a different BLM mutation. Thus, one of the daughter chromosomes has the recombined *BLM* gene with both point mutations, while it homologue has a normal copy of the gene. Mapping the presumptive crossover point led to isolation and cloning of the *BLM* gene (Ellis et al., 1995). Interestingly, the 1417-amino-acid protein product predicted by the DNA sequence of *BLM* belongs to the RecQ family of DNA helicases (unwinding enzymes), and this may account for the well-known retardation of DNA chain elongation during replication in BS cells. Another of these helicases is mutated in Werner syndrome (WS), as described below.

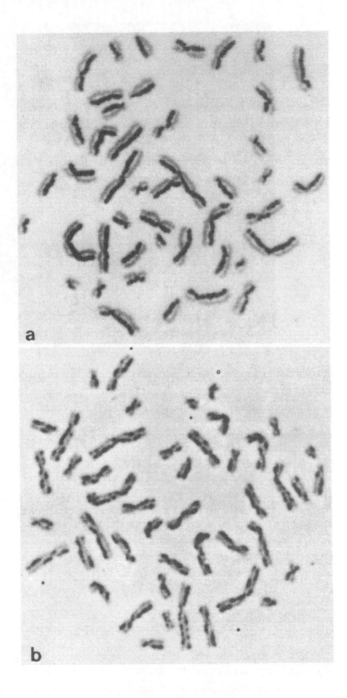

Figure 24.1. (a) Sister chromatid exchanges (SCEs) in a normal lymphocyte. (b) Highly increased number of SCEs in a lymphocyte from a patient with Bloom syndrome (courtesy of RSK Chaganti).

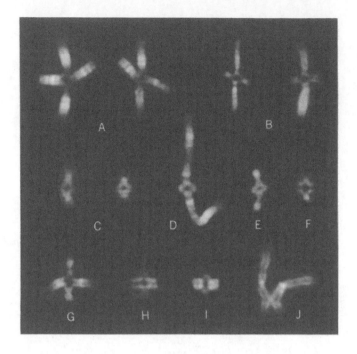

Figure 24.2. Quadriradial configurations in chromosomes 1q(A), 6p (B), 19 (C), 1q (D), 20q (E), 22q (F), 12q (G), 18q (H), centric region of 18 (I), and 3p (J), indicative of mitotic chiasmata in Bloom syndrome (Kuhn, 1976).

Mitotic Recombination or Crossing Over

Chiasmata and genetic recombination occur at high frequency in meiosis (Chapter 9) but far less often in mitosis. The most striking feature of the chromosome instability in Bloom syndrome is the marked tendency for mitotic chiasmata to form between homologous chromosomes to give distinctive quadriradial configurations (Figs. 20.1, 24.2). An interesting example of mitotic recombination (MR) arising by exchange between homologues is the transfer of a distinctive Q-bright satellite from one acrocentric chromosome to another (Therman et al., 1981). Cytological evidence of MR in BS cells (Chaganti et al., 1974) has been confirmed by molecular analyses (Groden et al., 1990). MR occurs far more frequently in Bloom syndrome than in normal cells, in which the rate of mitotic recombination is generally about 1–10 per 100,000 cells (Gupta et al., 1997), with marked individual variation of unknown origin (Holt et al., 1999). Mitotic crossing over leads to LOH (homozygosity) for all loci

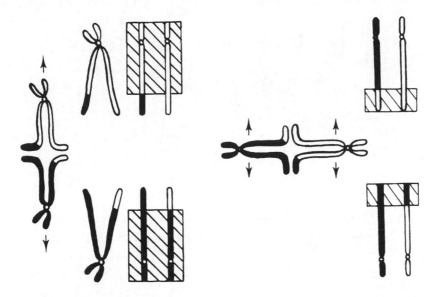

Figure 24.3. Segregation after crossing over in a meiotic bivalent (left) and a mitotic chiasma (right). The chromatids distal to a meiotic chiasma are nonsisters, but those distal to a mitotic chiasma are sister chromatids. Note the loss of heterozygosity (LOH) with homozygosity of markers distal to the mitotic crossover point (Therman and Kuhn, 1981).

distal to the breakpoint (Fig. 24.3), as seen in BS lines that have reverted to a normal SCE frequency. Loss of heterozygosity is a major mechanism of carcinogenesis in general (Chapter 28), and it may account for the high incidence of cancer in BS, because many chromosome segments can become homozygous via the abundant SCEs (Therman and Kuhn, 1981). In this regard, the highly nonrandom distribution of mitotic chiasmata in BS may be relevant. Although about one-sixth of the chiasmata are in centromeric heterochromatin, there are also specific hotspots in distal 1p, 3p21, 6p21, 11q13, 12q13, 17q12, and distal 22q (Kuhn, 1976), some near tumor suppressor genes.

Fanconi Anemia

This autosomal recessive disease is marked by bone marrow failure (pancytopenia), skeletal anomalies (such as absence of the thumb or radius), hyperpigmentation of the skin, reduced fertility, and an increased incidence of cancer (Alter

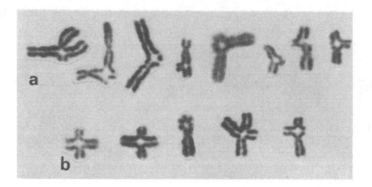

Figure 24.4. Chromosome abnormalities in lymphocytes of a patient with Fanconi anemia. (a) Adjacent quadriradials. (b) Alternate quadriradials (courtesy of EM Kuhn).

and Potter, 1983). Cultured FA cells have a decreased rate of DNA replication, a prolonged G2 phase, increased spontaneous chromosome breakage, and increased sensitivity to bifunctional alkylating agents like diepoxybutane (DEB) and mitomycin C (MMC) because the DNA interstrand crosslinks they induce cannot be repaired. The most characteristic chromosome anomalies in FA are translocations between nonhomologous chromosomes (Fig. 24.4), which were first described by Schroeder et al. (1964). Although the proportion of lymphocytes with aberrations may be as high as 30%, there is no increase in intrachromosomal interchanges, as indicated by a normal SCE frequency (Chaganti et al., 1974). The greatly increased chromosome breakage induced by DEB facilitates both prenatal and postnatal diagnosis (Auerbach et al., 1989), but more effective molecular diagnostic tests are becoming available as the various mutant genes causing FA are identified.

Pairwise fusion of FA cells from unrelated patients produces hybrid heterokaryons (Chapter 23) whose sensitivity to a crosslinking agent is reduced to normal only if different mutant genes are present in the two patients, so that the fused cell contains some normal product of each gene. Such analyses have shown that cells from unrelated patients with FA fall into at least eight complementation groups (FA-A to FA-H), indicating that mutations in as many as eight different genes may cause FA (Joenje et al., 1997). The proteins of all eight complementation groups may bind to each other to form a giant complex that carries out an important but still unknown functiion in DNA replication or repair. Three of these genes have been cloned. The normal FAA and FAC protein prod-

Table 24.1. The Genetic Basis of Some Chromosome Instability Syndromes

Syndrome	Gene	Location	Product or function
Bloom	*BLM*	15q26	RecQ DNA helicase
Fanconi A	*FANCA*	16q24.3	DNA repair
Fanconi C	*FANCC*	9q22.3	DNA repair
Fanconi D	*FANCD*	3p22–p26	DNA repair
Fanconi E	*FANCE*	6p21–p22	DNA repair
Fanconi G	*FANCG*	9p13	Postreplication repair
Ataxia telangiectasia	*ATM*	11q22–q23	Cell cycle checkpoint
Nijmegen breakage	*NBS1*	8q21.3	Nibrin: DNA DSB repair
Werner	*WRN*	8p11–p12	RecQ DNA helicase
Roberts	*RS*		DNA repair?

ucts of two of these genes, *FANCA* and *FANCC*, bind to each other, but mutant FAC fails to bind to FAA, suggesting a mechanism for the phenotypic effect of the mutation (Kupfer et al., 1997). The third cloned gene, *FANCG*, maps to 9p13 and is identical to *XRCC9*, which may be involved in DNA postreplication repair or cell cycle checkpoint control (Winter et al., 1998). *FANCD* has been mapped to 3p22–p26. *FANCE* has been mapped by *homozygosity mapping* (Table 24.1). This technique involved screening the entire genome with closely spaced, highly polymorphic (and thus informative) genetic markers. The affected individuals in three families showed a single 18.2-cM region of 6p21–p22 in which all the marker alleles were homozygous, thus mapping the gene to this region (Waisfisz et al., 1999). About 40% of the mutations in the large *FANCA* gene are large intragenic deletions that remove up to 31 of the 43 exons (Morgan et al., 1999).

Ataxia Telangiectasia: A Cell Cycle Checkpoint Disorder

This autosomal recessive disease is characterized by progressive cerebellar ataxia, telangiectasia of eyes and skin, growth retardation, elevated serum alpha-fetoprotein, severe immunodeficiency, premature aging, and a greatly increased risk of developing solid or lymphoid tumors (Shiloh, 1995). Surprisingly, even heterozygotes show some increased risk of cancer (Swift et al., 1991). Ataxia develops between 3 and 6 years of age, and the patients usually die of pulmonary infections or cancer. AT patients have a greatly increased sensitivity to ionizing radiation and radiomimetic chemicals (Gatti and Hall, 1983).

Chromosome aberrations are less frequent in AT than in BS or FA. Random breakage often leads to cell clones with a translocation, most commonly involving chromosome 14 (Kaiser-McCaw et al., 1975). SCEs are not increased. There is, however, a striking finding: frequent telomere-telomere fusions. These, like the premature aging, may be due to the greatly elevated rate of telomere shortening seen in AT cells (Metcalfe et al., 1996). There is also a marked increase in intrachromosomal recombination; this could account for the abnormal rearrangements of immunoglobulin and T-cell receptor genes that lead to immunodeficiency in these patients (Meyn, 1993).

The AT gene, called *ATM*, maps to 11q22–q23 (Gatti et al., 1988). It has been cloned and sequenced (Savitsky et al., 1995). Its deduced protein product is similar to the signal-transducing phosphatidylinositol-3 kinases (PI-3Ks). The ATM protein is involved in meiotic recombination (Keegan et al., 1996). It also functions as an early signal in the activation of the G1 checkpoint, which blocks entry of cells into S until any DNA damage is repaired (Chapter 2). AT cells can move into S despite DNA damage, resulting in unrepaired DNA strands, chromosome breaks and rearrangements, or cell death. Normally, in cells exposed to a DNA-damaging agent such as ionizing radiation, a phosphate group is first removed from the serine residue in position 376 of the p53 protein, which is then able to bind 14-3-3 proteins and become active (Waterman et al., 1998). There is a coordinate rise in both the p53 tumor suppressor protein and the growth arrest and DNA damage inducible protein 45 (GADD45) protein, which both act in the signal-transducing pathway that activates the G1 checkpoint discussed in Chapter 2. However, in AT cells, no increase in either protein occurs in response to radiation, indicating that the ATM protein is essential for cell cycle arrest following DNA damage (Kastan et al., 1992). The importance of this for carcinogenesis is discussed in Chapter 26.

Related Disorders with Chromosome Instability

Nijmegen breakage syndrome (NBS) is a rare autosomal recessive disorder that was initially considered to be a variant of ataxia telangiectasia because of their virtually identical phenotypes: growth retardation, hypersensitivity to ionizing radiation, chromosomal translocations arising at sites of V(D)J rearrangements, immunodeficiency, gonadal failure, and increased risk of cancer. However, the gene for NBS is at a different location: 8q21.3 rather than 11q22–q23. Positional cloning has identified the *NBS1* gene, which is transcribed in all tissues (Matsuura et al., 1998; Varon et al., 1998). A shorter transcript is expressed at high levels in the testis, suggesting that the NBS1 protein is involved in meiotic recombination. If this is also true in the female germline, it could account for the presence of ovarian failure in NBS. Mutations of the gene are often submicroscopic deletions, identifiable by molecular analysis; a 5-bp deletion is common. The *NBS1* gene product is nibrin (Varon et al., 1998), part of a multiprotein complex that also contains the human MRE11, RAD50, and BRCA1 proteins, in keeping with its role in DSB repair during meiotic recombination (Carney et al., 1998; Zhong et al., 1999; see also Chapter 10). MRE11 maps to 5q23–q31, a region that is frequently altered in myeloid leukemia.

Werner syndrome (WS) is a rare autosomal recessive disorder marked by growth retardation and premature aging. Both chromosome structural aberrations and the risk for cancer are increased (Salk et al., 1981). Gene mutations also occur at a high frequency. Three-fourths of them are due to submicroscopic deletions, with half of the deletions more than 20 kb long (Fukuchi et al., 1989). The initiation of DNA replication is impaired in WS cells, and the locus-specific mutation rate is increased 50-fold. The WS gene, called *WRN*, has been cloned and sequenced (Yu et al., 1996). The gene product has been shown to be an active DNA helicase of the RecQ family (Gray et al., 1997). Helicases are involved in DNA replication and repair but also in transcription. Molecular studies have shown that the gene is expressed only in nonepithelial cells, which is probably the reason that only nonepithelial tumors are common in this disorder.

The xeroderma pigmentosum (XP) group of autosomal recessive diseases is not associated with an increased rate of spontaneous chromosome breakage but shows very high susceptibility to sunlight (UV)-induced skin cancers. Complementation analyses (Chapter 23) and follow-up studies have shown that mutations of at least eight different genes, *XPA–XPG* and *XPV* (variant), produce the XP phenotype, although only those of *XPA*, *XPB*, *XPD*, and *XPG* are associated with progressive neurological degeneration (Cleaver and Kraemer, 1995). *XPV* encodes DNA polymerase η, which carries out error-free replication of UV-damaged DNA by bypassing UV-induced thymine dimers (Masutani et al., 1999). The XPB and XPD proteins, like BLM and WRN, are RECQ DNA heli-

Table 24.2. Genes for Eight Complementation Groups of Xeroderma Pigmentosum

Gene	Location	Function in excision repair of UV-damaged DNA
XPA	9q34.1	Binds UV-damaged DNA preferentially
XPB	2q21	RecQ DNA helicase
XPC	3p25.1	?
XPD	19q13.2*	RecQ DNA helicase
XPE	?	?
XPF	16p13	Component of a nuclease[†]
XPG	13q32–q33	Endonuclease
XPV	?	DNA polymerase η

*The ERCC1 gene is also at 19q13.2

[†]ERCC1 and XPF encode components of the same nuclease

cases. All four of these proteins bind to the p53 protein and reduce p53-mediated cell killing, or *apoptosis* (reviewed by Spillare et al., 1999). This contributes to the mutator phenotype (see below) of these disorders. More generally important in XP is the error-prone replication of DNA in individuals homozygous for a mutation in any one of the XP genes. All eight of them are required for excision repair of DNA that has been damaged by UV or certain other agents (Table 24.2).

Bloom, Werner, and xeroderma pigmentosum syndromes are good examples of *mutator phenotypes*: Inactivation of a mutator locus by deletion or mutation greatly increases the risk of mutation at other loci throughout the genome. The importance of mutator phenotypes as causes of cancer is discussed further in Chapter 26. Given the complexity of DNA replication and repair and the still incomplete understanding of the many proteins (and thus genes) involved, it would not be surprising if other genotypes are found that predispose to increased chromosome aberrations. In fact, a number have already been reported. One intriguing report involved a girl with craniosynostosis, microcephaly, birdlike facies, and mental retardation. Her lymphocytes showed a high frequency of spontaneous chromosome breakage, endomitosis, endoreduplication, and hypersensitivity to alkylating and radiomimetic agents. Her parents were consanguineous, supporting the notion that this may be a newly recognized rare recessive disorder (Tommerup et al., 1993).

The disorders discussed thus far in this chapter are marked by increased rates of both spontaneous and induced chromosome damage. In addition, a few conditions show hypersensitivity to DNA-damaging agents despite the absence of a high rate of spontaneous chromosome damage. One of these is Roberts syndrome (RS), a rare autosomal recessive disease characterized by severe pre- and postnatal growth retardation and symmetrical limb reduction deformities (tetraphocomelia). About half the cases show a striking abnormality (the RS effect): "puffing," or decondensation, of the constitutive heterochromatin and premature separation of the centromeres so that some sister chromatids look straight, or parallel, with no primary constriction (Tomkins et al., 1979). Because a *Drosophila* mutant with a defect in mitotic condensation of heterochromatin was reported to be hypersensitive to cell killing by mutagens, Burns and Tomkins (1989) looked for, and found, a similar hypersensitivity of RS cells to cell killing by mitomycin C (MMC). MMC induces chromosome breaks mostly in the heterochromatin (Shaw and Cohen, 1965). Such breakage, if unrepaired, would produce multiple acentric chromosome fragments, whose loss in succeeding mitoses would lead to deletion of essential genes and thus to cell death. This suggests that there is a defect in the repair of this kind of DNA damage in RS patients. The mechanism of the undercondensation of heterochromatin in RS remains unclear.

References

Alter BP, Potter NU (1983) Long-term outcome in Fanconi's anemia: description of 26 cases and review of the literature. In: German J (ed) Chromosome mutation and neoplasia. Liss, New York, pp 43–62

Auerbach AD, Ghosh R, Pollio PC, et al. (1989) Diepoxybutane test for prenatal and postnatal diagnosis. In: Schroeder-Kurth TM, Auerbach AD, Obe G (eds) Fanconi anemia: clinical, cytogenetic and experimental aspects. Springe Verlag, Berlin, pp 71–82

Burns MA, Tomkins DJ (1989) Hypersensitivity to mitomycin C cell-killing in Roberts syndrome fibroblasts with, but not without, the heterochromatin abnormality. Mut Res 216:243–249

Carney JP, Maser RS, Olivares H, et al. (1998) The hMre11/hRad50 protein complex and Nijmegen breakage syndrome: linkage of double-strand break repair to the cellular DNA damage response. Cell 93:477–486

Chaganti RSK, Schonberg S, German J (1974) A manyfold increase in sister chromatid exchanges in Bloom's syndrome lymphocytes. Proc Natl Acad Sci USA 71:4508–4512

Cleaver JE, Kraemer KH (1995) Xeroderma pigmentosum and Cockayne syndrome. In: Scriver CR, Beaudet AL, Sly WS, et al. (eds) The metabolic and molecular bases of inherited disease, 7th edn. McGraw-Hill, New York, pp 4393–4419

Ellis NA, Groden J, Ye T-Z, et al. (1995) The Bloom's syndrome gene product is homologous to RecQ helicases. Cell 83:655–666

Fukuchi K, Martin GM, Monnat RJ Jr, et al. (1989) Mutator phenotype of Werner syndrome is characterized by extensive deletions. Proc Natl Acad Sci USA 86:5893–5897

Gatti RA, Hall K (1983) Ataxia-telangiectasia: search for a central hypothesis. In: German J (ed) Chromosome mutation and neoplasia. Liss, New York, pp 23–41

Gatti RA, Berkel I, Boder E, et al. (1988) Localization of an ataxia-telangiectasia gene to chromosome 11q22–23. Nature 336:577–580

German J (1993) Bloom syndrome: a mendelian prototype of somatic mutational disease. Medicine 72:393–406

Gray MD, Shen J-C, Kamath-Loeb AS, et al. (1997) The Werner syndrome protein is a DNA helicase. Nat Genet 17:100–103

Groden J, Nakamura Y, German J (1990) Molecular evidence that homologous recombination occurs in proliferating human somatic cells. Proc Natl Acad Sci USA 87:4315–4319

Gupta PK, Sahota A, Boyadjiev SA, et al. (1997) High frequency *in vivo* loss of heterozygosity is primarily a consequence of mitotic recombination. Cancer Res 57:1188–1193

Holt D, Dreimanis M, Pfeiffer M, et al. (1999) Interindividual variation in mitotic recombination. Am J Hum Genet 65:1423–1427

Joenje H, Oostra AB, Wijker M, et al. (1997) Evidence for at least eight Fanconi anemia genes. Am J Hum Genet 61:940–944

Kaiser-McCaw B, Hecht F, Harnden DG, et al. (1975) Somatic rearrangement of chromosome 14 in human lymphocytes. Proc Natl Acad Sci USA 72:2071–2075

Kastan MB, Zhan Q, El-Deiry WS, et al. (1992) A mammalian cell cycle checkpoint pathway utilizing p53 and GADD45 is defective in ataxia-telangiectasia. Cell 71:587–597

Keegan KS, Holtzman DA, Plug AW, et al. (1996) The Atr and Atm protein kinases associated with different sites along meiotically paired chromosomes. Genes Dev 10:2423–2437

Kuhn EM (1976) Localization by Q-banding of mitotic chiasmata in cases of Bloom's syndrome. Chromosoma 57:1–11

Kupfer GM, Naf D, Suliman A, et al. (1997) The Fanconi anaemia proteins, FAA and FAC, interact to form a nuclear complex. Nat Genet 17:487–490

Masutani C, Kusumoto R, Yamada A, et al. (1999) The *XPV* (xeroderma pigmentosum variant) gene encodes human DNA polymerase η. Nature 399:700–704

Mathew S, Murty VVVS, German J, et al. (1993) Confirmation of 15q26.1 as the site of the FES protooncogene by fluorescence in situ hybridization. Cytogenet Cell Genet 63:33–34

Matsuura S, Tauchi H, Nakamura A, et al. (1998) Positional cloning of the gene for Nijmegen breakage syndrome. Nat Genet 19:179–181

McDaniel LD, Schultz RA (1992) Elevated sister chromatid exchange phenotype of Bloom syndrome cells is complemented by human chromosome 15. Proc Natl Acad Sci USA 89:7968–7972

Metcalfe JA, Parkhill J, Campbell L, et al. (1996) Accelerated telomere shortening in ataxia telangiectasia. Nat Genet 13:350–353

Meyn MS (1993) High spontaneous intrachromosomal recombination rates in ataxia-telangiectasia. Science 260:1327–1330

Morgan NV, Tipping AJ, Joenje H, et al. (1999) High frequency of large intragenic deletions in the Fanconi anemia group A gene. Am J Hum Genet 65:1330–1341

Salk D, Au K, Hoehn H, et al. (1981) Cytogenetics of Werner's syndrome cultured skin fibroblasts: variegated translocation mosaicism. Cytogenet Cell Genet 30:92–107

Savitsky P, Bar-Shira A, Gilad S, et al. (1995) A single ataxia telangiectasia gene with a product similar to PI-3 kinase. Science 268:1749–1753

Schroeder TM, Anschütz F, Knopp A (1964) Spontane Chromosomenaberrationen bei familiärer Panmyelopathie. Humangenetik 1:194–196

Shaw MW, Cohen MM (1965) Chromosome exchanges in human leukocytes induced by mitomycin C. Genetics 51:181–190

Shiloh Y (1995) Ataxia-telangiectasia: closer to unraveling the mystery. Eur J Hum Genet 3:116–138

Spillare EA, Robles AI, Wang ZW, et al. (1999) p53-mediated apoptosis is attenuated in Werner syndrome cells. Genes Dev 13:1355–1360

Swift M, Morrell D, Massey RB, et al. (1991) Incidence of cancer in 161 families affected by ataxia-telangiectasia. N Engl J Med 325:1831–1835

Therman E, Kuhn EM (1981) Mitotic crossing-over and segregation in man. Hum Genet 59:93–100

Therman E, Kuhn EM (1985) Incidence and origin of symmetric and asymmetric dicentrics in Bloom's syndrome. Cancer Genet Cytogenet 15:293–301

Therman E, Otto PG, Shahidi NT (1981) Mitotic recombination and segregation of satellites in Bloom's syndrome. Chromosoma 82:627–636

Tomkins DJ, Hunter A, Roberts M (1979) Cytogenetic findings in Roberts-SC phocomelia syndrome(s). Am J Med Genet 4:17–26

Tommerup N, Mortensen E, Nielsen MH, et al. (1993) Chromosomal breakage, endomitosis, endoreduplication, and hypersensitivity toward radiomimetic and alkylating agents: a possible new autosomal recessive mutation in a girl with craniosynostosis and microcephaly. Hum Genet 92:339–346

Varon R, Vissinga C, Platzer M, et al. (1998) Nibrin, a novel DNA double-strand break repair protein, is mutated in Nijmegen breakage syndrome. Cell 93:467–476

Waisfisz Q, Saar K, Morgan NV, et al. (1999) The Fanconi anemia group E gene, *FANCE*, maps to chromosome 6p. Am J Hum Genet 64:1400–1405

Waterman MJF, Stavridi ES, Waterman JLF, et al. (1998) ATM-dependent activation of p53 involves dephosphorylation and association with 14-3-3 proteins. Nat Genet 19:175–178

Winter JP de, Waisfisz Q, Rooimans MA, et al. (1998) The Fanconi anaemia group G gene *FANCG* is identical with *XRCC9*. Nat Genet 20:281–283

Yu CE, Oshima J, Fu Y-H, et al. (1996) Positional cloning of the Werner's syndrome gene. Science 272:258–262

Zhong Q, Chen C-F, Li S, et al. (1999) Association of BRCA1 with the hRad50-hMre11-p95 complex and the DNA damage response. Science 285:747–750

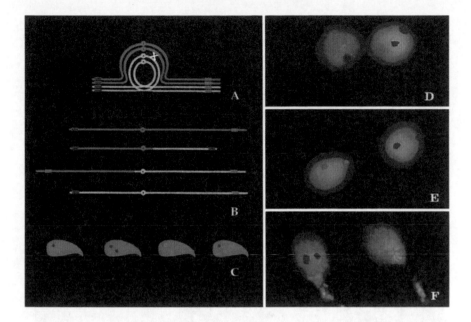

Figure 16.3. Recombination within a pericentric inversion, as visualized by two-color FISH. (A-C) Diagrams of a crossover within the inverted segment, the recombinant and nonrecombinant chromosomes, and nonrecombinant (green-red) and recombinant (two red or two green) sperm. (D-F) Images from two-color FISH showing recombinant and nonrecombinant sperm (reproduced from Jarrola et al., Am J Hum Genet 63:218–224, copyright 1998, American Society of Human Genetics, with permission of the University of Chicago Press).

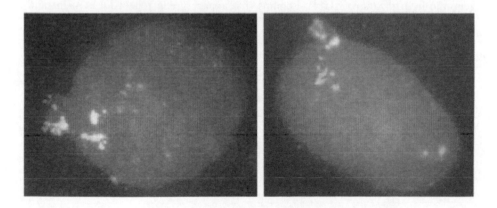

Figure 25.3. Preferential segregation of one amplified marker into nuclear blebs containing $P3C4$ only (left) or $AMPD3$ only (right) (reproduced from Toledo et al., EMBO J 11:2665–2673, 1992, with permission of Oxford University Press).

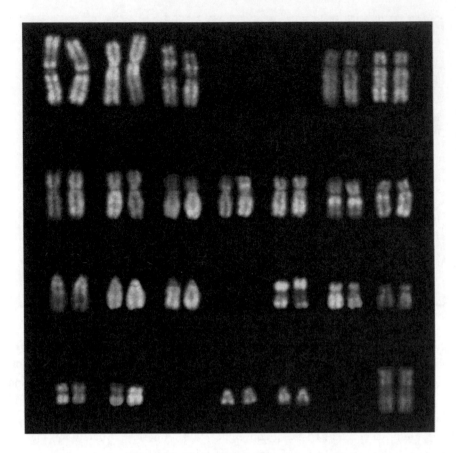

Figure 26.1. Analysis of chromosome changes in a breast cancer by comparative genomic hybridization (CGH). Regions that appear green reflect DNA gains and amplification in the cancer (e.g., 8q, 14, 17q22–q24, and 20q). Regions that appear red reflect DNA losses and deletions (e.g., 1p21–p31, 8p, 11q14–qter, and 16q12–q21) (reprinted from Forozan et al., Trends Genet, v13, Genome screening by comparative genomic hybridization, p 408, copyright 1997, with permission from Elsevier Science).

25

DNA and Gene Amplification

Double minutes (DMs) and homogeneously stained regions (HSRs) provide good examples of phenomena that were originally regarded as cytological oddities but have turned out to be expressions of a fundamental process, called DNA or gene *amplification*. They arise almost exclusively in transformed and malignant cells. DMs were first described in the 1960s. They occur in most types of solid tumors and many leukemias (Chapter 27). Biedler and Spengler (1976) first described HSRs in human neuroblastoma cell lines and methotrexate-resistant hamster cell lines. HSRs are rare in human tumors but more common in cultured tumor cell lines (Benner et al., 1991).

Double Minutes and Homogeneously Stained Regions

Double minutes are small paired chromatin structures that vary markedly in number and also in size, from very small double dots (Fig. 25.1a) to larger spherical structures and rings (Fig. 25.1b,c). They resemble tiny acentric chromosomes even by electron microscopy (Fig. 25.2; Jack et al., 1987). The lack of telomeric sequences and further molecular findings indicate that DMs contain circular DNA molecules (Lin et al., 1990; Von Hoff et al., 1990). DMs appear to replicate once in every mitotic cycle, during early S phase. Why, then, is their number so highly variable? The reason is that DMs lack centromeres, as shown, for example, by their lack of binding of antikinetochore antibodies (Haaf and Schmid, 1988). Consequently, their chromatids do not separate at anaphase. DMs tend to associate with centric chromosomes and travel at random towards either pole of the anaphase spindle, so one daughter cell may gain DMs and the other lose them. Other DMs lag behind, fail to be included in either daughter nucleus, and form micronuclei.

HSRs are interstitial or terminal additions or expansions of a chromosome arm that usually stain uniformly and rather lightly by any of the chromosome banding techniques (Fig. 25.1d). However, in rare cases an HSR has interspersed G- or C-bands, reflecting its heterogeneous origin (Cowell, 1982; Holden et al., 1987). HSRs, like DMs, generally replicate within a short period during early S phase. In most cases, they are located on the same chromosome arm as the native locus of the amplified DNA or gene, but sometimes they are present on a different chromosome or on multiple chromosomes. Since HSRs segregate normally in mitosis, they are more stable than DMs. However, they may increase or decrease in length and sometimes undergo other structural changes.

DMs and HSRs Are Expressions of Gene Amplification

Methotrexate (MTX) binds to and inactivates the enzyme dihydrofolate reductase (DHFR) and leads to cell death, especially of cancer cells. The development of resistance to MTX is accompanied by increased expression of DHFR and the appearance of HSRs. Observing this, Biedler and Spengler (1976) proposed that HSRs are expressions of gene amplification. Resistance to MTX or to any of

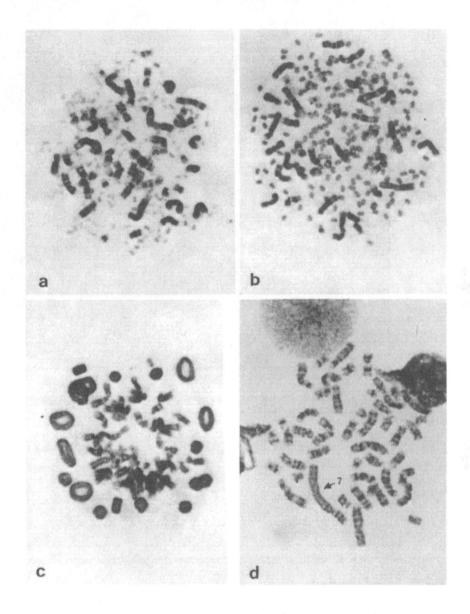

Figure 25.1. G-banded metaphase plates from a neuroblastoma cell line. (a) Small double minutes (DMs). (b) Medium-sized DMs. (c) Large rings. (d) Long homogeneously stained region on a chromosome 7 (Biedler JL, Ross RA, Shanske S, et al. 1980. Human neuroblastoma cytogenetics: search for significance of homogeneously stained regions and double minute chromosomes. In: Evans AE [ed] Advances in Neuroblastoma Research. Raven, New York, pp 81–96).

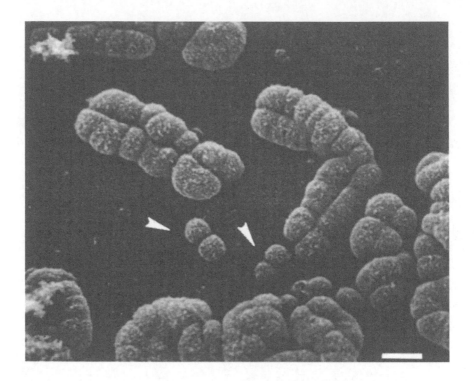

Figure 25.2. Scanning electron microscopy of normal and double minute (DM) chromosomes (arrows). The DMs vary in size, but their structure corresponds to that of normal chromosomes (reproduced from Jack et al., 1987; with permission of S Karger AG, Basel).

several other anticancer drugs, such as colchicine, vincristine, phosphonacetyl-L-aspartate (PALA), or actinomycin D, is usually associated with the development of DMs or HSRs containing multiple copies of the gene conferring drug resistance (Cowell, 1982). Oncogene amplification has been observed in many different kinds of cancer (Chapter 27). In most cases, the amplified genes are carried on extrachromosomal DMs or smaller submicroscopic circular extrachromosomal DNA molecules, called *episomes* or *amplisomes* (Pauletti et al., 1990). Amplified *MYC* oncogenes carried by episomes only a few hundred kb in size have been demonstrated by a variant of pulsed-field gel electrophoresis (PFGE) called field inversion gel electrophoresis (FIGE) in colon cancer and leukemia cell lines (Von Hoff et al., 1988). Standard gel electrophoresis separates, by size, fragments of DNA less than about 23 kb in size, but PFGE or FIGE will resolve fragments up to 1500 kb or larger (Cantor et al., 1988).

When cells containing DMs are treated with hydroxyurea, budding of the nuclear membrane and micronucleus formation occur. Acentric episomes and DMs are preferentially included in the micronuclei, leaving the cells with fewer DMs, fewer copies of the amplified oncogene, and reduced tumorigenicity. This also provides a mechanism for easy isolation and identification of an amplified gene (Shimizu et al., 1998). This can also be done by comparative genomic hybridization (Chapter 8), which has been used to detect and map amplified DNA sequences in breast cancer (Kallioniemi et al., 1994). An alternative approach is microdissection of the DMs and PCR amplification of the microdissected DM DNA to yield a probe for FISH or other molecular analysis. Findings with this approach have confirmed the origin of DMs in the HL-60 leukemia line from a single chromosomal site, 8q24.1–q24.2, where the *MYC* gene resides (Sen et al., 1994).

In some cases, DMs and HSRs are interconvertible. In the earliest passages of HL-60 leukemia cells, amplified MYC oncogenes are present on submicroscopic amplisomes. At later passages, these have been replaced by DMs, and at still later passages the amplified *MYC* genes have shifted to a chromosomal site (Von Hoff et al., 1990). Integration of DMs into random chromosomal sites to yield an HSR appears to be a common event. HSRs can sometimes give rise to DMs, and one mechanism for this has been identified, as described below (Toledo et al., 1992).

Normal cells rarely develop drug resistance by gene amplification (Livingstone et al., 1992). A 100-fold amplification of the CHE (cholinesterase) gene was observed in a farmer whose parents were exposed over many years to an organophosphate. A son and a grandchild inherited the CHE amplicon (Prody et al., 1989). A father and two of his sons, all rapid metabolizers of the antihypertensive drug debrisoquine had a 12-fold amplification of the cytochrome P450 *CYP2D* gene, whose product metabolizes the drug (Johansson et al., 1993). HSRs containing amplified rRNA genes (Miller et al., 1978; Cowell, 1982) or GACA repeats (Schmid et al., 1994) have been observed in a number of individuals or families. Some fragile sites may arise by amplification of minisatellites. For example, the *FRA3B* fragile site is an amplified AT-rich minisatellite (Chapter 20).

Why should DNA amplification be so rare in normal cells? One explanation is that there are genes that suppress amplification. Cancer cells support a high level of amplification, but when they are fused with normal cells, the hybrids are unable to amplify, suggesting that the cancer cells lack the product of a gene that suppresses amplification (Tlsty et al., 1992). The *TP53* protooncogene is a likely candidate, because cancer cells often have nonfunctional p53 protein due to *TP53* mutations (Chapter 27). Cells homozygous for a nonfunctional *TP53* allele are capable of amplifying a gene, whereas cells heterozygous for the mutant allele are, like normal cells, incapable of gene amplification (Livingstone et al., 1992). The introduction of wild-type p53 into cells with nonfunctional p53 alleles inhibits gene amplification (Yin et al., 1992). Thus, the normal *TP53* gene product suppresses amplification. To understand how p53 prevents amplification, one must first learn how amplification is brought about.

Mechanisms of Gene Amplification

Gene amplification arises by a variety of mechanisms, most of them poorly understood. Extrachromosomal copies of a gene may arise either by double-strand breakage in a stalled replication bubble, followed by recombination to form a circular molecule (Windle and Wahl, 1992), or by looping out and excision from one homologue of a DNA segment that includes the gene (Coquelle et al., 1997). The breakpoints and amplisome size depend upon the local chromatin environment of the gene. The breakpoints that lead to *DHFR*-containing DMs tend to occur 500 kb upstream and 200 kb downstream of the *DHFR* gene, very close to unmethylated CpG clusters, or *islands* (Foureman et al., 1998). CpG islands are usually found at the promoter region of housekeeping genes and are in an open chromatin conformation that is readily accessible both to transcription factors and to nucleases. Nuclease-sensitive sites are implicated in *MYC* gene amplification (Razin et al., 1995). The p53 protein is essential for DNA damage checkpoint function (Chapter 26). Therefore, *TP53* mutations would greatly increase the probability that the breaks would not be repaired and an extrachromosomal amplisome could arise. Initially, a single copy of the amplified gene is carried by each amplisome, but DMs tend to have dimers or larger numbers of the gene, arranged as either head-to-tail direct repeats or head-to-head inverted repeats (Fakharzadeh et al., 1993). Further increases in copy number are brought about by the unequal random segregation of DMs at anaphase.

Intrachromosomal amplification may arise at a single step by multiple cycles of unscheduled DNA replication at a single locus, followed by multiple recombination events that link individual units together and to the chromosome (Roberts et al., 1983). Windle and Wahl (1992) favor a modified version of this onionskin model in which the initiating event is the occurrence of double-strand breaks (DSBs) in stalled replication bubbles. This is followed by amplification and recombination, leading to extrachromosomal DMs or intrachromosomal HSRs. The DMs appear to be able to integrate almost at random into any chromosome. This mechanism may account for the presence of the tandemly repeated copies of the N-MYC oncogene observed in HSRs in neuroblastomas, in which the HSRs are on different chromosomes from the one with the native N-MYC locus (Amler et al., 1992). However, it is unlikely that HSRs generally arise by integration of DMs or smaller episomal amplicons. In most cases HSRs are found on the same chromosome arm as the native site of the unamplified gene, and the unit of intrachromosomal amplification is usually much larger than the size of DMs, especially early in the process.

A second mechanism for intrachromosomal amplification is unequal homologous exchange between sister chromatids, which is also responsible for the well-known variations in the number of tandemly repetitive rRNA genes and satellite DNAs. It probably accounts for the occasional amplification of rRNA genes (Miller et al., 1978). In fact, unequal SCEs have been observed in a melanoma cell line in which the HSRs contained both amplified ribosomal RNA genes and another tandemly repetitive element (Holden et al., 1987). While unequal homologous exchanges increase or decrease the copy number of tandem repeats, they do not account for DMs or for the initial amplification event needed to produce the tandem repeats of a single-copy gene necessary for unequal exchange to operate.

Cowell and Miller (1983) proposed a novel explanation for intrachromosomal amplification, suggested by their study of a rapidly evolving HSR. Their hypothesis was that the initial event leading to DNA amplification is telomere loss, leading to the formation of a dicentric chromosome, which undergoes anaphase bridge-breakage-fusion-bridge cycles, leading to amplification. This mechanism gained support from the finding that the generation of a large inverted duplication is an early event in the amplification process (Ford and Fried, 1986). Furthermore, it provides an explanation for the later observation that the unit of amplification is initially quite large but decreases with time (see, for example, Toledo et al., 1992).

The telomere loss/bridge-breakage-fusion-bridge cycle hypothesis has gained further support from observations on telomere length and chromosome instability. In sperm, the TTAGGG repeats at the telomeres of the various chromosomes are at their maximum length, 10–14 kb. In somatic cells, which lack telomerase, the telomeres are several kb shorter and quite heterogeneous in length. In primary tumors and some tumor cell lines, telomeres are even shorter, down to as little as 5 kb or so (de Lange et al., 1990). When cultured diploid cells are infected with a transforming DNA virus, such as SV40, the telomeres get even shorter, down to an average of 1.5 kb, and dicentric chromosomes become abundant, setting the stage for the bridge-breakage-fusion-bridge cycles that can lead to amplification. If the cells turn on their telomerase gene, telomere length stabilizes and the cells are immortalized (Counter et al., 1992). Ironically, amplification is one mechanism by which this is achieved. Thus, amplification may play a general and very important role in carcinogenesis, since most cancers show amplification of the gene for the telomerase RNA gene (Soder et al., 1997).

Telomere loss can also occur as a result of chromosome breakage, and that is probably why fragile sites play such an important role in intrachromosomal amplification: They are hotspots for the chromosome breakage that is involved in generating amplifiable structures (Coquelle et al., 1997). Some environmental agents may be carcinogenic because they induce breakage at fragile sites. Cigarette smokers tend to show an increase in breaks at several fragile sites and also at the sites of several oncogenes, such as *BCL1* (Kao-Shan et al., 1987). The various drugs that induce drug resistance through intrachromosomal amplification also break chromosomes and, through the formation of dicentric chromosomes, trigger bridge-breakage-fusion-bridge cycles that lead to amplification. Intrachromosomal amplification of *DHFR* and *AMPD2* (adenylate [adenosine mono phosphate] deaminase 2) genes is associated with the presence of breaks at nearby fragile sites telomeric to the locus of each gene. In some cases, the amplified region is flanked by two fragile sites. The telomeric one is important in generating a dicentric chromosome, while the one on the centromeric side, which arises in a subsequent bridge-breakage-fusion-bridge cycle, defines the size of the unit of amplification; this may include several genes from a region. For example, fragile sites *FRA11F* at 11q14.2 and *FRA11A* at 11q13.3 flank the *BCL1* (B cell CLL/lymphoma), *cyclin D1*, and fibroblast growth factor genes *FGF3* and *FGF4* in 11q13, which are coamplified in breast, head and neck, and other carcinomas (Coquelle et al., 1997). As the names imply, several of the coamplified genes are oncogenic, and their joint amplification may add to the malignancy of the cells.

The most exciting new insight into gene amplification has come from a two-color FISH study. Toledo et al. (1992) coamplified the *AMPD2* gene and the *P3C4* marker, 5 Mb away. Analysis of metaphase chromosomes showed an HSR containing long arrays of inverted repeats, with copies of the two genes interspersed throughout. This is best explained by bridge-breakage-fusion-bridge

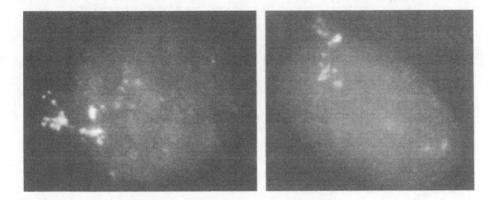

Figure 25.3. Preferential segegation of one amplified marker into nuclear blebs containing *P3C4* only (left) or *AMPD3* only (right) (reproduced from Toledo et al., EMBO J 11:2665–2673, 1992, with permission of Oxford University Press) (See color insert).

cycles. The really novel findings came from an examination of interphase cells. In these, the amplified *AMPD2* genes were clustered in one nuclear domain and the *P3C4* genes in a different domain, at a different point at the nuclear envelope. Sometimes one of the clusters was extruded from the nucleus, forming a micronucleus. Surprisingly, each micronucleus usually contained only one of the two types of genes (Fig. 25.3; see color insert). This clearly indicates that multiple chromosome breaks have taken place (presumably in interphase), with selective loss of one type of gene from the amplified region and retention of the other (Fig. 25.4). Micronucleus formation of this type has been seen in at least one other amplifying line (Smeets et al., 1994), so this may be a general mechanism for shortening the unit of amplification while retaining a selectable drug-resistance or oncogenic marker. As a corollary, one might predict that genes that continue to be coamplified must confer a selective growth advantage over those containing a single amplified gene (Coquelle et al., 1997).

Nuclear protrusions, or blebs, have long been observed in cancer cells and were thought to depend upon the presence of a very long chromosome arm that simply trailed behind at anaphase and protruded at interphase. However, their formation most often reflects the presence of an amplified region that is undergoing reorganization. The amplified sequences enclosed by such projections are frequently detected in the micronuclei that arise from them (Pedeutour et al., 1994). The findings of Toledo et al. (1992) show how complex the evolution of HSTs and DMs can be. The massive breakage associated with the selective

Figure 25.4. (A) Metaphase diagram showing two amplified interspersed markers, P_3C_4 and $AMPD_3$. (B) Diagrams showing preferential clustering of these markers into separate nuclear areas, followed by formation of a nuclear bleb containing the extra copies of one marker and their sequestration into a micronucleus. After repair of the multiple breaks, the remaining nuclear amplified unit is shorter and enriched for the other marker (reproduced from Toledo et al., EMBO J 11:2665–2673, 1992, with permission of Oxford University Press).

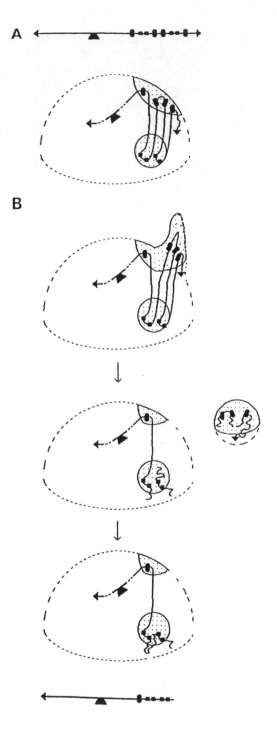

extrusion of amplified copies of an interspersed marker leaves the amplified gene copies on short DNA fragments. This could lead to the formation of either an HSR in which the copies of the gene are now considerably closer to each other than they were before, or DMs. This may be the mechanism by which an HSR is replaced by DMs.

How Does Gene Amplification Lead to Cancer?

In many cancer cells, one or more cellular oncogenes are amplified and the gene product is overexpressed. This leads to unregulated cell proliferation and plays an important role in carcinogenesis, as described in Chapter 27. A second mechanism is blocking of programmed cell death (apoptosis). Many cancer cells produce antigens that are recognized by the body's immune system. This leads to cell killing by activated T lymphocytes or natural killer lymphocytes, which produce a protein called Fas ligand (FasL). FasL induces apoptosis in the target cells by binding to the cell death receptor Fas. Nearly half of 35 lung and colon cancers were found to have 2- to 16-fold amplification of the gene for a decoy receptor, called DcR3, that binds FasL and prevents it from binding to Fas. This blocks cancer cell killing by the body's immune surveillance mechanism (Pitti et al., 1998).

References

Amler LC, Shibasaki Y, Savelyeva L, et al. (1992) Amplification of the *N-myc* gene in human neuroblastomas: tandemly repeated amplicons within homogeneously staining regions on different chromosomes with the retention of single copy gene at the resident site. Mutat Res 276:291–297

Benner SE, Wahl GM, Von Hoff DD (1991) Double minute chromosomes and homogeneously stained regions in tumors taken directly from patients versus in vitro human tumor cell lines. Anticancer Drugs 2:11–25

Biedler JL, Spengler BA (1976) Metaphase chromosome anomaly: association with drug resistance and cell-specific products. Science 191:185–187

Biedler JL, Ross RA, Shanske S, et al. (1980) Human neuroblastoma cytogenetics: search for significance of homogeneously stained regions and double minute chromosomes. In: Evans AE (ed) *Advances in Meuroblastoma research*. Raver, New York, pp 81–96

Cantor CR, Smith CL, Mathew MK (1988) Pulsed-field gel electrophoresis of very large DNA molecules. Annu Rev Biophys Biophys Chem 17:287–304

Coquelle A, Pipiras E, Toledo F, et al. (1997) Expression of fragile sites triggers intrachromosomal mammalian gene amplification and sets boundaries to early amplicons. Cell 89:215–225

Counter CM, Avilion AA, LeFeuvre C, et al. (1992) Telomere shortening associated with chromosome instability is arrested in immortal cells which express telomerase activity. EMBO J 11:1921–1929

Cowell JK (1982) Double minutes and homogeneously staining regions: gene amplification in mammalian cells. Annu Rev Genet 16:21–59

Cowell JK, Miller OJ (1983) Occurrence and evolution of homogeneously staining regions may be due to breakage-fusion-bridge cycles following telomere loss. Chromosoma 88:216–221

de Lange T, Shiue L, Myers RM, et al. (1990) Structure and variability of human chromosome ends. Mol Cell Biol 10:518–527

Fakharzadeh SS, Rosenbloom-Vos L, Murphy M, et al. (1993) Structure and organization of amplified DNA in double minutes containing the mdm2 oncogene. Genomics 15:283–290

Ford M, Fried M (1986) Large inverted duplications are associated with gene amplification. Cell 45:425–430

Foureman P, Winfield JA, Hahn PJ (1998) Chromosome breakpoints near CpG islands in double minutes. Gene 218:121–128

Haaf T, Schmid M (1988) Analysis of double minutes and double minute–like chromatin in human and murine tumor cells using antikinetochore antibodies. Cancer Genet Cytogenet 30:73–82

Holden JJA, Hough MR, Reimer DL, et al. (1987) Evidence for unequal crossing-over as the mechanism for amplification of some homogeneously staining regions. Cancer Genet Cytogenet 29:139–149

Jack EM, Waters JJ, Harrison CJ (1987) A scanning electron microscopy study of double minutes from a human tumour cell line. Cytogenet Cell Genet 44:49–52

Johansson I, Lundquist E, Bertilsson L, et al. (1993) Inherited amplification of an active gene in the cytochrome P450 CYP2D locus as a cause of ultrarapid metabolism of debrisoquine. Proc Natl Acad Sci USA 90:11825–11829

Kallioniemi A, Kallioniemi O-P, Piper J, et al. (1994) Detection and mapping of amplified DNA sequences in breast cancer by comparative genomic hybridization. Proc Natl Acad Sci USA 91:2156–2160

Kao-Shan CS, Fine RL, Whang-Peng J, et al. (1987) Increased fragile sites and sister chromatid exchanges in bone marrow and peripheral blood of young cigarette smokers. Cancer Res 47:6278–6282

Lin CC, Meyne J, Sasi R, et al. (1990) Apparent lack of telomere sequences on double minute chromosomes. Cancer Genet Cytogenet 48:271–274

Livingstone LR, White A, Sprouse J, et al. (1992) Altered cell cycle arrest and gene amplification potential accompany loss of wild-type p53. Cell 70:923–935

Miller DA, Breg WR, Warburton D, et al. (1978) Regulation of rRNA gene expression in a human familial 14p+ marker chromosome. Hum Genet 43:289–297

Pauletti G, Lae E, Attardi G (1990) Early appearance and long-term persistence of the submicroscopic extrachromosomal elements (amplisomes) containing the amplified DHFR genes in human cell lines. Proc Nat Acad Sci USA 87:2955–2959

Pedeutour F, Suijkerbuijk RF, Forus A, et al. (1994) Complex composition and co-amplification of SAS and MDM2 in ring and giant marker chromosomes in well-differentiated liposarcoma. Genes Chrom Cancer 10:85–94

Pitti RM, Marsters SA, Lawrence DA, et al. (1998) Genomic amplification of a decoy receptor for Fas ligand in lung and colon cancer. Nature 396:699–703

Prody CA, Dreyfus P, Zamir R, et al. (1989) De novo amplification within a "silent" human cholinesterase gene in a family subjected to prolonged exposure to organophosphorus insecticides. Proc Natl Acad Sci USA 86:690–694

Razin SV, Gromova H, Iarovian OV (1995) Specificity and functional significance of DNA interaction with the nuclear matrix: new approaches to clarify the old question. Int Rev Cytol 162B:405–408

Roberts JM, Buck LB, Axel R (1983) A structure for amplified DNA. Cell 33: 53–63

Schmid M, Nanda I, Steinlein C, et al. (1994) Amplification of (GACA)n simple repeats in an exceptional 14p+ marker chromosome. Hum Genet 93: 375–382

Sen S, Sen P, Mulac-Jericevic B, et al. (1994) Microdissected double minute DNA detects variable patterns of chromosomal localization and multiple abundantly expressed transcripts in normal and leukemic cells. Genomics 19:542–555

Shimizu N, Itoh N, Utiyama H (1998) Selective entrapment of extrachromosomally amplified DNA by nuclear budding and micronucleation during S phase. J Cell Biol 140:1307–1320

Smeets DFCM, Moog U, Weemaes CMR, et al. (1994) ICF syndrome: a new case and review of the literature. Hum Genet 94:240–246

Soder AI, Hoare SF, Muir S, et al. (1997) Amplification, increased dosage and in situ expression of the telomerase RNA gene in human cancer. Oncogene 14:1013–1021

Tlsty TD, White A, Sanchez J (1992) Suppression of gene amplification in human cell hybrids. Science 255:1425–1427

Toledo F, Le Roscouet D, Buttin G, et al. (1992) Co-amplified markers alternate in megabase long chromosomal inverted repeats and cluster independently in interphase nuclei at early steps of mammalian gene amplification. EMBO J 11:2665–2673

Von Hoff DD, Needham-Van Devanter J, Yucel J, et al. (1988) Amplified human MYC oncogenes localized to replicating submicroscopic circular DNA molecules. Proc Natl Acad Sci USA 85:4804–4808

Von Hoff DD, Forseth B, Clare CN, et al. (1990) Double minutes arise from circular extrachromosomal DNA intermediates which integrate into chromosomal sites in human HL-60 leukemia cells. J Clin Invest 85: 1887–1895

Windle BE, Wahl GM (1992) Molecular dissection of mammalian gene amplification: new mechanistic insights revealed by analysis of very early events. Mutat Res 276:199–224

Yin Y, Tainsky MA, Bischoff FZ, et al. (1992) Wild-type p53 restores cell cycle control and inhibits gene amplification in cells with mutant p53 alleles. Cell 70:937–948

26

Genome Destabilization and Multistep Progression to Cancer

Neoplasia is a broad group of diseases in which one or more cell lineages have escaped from regulators of cell proliferation as the result of inherited or acquired genetic changes. Neoplasms can be caused by exogenous or endogenous carcinogens. Initially, this leads to benign tumors, such as dysplasia or polyps, with limited growth potential. Further genetic changes, which may take years to occur, lead to escape from additional regulators. The result is tumor progression and finally a cancer that can infiltrate and destroy adjacent normal tissues. Some cancers eventually gain the ability to spread to new sites via blood and lymphatic vessels; that is, they can metastasize.

Immortality of Transformed and Malignant Cells

A key feature of cancer cells and cells transformed by DNA tumor viruses or chemicals is their ability to divide indefinitely (escape senescence): They are immortal. Diploid cells, on the other hand, senesce after a limited number of cell doublings: They are mortal. A consideration of the key regulators of the cell cycle (Chapter 2) suggests two obvious mechanisms for cell senescence: activation of a gene for a CDK inhibitor, such as p21 (*CIP1*) or p16 (*INK4A*) and inactivation of a tumor suppressor gene such as *TP53* or *RB1*. Disruption of the *CIP1* gene by two sequential rounds of targeted homologous recombination made diploid human fibroblasts immortal and prevented the cells from arresting the cell cycle in response to DNA damage (Brown et al., 1997). Escape from senescence might be brought about by mutation or loss of any one of these classes of genes, or by inactivation of the *TP53* or *RB1* gene products by binding to a viral oncoprotein (Chapters 27 and 28). Loss or mutation of a sequence on the long arm of chromosome 6 is sometimes involved, because microcell-mediated introduction of a normal chromosome 6 suppresses growth and causes senescence of SV40-transformed cells (Sandhu et al., 1994). Interestingly, a variety of cancers, including breast cancer, ovarian cancer, and non-Hodgkin's lymphoma, have deletions of 6q26–q27, as shown by loss of heterozygosity (LOH) of genetic markers in the region. This region is also deleted in SV40-transformed cell lines. A gene called *SEN6* (senescence-inducing gene on chromosome 6) is thought to be responsible (Banga et al., 1997).

Another cause of cell senescence is defective telomere function. This can be the result of a deficiency of the TRF2 (telomere-repeat binding) protein. The now unprotected telomeres (chromosome ends) act like double-stranded breaks and trigger the DNA damage checkpoint and cell death by p53-dependent apoptosis (Karlseder et al., 1999). The same endpoint can be reached in cell aging because telomerase activity is low or absent in most somatic cells after embryogenesis is completed. Telomeres thus get shorter with each mitosis, because their synthesis requires telomerase. In contrast, transformed and malignant cells have reactivated their telomerase gene, and telomerase activity is actually elevated in most cancers. Interphase cytogenetic analysis, using comparative genomic hybridization (Chapter 8), has shown that 97% of cancers show a small increase in copy number of the telomerase RNA gene located at 3q26.3. In a few cancers, there is even greater amplification of this gene (Chapter 25).

Genetic Basis of Cancer: Sequential Chromosome or Gene Mutations

It has been known for many years that multiple chromosome changes, both numerical and structural, are extremely common in cancer cells (reviewed by Heim and Mitelman, 1995). Comparative genomic hybridization is a powerful tool for rapidly analyzing these changes (Fig. 26.1; see color insert). These karyotypic changes lead to tumorigenesis and progression by causing gain, activation, loss, or inactivation of specific genes (Chapters 26–28). Mutant genes that cause enhanced susceptibility to malignant disease can also be inherited. Mutations of some genes produce an autosomal dominant pattern of cancer susceptibility. The Li-Fraumeni syndrome is due to a mutation of the *TP53* gene (Malkin et al., 1990). The familial breast and ovarian cancer syndrome is due to a mutation of the *BRCA1* gene (Mik et al., 1994). Inherited mutations of other genes produce an autosomal recessive pattern of tumor susceptibility, as in the chromosome instability syndromes discussed in Chapter 24. More than 20 hereditary cancer susceptibility syndromes are known that are due to specific germline-transmitted mutant genes. These account for only about 1% of all cancers; most of the remaining 99% are due to *somatic mutations* at the chromosome or gene level (Fearon, 1997).

The number of acquired genetic changes needed to produce cancer was estimated over 30 years ago by mathematical calculations based on the marked age dependence of cancer incidence and mortality. Armitage and Doll, in 1954, arrived at estimates of three to seven for different individual types of cancer, and this was supported by the more extensive analysis of Ashley (1969). These rather accurate estimates strongly suggested that some of the numerous genetic changes seen in cancer were causal. Burch, in 1962, stressed the importance of inherited mutations in accounting for childhood cancers and pointed out that somatic mutations sometimes cause cancers that are usually familial, and vice versa. Knudsen (1971) was the first to present convincing evidence that as few as two mutations were enough to cause at least one type of cancer, the childhood tumor retinoblastoma (Chapter 28). However, most cancers require more genetic changes. Some of these are point mutations of particular genes, but more are chromosome rearrangements, deletions, or amplifications. However, these chromosomal changes are also important because they affect the expression of specific genes.

Three main types of genes play a role in cancer induction. The first are those that block senescence, increase mutation rates, or lead to chromosome instabil-

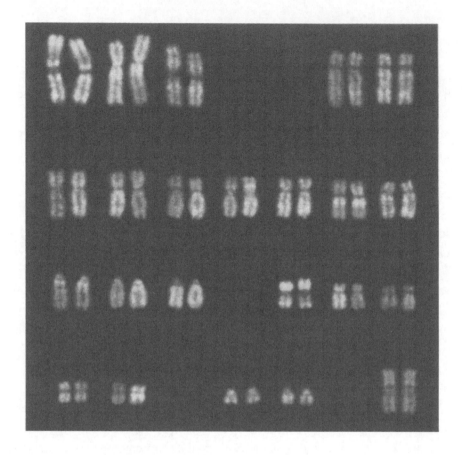

Figure 26.1. Analysis of chromosome changes in a breast cancer by comparative genomic hybridization (CGH). Regions that appear green reflect DNA gains and amplification in the cancer (e.g., 8q, 14, 17q22–q24, and 20q). Regions that appear red reflect DNA losses and deletions (e.g., 1p21–p31, 8p, 11q14–qter, and 16q12–q21) (reprinted from Forozan et al., Trends Genet, v13, Genome screening by comparative genomic hybridization, p 408, copyright 1997, with permission from Elsevier Science) (See color insert).

ity (see DNA Damage Checkpoint and Structural Instability of Chromosomes). The second are growth-stimulating genes, called *cellular oncogenes* or *protooncogenes*, whose normal alleles may be activated or mutate into *oncogenes* (Chapter 27). The third are the *tumor suppressor genes*, sometimes called *antioncogenes*, whose normal function is to regulate cell proliferation or produce differentiation of cells. When these genes are lost or inactivated by mutation, the absence of their products allows malignant growth to occur (Chapter 28).

Clonal Origin, Selection, and Multistep Tumor Progression

The extremely wide range of chromosome constitutions found in cancer cells suggests that karyotypic changes occur almost at random and reflect marked genomic instability. Marked variation in nuclear size from cell to cell, a well-known characteristic of many cancers, is a direct result of corresponding variations in chromosome number. Endoreduplication and endomitosis are more frequent in malignant than in normal cells, and the resulting endopolyploidy may reach high levels. Selection favors the fastest-growing cell type, which proliferates into a clone of cells. However, new chromosome constitutions are continually generated in these genomically unstable transformed cells, and continued selection leads to new clones (formerly called stemlines) that have accumulated more cancer-promoting genetic changes. Benign tumors are not always monoclonal. One patient with familial adenomatous polyposis was also an XO/XY mosaic. In situ hybridization with Y-specific probes showed that many of his colon adenomas contained some cells with a Y chromosome and other cells without a Y; that is, they were polyclonal in origin, arising from more than one cell (Novelli et al., 1996). The expectation is that a more rapidly growing (malignant) clone will arise in one adenoma and in turn be supplanted by still more rapidly growing subclones, and the resultant cancer will then be monoclonal.

Despite the bewildering complexity of karyotypic changes seen in cancer cells, it has been shown that the stepwise progression in the degree of malignancy of some tumors is accompanied by specific changes in the karyotype. Figure 26.2 illustrates this in a breast cancer that was analyzed by comparative genomic hybridization. One of the first diseases to show clear-cut sequential changes was chronic myelogenous leukemia (CML), in which the primary aberration is t(9q;22q); other changes appear during the progression of the disease: a second 22q−, trisomy 8, and an i(17q) (Mitelman et al., 1976). Burkitt's lymphoma may require three or four steps, including an infection with Epstein-Barr virus in early childhood and activation of two oncogenes (Land et al., 1983). A particularly complex disease is colon cancer, in which at least seven genetic changes contribute to an increasingly malignant phenotype (Fig. 26.3). The clinical behavior of a cancer reflects these changes; a slow-growing cancer may suddenly become more malignant or, after responding satisfactorily to radiation or chemotherapy, may suddenly become resistant to treatment.

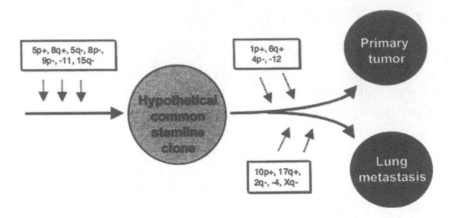

Figure 26.2. CGH analysis comparing chromosome changes in a primary breast cancer with its distant metastasis occurring one year later. Aberrations shared between the two are likely to reflect an early stemline clone from which both primary and metastatic tumors arose by independent clonal evolution. Solid arrows indicate the pathways from normal cell to stemline clone and from the stemline clone to either primary tumor or lung metastasis, short dotted arrows indicate where, in these pathways, various chromosome changes are thought to act (reprinted from Forozan et al., Trends Genet, v13, Genome screening by comparative genomic hybridization, pp 405–409, copyright 1997, with permission from Elsevier Science).

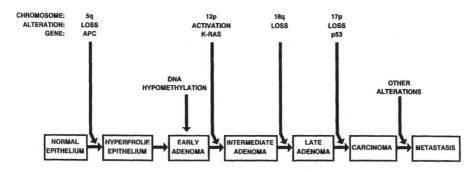

Figure 26.3. Multistep development of colon cancer (Kinzler and Vogelstein, copyright 1995, McGraw-Hill, with permission of the McGraw-Hill Companies).

Spindle Aberrations and Instability of Chromosome Number

Mitotic spindle abnormalities (Fig. 26.4) were first observed in cancer cells over a hundred years ago by Arnold and others. This observation, plus experimental analysis of the effects of multipolar spindles, led Boveri to suggest in 1914 that multipolar spindles and the resultant large variations in chromosome number among the daughter cells (*mixoploidy*) provide the genetic variation needed for the clonal selection of cells capable of uncontrolled growth. There is overwhelming evidence of abnormal spindles in tumor cells, such as the

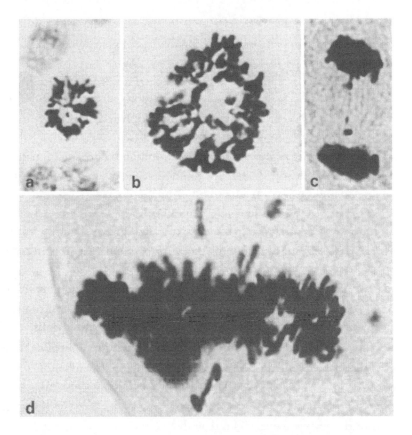

Figure 26.4. Mitotic stages from normal placenta (a) and cervical cancer (b–d). (a) Diploid metaphase (polar view). (b) Metaphase in octaploid range (polar view). (c) Anaphase with a bridge (side view). (d) Giant metaphase (side view) (Therman et al., 1984).

presence of multipolar mitoses (Fig. 26.5), which are practically nonexistent in normal cells (Therman and Kuhn, 1989).

Spindle abnormalities can lead to faulty alignment of the chromosomes in metaphase or anaphase, as seen in most cancers. In C-*mitosis*, named for its resemblance to colchicine-induced metaphase arrest, the chromosomes behave normally through prophase and up until metaphase. A defective or absent mitotic spindle prevents movement of the chromosomes to the metaphase plate or towards the spindle poles. The sister chromatids ultimately separate and form either a tetraploid restitution nucleus or a number of micronuclei of variable sizes. Such micronuclei are usually unable to divide further. C-mitoses and micronuclei are rare in normal cells but occur fairly often in cancer cells, with their abnormal spindles. The induction of micronuclei by spindle poisons such as the colchicine derivative diacetylmethyl (Colcemid) is now widely used for the production of microcell hybrids containing a single human chromosome (Chapter 23).

The centrosome is the chief microtubule organizing center (MTOC) of both interphase and dividing cells (Kellogg et al., 1994), and aberrations of the centrosome are the cause of the enlarged and multipolar spindles seen in many cancer cells. Centrosome duplication usually occurs only once in each cell cycle. However, in the absence of the p53 protein, multiple copies of functionally competent centrosomes can be generated in a single cell cycle (Fukusawa et al., 1996). This may be a major mechanism underlying the carcinogenic effect of mutations in the *TP53* gene, which encodes p53. Amplification or overexpression of the *STK15* gene, which encodes a centrosome-associated kinase, is another cause of centrosome duplication and aneuploidy. It occurs in about 12% of breast cancers and in a wide range of other cancer cell lines (Zhou et al., 1998).

Lengauer et al. (1997) carried out a molecular analysis of aneuploidy in several cloned colorectal cancer cell lines, using multicolor FISH with nine chromosome-specific probes. Some of the cancers showed dramatic variation in the number of copies of each chromosome from cell to cell within individual clones, with loss or gain occurring at a rate of about 1% per cell generation for each chromosome. This chromosomal instability (CIN) phenotype was not secondary to the malignant phenotype, because other cancers of similar phenotype did not show it. Instead, they were near-diploid, with an invariant chromosome number, but showed a microsatellite instability (MIN) phenotype as the result of defective DNA mismatch repair (see below). Nearly 85% of colorectal cancers have the CIN phenotype and 15% the MIN phenotype. CIN is dominant: Fusion of CIN with non-CIN cells produces hybrids with a CIN phenotype (Cahill et al.,

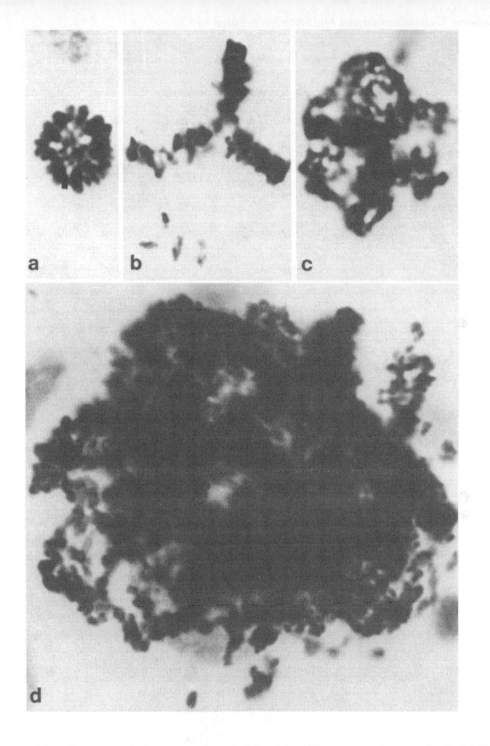

Figure 26.5. Metaphase spreads from normal trophoblast (a) and from cervical cancer (b–d). (a) Diploid metaphase in polar view. (b) Side-view of a tripolar metaphase with laggards. (Therman et al., 1984). (c) Six-polar metaphase. (d) Highly polyploid metaphase with numerous poles (Therman and Kuhn, 1989, reprinted with permission from CRC Press Inc.).

1998). Microcell-mediated transfer of single chromosomes from CIN cells to non-CIN cells might be useful in identifying the chromosome, and ultimately the gene, whose mutation is responsible for the CIN phenotype.

The Major Mechanism of Spindle Aberrations and Heteroploidy

Genome integrity depends upon checkpoint mechanisms that arrest cell cycle progression until damage produced by genotoxic agents or microtubular toxins is repaired or cell death by apoptosis is initiated (Hartwell and Kastan, 1994). In Chapter 2, we discussed the *spindle assembly checkpoint*, which prevents a cell from entering anaphase until a spindle is fully assembled and the kinetochore of every chromosome has a bipolar attachment to the spindle poles. The product of the *TP53* gene, p53, is required for proper function of this checkpoint. Information is growing on just how this works and on the other genes involved. A mutation of *TP53* is present in more than half of all cancers (Hainaut et al., 1997), making it the most common cause of the spindle abnormalities and heteroploidy seen in cancer. The absence of functional p53 protein can be the result of either mutation or loss of both copies of the gene. However, in many cases, a single dominant negative mutation is found in which the abnormal p53 protein from the mutant allele inhibits the product of the normal allele (Gualberto et al., 1998).

Several other genes have recently been identified whose products are required for the spindle assembly checkpoint to function. These include *MAD2* and *BUB1*. The expression of *MAD2* is reduced in some breast cancer cells with abnormal checkpoint function. Mutations in *BUB1* are fairly common in some cancers with chromosome instability and aneuploidy due to deficient checkpoint function (Cahill et al., 1998). Sarcomas and other tumors without a *TP53* mutation may show overexpression of the *MDM2* (murine double minute 2) gene. *MDM2* binds to the p53 protein and targets it for rapid degradation. Overexpression, which is sometimes the result of amplification of the *MDM2* gene, may thus be oncogenic because it reduces the level of p53 protein (Momand and Zambetti, 1996).

Structural Chromosome Changes

The presence of large marker chromosomes in some cancer cells was noted many years ago. As cytological methods improved, it became clear that there were structurally altered chromosomes of all sizes, going all the way down to double minute chromosomes (DMs). The introduction of banding techniques made it possible to identify dozens, and eventually hundreds, of specific translocations, inversions, and deletions, as well as a new type of marker chromosome with a nonbanding, homogeneously staining region (HSR). A single tumor, or a cell line derived from it, usually has several structurally rearranged, or *marker*, chromosomes, indicating a general breakdown of the mechanisms for maintaining chromosome integrity. This is illustrated by the high frequency of ectopic nucleolus organizer regions in cancer cells, the result of frequent recombination events (Atkin and Baker, 1995). Although some of the rearranged chromosomes in cancer cells remain unchanged over many years of growth in culture, as expected if some of the rearrangements play a role in carcinogenesis or tumor progression, cancer cells also continue to generate new marker chromosomes. Miller (1973) developed a simple method to estimate the heterogeneity of any cell population from the karyotype analysis of as few as 5–10 cells. Comparative genomic hybridization and FISH with chromosome painting or microdissection-PCR probes (Chapter 8) are powerful methods for identifying the origin and make-up of rearranged chromosomes.

DNA Damage Checkpoint and Structural Instability of Chromosomes

A breakdown of the G1 DNA damage checkpoint is the major mechanism underlying the structural instability of chromosomes that is so important in tumorigenesis and tumor progression. This checkpoint arrests the cell cycle in G1 when there is DNA damage and blocks entry into S until the DNA damage has been repaired or the cell dies (Chapter 2). This is important because nucleotide substitutions and DNA strand breakage are extremely common events. Spontaneous DNA breaks are induced by the oxidants that are produced in such abundance in our cells by normal metabolic processes (Ames, 1989). The cellular DNA repair machinery normally repairs all but the minute fraction of these that account for most of the background mutation rate. If the damage is not repaired before DNA replication, the result is a mutation or a double-strand break (DSB). DSBs can lead to chromosome rearrangements (Chapter 14); they are also essential for gene amplification to occur (Chapter 25).

The G1 DNA damage checkpoint function requires the normal protein products of the *ATM* (ataxia telangiectasia) and *TP53* genes (Chapter 2). *ATM* mutations are carcinogenic, though rare (Chapter 24), but *TP53* mutations are present in more than half of all cancers, as noted above. *TP53* mutations act as a "mutator phenotype" by allowing cells to live (since apoptosis requires p53) and the S phase to proceed despite DNA damage. This leaves unrepaired DNA and leads to chromosome breaks and point mutations. Because they disable the checkpoint and are so abundant, *TP53* mutations are the major cause of the structural as well as the numerical chromosome changes seen in cancer. This includes the gene amplifications responsible for DMs and HSRs, because of the role DSBs play in their origin. The introduction of wild-type p53 restores cell cycle checkpoint control and inhibits gene amplification in cells with nonfunctional mutant *TP53* alleles (Yin et al., 1992).

Why should *TP53* mutations be so prevalent in cancer cells? One reason is that many of the *TP53* mutations produce a dominant negative phenotype; that is, a *single* mutant allele results in the absence of any functional p53 protein in the cell (Gualberto et al., 1998). For other tumor suppressor genes, mutation or loss of *both* copies of the gene is required, and this is much less common (the probability of two independent mutations versus the probability of one mutation). Another reason is that *TP53* has several mutational hotspots. Of more than 5000 somatic mutations of *TP53*, nearly 20% occurred in only five codons: those for amino acids 175, 245, 248, 249, and 273 in the p53 protein (Hainaut et al., 1997).

A striking feature of these hotspots is that they vary from cancer to cancer, reflecting the different environmental agents involved in their causation. For example, in skin cancers, where exposure to sunlight is the carcinogenic agent, the hotspots are adjacent pyrimidines, because ultraviolet light induces the cyclobutane ring type of pyrimidine dimers from adjacent TT, TC, and CC pairs. *TP53* mutations are seen in 90% of squamous cell carcinomas but in only 50% of basal cell carcinomas, which arise from a deeper layer of the skin, although they still show the mutation signature of UV damage. In contrast, in lung cancer, there are three different hotspots, at the codons for amino acid positions 157, 248, and 273 in the p53 protein. These are the sites of selective strong binding of the metabolic product of benzpyrene, the major carcinogen in tobacco smoke, providing a direct etiologic link between a carcinogen in tobacco smoke and lung cancer. All three codons contain a methylated CpG dinucleotide, and the benzpyrene metabolite does not bind if these sites are experimentally demethylated (Denissenko et al., 1997).

Other Causes of Structural Instability of Chromosomes

Mutations of several poorly characterized genes that are involved in spindle assembly and DNA damage checkpoints account for some cases of chromosome instability. One of these genes, *RAD1*, is an essential component of the cell cycle checkpoint that is activated by DNA damage. *RAD1* maps to 5p13.2, a region that is frequently altered in small cell lung cancer, bladder cancer, squamous cell carcinoma, and adenocarcinoma (Marathi et al., 1998). Malfunction of the DNA damage checkpoint is not the only cause of structural instability. Failure to repair damaged DNA can also lead to chromosome changes and an increased risk of cancer.

There are multiple DNA repair pathways, each of them dependent upon the normal function of dozens of genes. Mutations of some of these occur, but are rare (Chapter 24). However, some are considerably more common, for example, genes in the DNA mismatch repair pathway that is initiated when a nucleotide sequence in one strand of the DNA ends up with an illegitimate pairing partner. The altered conformation triggers a repair pathway that involves nicking of one strand about 100 bp both 5′ and 3′ of the mismatch, excising the mismatched region and filling in the gap by using the other strand as template. Mutations in four of the five genes known to be required for this *long patch repair* have been seen in hereditary nonpolyposis colon cancer (HNPCC), a family of autosomal dominant disorders (Kolodner, 1995). Female carriers of a mutant *MSH6* gene are even more likely to develop uterine endometrial hyperplasia and carcinoma than colon carcinoma (Wijnen et al., 1999). Mutations in the related genes *MSH2*, *MLH1*, *PMS2*, and *PMS1* produce a mutator phenotype, with elevated frequencies of mutations of many other genes and of microsatellite repeats, such as dinucleotide or trinucleotide repeats. Methylation of the *MLH1* promoter silences the gene and is the primary cause of the microsatellite instability seen in sporadic endometrial cancers (Simkins et al., 1999).

Double-strand breaks in DNA are repaired by a pathway that involves using an undamaged homologous DNA molecule as the template for repair and is called *recombinational DNA repair* (Chapter 14). Two of the proteins in the DSB repair complex are RAD50 and MRE11. *RAD50* maps to 5q31.1, the region most often deleted in acute myeloid leukemia, and *MRE11* maps to 11q21, a region frequently altered in various cancers (Dolganov et al., 1996). The two major breast cancer susceptibility genes, *BRCA1* and *BRCA2*, are probably also involved in DSB repair. Their products are associated with the RAD51 protein, another member of the DSB repair complex (Patel et al., 1998).

Changes in DNA methylation may be an important cause of structural changes, as many cancers show genome-wide demethylation. Demethylation affects all classes of transposable elements, including the very abundant L1 LINE retrotransposons, which are the major sites of methylation throughout the genome. Demethylation of their genes, including the reverse transcriptase gene, may be the reason for the much greater abundance of their transcripts in cancer cells than in normal cells (Jürgens et al., 1996). If this leads to increased transpositional activity, it increases genome instability, because movement of transposons is associated with chromosome breakage, as pointed out many years ago by McClintock. The induction of multiradials by demethylation of satellite DNAs (Chapter 22) may be due to the activation of L1 LINE retrotransposons in heterochromatin. The generally high level of methylation of transposable elements, coupled with the well-known silencing of gene expression by methylation, has led to the hypothesis that DNA methylation developed as a mechanism for inactivating incoming DNA sequences, such as retroviruses, that integrate into the host's genome (Yoder et al., 1997).

Demethylation can lead to the relaxation or loss of imprinting of specific genes frequently seen in cancers. Two-thirds of Wilms tumors that are not associated with LOH of markers on 11p show biallelic expression (loss of imprinting) of the *H19* and *IGF2* (insulin-like growth factor 2) genes at 11p15.5, leading to overexpression of the proliferation-promoting *IGF2* product (Rainier et al., 1993).

Environmental Causes of Cancer

The increasing molecular evidence of the important role mutations play in tumorigenesis and progression comes as no surprise to those who learned long ago of the importance of environmental agents as causes of cancer, because most carcinogens are mutagens, such as X-rays, other types of radiation, chemicals, and viruses. A few, such as the phorbol esters, are not. Instead, they act as tumor promoters, primarily by stimulating cell proliferation. Many mutations arise as errors of DNA replication, so tumor promoters increase the chance that another mutation can arise in a cell lineage that already harbors one or more relevant mutations. DNA tumor viruses such as polyoma, SV40, human papilloma (HPV), and adenovirus, can transform diploid fibroblasts into immortal cells that can ultimately become malignant.

Virus-induced cell transformation is a multistep process that requires the continued presence of a single viral protein: large T for polyoma or SV40, E1A for adenovirus, or E7 for HPV. These proteins bind, and inactivate, the RB1 tumor suppressor protein (Ko and Prives, 1996); HPV E7 also binds the p53 tumor suppressor protein and induces its degradation (White et al., 1994). The normal p53 protein is polymorphic, sometimes with proline in position 72 and sometimes with arginine. Although both alleles are functional, arg72-p53 is more readily degraded than pro72-p53. Patients with HPV-associated cervical cancers are much more likely to be homozygous for the arginine-encoding allele than expected from the allele frequencies. In fact, individuals homozygous for the arg-encoding allele have a seven-fold greater risk of developing cervical cancer than heterozygotes do but no increased risk for developing non-HPV-associated cancers (Storey et al., 1998).

Just how important are environmental mutagens and tumor promoters in comparison to inherited mutations or other endogenous risk factors? They may be extremely important. Different populations show striking differences, up to 30-fold or greater, in the incidence of many different types of cancer. This is far more than could be accounted for by genetic differences between the populations. Furthermore, a specific environmental factor has been demonstrated in some cases, such as tobacco smoking in lung cancer, Epstein-Barr virus in nasopharyngeal carcinoma and Burkitt lymphoma, papilloma virus in cervical cancer, and sunlight (UV) exposure in skin cancer. As many as 15% of all cancers in developed countries may be virus-related and 25% in developing countries (Zur Hausen, 1997).

References

Ames B (1989) Endogenous DNA damage as related to cancer and aging. Mutat Res 250:3–16

Ashley DJB (1969) The two "hit" and multiple "hit" theories of carcinogenesis. Br J Cancer 23:313–328

Atkin NB, Baker MC (1995) Ectopic nucleolus organizer regions. A common anomaly revealed by Ag-NOR staining of metaphases from nine cancers. Cancer Genet Cytogenet 85:129–132

Banga SS, Kim S-H, Hubbard K, et al. (1997) *SEN6*, a locus for SV40-mediated immortalization of human cells, maps to 6q26–27. Oncogene 14:313–321

Brown JP, Wei W, Sedivy JM (1997) Bypass of senescence after disruption of p21(*CIP1/WAF1*) gene in normal diploid human fibroblasts. Science 277:831–834

Cahill DP, Lengauer C, Yu J, et al. (1998) Mutations of mitotic checkpoint genes in human cancers. Nature 392:300–303

Denissenko MF, Chen JX, Tang MS, et al. (1997) Cytosine methylation determines hot spots of DNA damage in the human *P53* gene. Proc Natl Acad Sci USA 94:3893–3898

Dolganov GM, Maser RS, Novikov A, et al. (1996) Human Rad50 is physically associated with human Mre11: identification of a conserved multiprotein complex implicated in recombinational DNA repair. Mol Cell Biol 16:4832–4841

Fearon ER (1997) Human cancer syndromes: clues to the origin and nature of cancer. Science 278:1043–1050

Forozan F, Kahru R, Kononen J, et al. (1997) Genome screening by comparative genomic hybridization. Trends Genet 13:405–409

Fukusawa K, Choi T, Kuriyama R, et al. (1996) Abnormal centrosome amplification in the absence of p53. Science 271:1744–1747

Gualberto A, Aldape K, Kozakiewicz K, et al. (1998) An oncogenic form of p53 confers a dominant, gain of function phenotype that disrupts spindle checkpoint control. Proc Natl Acad Sci USA 95:5166–5171

Hainaut P, Soussi T, Shomer B, et al. (1997) Database of p53 gene somatic mutations in human tumors and cell lines: updated compilation and future prospects. Nucleic Acids Res 25:151–157

Hartwell LH, Kastan MB (1994) Cell cycle control and cancer. Science 266:1821–1828

Heim S, Mitelman F (1995) Cancer cytogenetics, 2nd edn, Wiley-Liss, New York

Jürgens B, Schmitz-Dräger BJ, Schulz WA (1996) Hypomethylation of L1 LINE sequences prevailing in human urothelial carcinomas. Cancer Res. 56:5698–5703

Karlseder J, Broccoli D, Dai Y, et al. (1999) p53- and ATM-dependent apotosis induced by telomeres lacking TRF2. Science 283:1321–1325

Kellogg D, Moritz M, Alberts B (1994) The centrosome and cellular organization. Annu Rev Biochem 63:639–674

Kinzler KW, Vogelstein B (1995) Colorectal tumors. In: The metabolic and molecular bases of inherited disease, 7th ed. Scriver CL, Beaudet AL, Sly WS, Valle D (eds), McGraw-Hill, New York, pp 643–663

Knudsen AG (1971) Mutation and cancer: statistical study of retinoblastoma. Proc Natl Acad Sci USA 68:820–823

Ko LJ, Prives C (1996) p53: puzzle and paradigm. Genes Dev 10:1054–1072

Kolodner RD (1995) Mismatch repair mechanisms and relationship to cancer susceptibility. Trends Biochem Sci 20:397–401

Land H, Parada LF, Weinberg RA (1983) Cellular oncogenes and multistep carcinogenesis. Science 222:771–778

Lengauer C, Kinzler KW, Vogelstein B (1997) Genetic instability in colorectal cancers. Nature 386:623–627

Malkin D, Li FP, Strong LC, et al. (1990) Germ line p53 mutations in a familial syndrome of breast cancer, sarcomas, and other neoplasms. Science 250:1233–1238

Marathi UK, Dahlen M, Sunnerhagen P, et al. (1998) *RAD1*, a human structural homolog of the *Schizosaccharomyces pombe RAD1* cell cycle checkpoint gene. Genomics 54:344–347

Mik Y, Swensen J, Shattuck-Eidens D, et al. (1994) Isolation of a strong candidate for the 17q-linked breast and ovarian cancer susceptibility gene, *BRCA1*. Science 266:66–71

Miller OJ (1973) Analysis of heterogeneity. In: Caspersson T, Zech L (eds) Nobel Symposium 23. Chromosome identification—technique and applications in biology and medicine. Academic, New York, pp 177–178

Mitelman F, Levan G, Nilsson P, et al. (1976) Non-random karyotypic evolution in chronic myeloid leukemia. Int J Cancer 18:24–30

Momand J, Zambetti GP (1996) Analysis of the proportion of p53 bound to mdm-2 in cells with defined growth characteristics. Oncogene 12: 2279–2289

Novelli MR, Williamson JA, Tomlinson IPM, et al. (1996) Polyclonal origin of colonic adenomas in an XO/XY patient with FAP. Science 272:1187–1190

Patel KJ, Yu VPCC, Lee H, et al. (1998) Involvement of BRCA2 in DNA repair. Mol Cell 1:347–357

Rainier S, Johnson LA, Dobry CJ, et al. (1993) Relaxation of imprinted genes in human cancer. Nature 362:747–749

Sandhu AK, Hubbard K, Kaur GP, et al. (1994) Senescence of immortal human fibroblasts by the introduction of normal human chromosome 6. Proc Natl Acad Sci USA 91:5498–5502

Simkins SB, Bocker T, Swisher EM, et al. (1999) *MLH1* promoter methylation and gene silencing is the primary cause of microsatellite instability in sporadic endometrial cancers. Hum Mol Genet 8:661–666

Storey A, Thomas M, Kalita A, et al. (1998) Role of a p53 polymorphism in the development of human papilloma-virus-associated cancer. Nature 393:229–234

Therman E, Kuhn EM (1989) Mitotic modifications and aberrations in cancer. CRC Crit Rev Oncogen 1:293–305

Therman E, Buchler DA, Nieminen U, et al. (1984) Mitotic modifications and aberrations in human cervical cancer cells. Cancer Genet Cytogenet 11:185–197

White AE, Livanos EM, Tlsty TD (1994) Differential disruption of genomic integrity and cell cycle regulation in normal human fibroblasts by the HPV oncoprotein. Genes Dev 8:666–677

Wijnen J, de Leeuw W, Vasen H, et al. (1999) Familial endometrial cancer in female carriers of MSH6 germline mutations. Nat Genet 23:142–144

Yin Y, Tainsky MA, Bischoff FZ, et al. (1992) Wild-type p53 restores cell cycle control and inhibits gene amplification in cells with mutant p53 alleles. Cell 70:937–948

Yoder JA, Walsh CP, Bestor TH (1997) Cytosine methylation and the ecology of intragenomic parasites. Trends Genet 13:335–340

Zhou H, Kuang J, Zhong L, et al. (1998) Tumour amplified kinase STK15/BTAK induces centrosome amplification, aneuploidy and transformation. Nat Genet 20:189–193

Zur Hausen H (1997) Viruses in human tumors—reminiscences and perspectives. Adv Cancer Res 68:1–22

27

Chromosomes and Cancer: Activation of Oncogenes

A large number of hematological (blood cell) and solid tumors of various types show consistent chromosome abnormalities, and there is overwhelming evidence that the chromosome changes are essential for the malignant phenotype. Almost every chromosome band is involved, indicating the large number of genes that can play a role in oncogenesis (Mitelman et al., 1997). Many of these rearrangements lead to cancer by activating a cellular oncogene (*protooncogene*). Protooncogenes are normal genes present in all metazoan cells. Genes homologous to protooncogenes are found in the retroviruses known to cause cancer in various animal species. They transform cells, either by being inserted into the host genome or by being present in multiple copies in the host cell (Bishop, 1983). The retroviruses originally acquired these oncogenes from the metazoan cells they infected. The oncogene of the Rous sarcoma virus is called *v-src*, and its homologue in the normal human genome is *c-SRC*, or *SRC*. More than 80 human protooncogenes have been localized to a specific chromosome or chromosome band. A normal cell can be transformed by activating one or more oncogenes in it. This most often occurs through chromosomal mechanisms such as translocation or amplification. In leukemias and lymphomas, these are mostly balanced reciprocal translocations; in solid tumors, deletions and trisomies are also common (Cobaleda et al., 1998; Helm and Mitelman, 1995).

Mechanisms of Oncogene Action

The normal functions of cellular oncogenes are quite important. Many play a role in the cell cycle and its checkpoints. Many stimulate cell proliferation through their role as growth factors, growth factor receptors, signal transducers from cytoplasm to nucleus, or activators of transcription or replication (Table 27.1). Other oncogenes enhance metastasis. Activated cellular oncogenes are dominant in their effects (a single copy is oncogenic), as shown by transfection experiments. Purified DNA fragments from a variety of cancers, when transferred to the DNA of recipient cells, a process called *transfection* (Chapter 23), transform nonneoplastic cells with high efficiency. DNA fragments from normal cells also accomplish transformation, although with a very low frequency, presumably by activating a cellular oncogene.

Another mechanism by which oncogenes act is blocking of cell differentiation. Overexpression of the MDM2 cell cycle regulator occurs in nearly 30% of sarcomas. It inhibits Myo-D and p53, thus blocking muscle cell differentiation as well as inducing aneuploidy. Thus, when overproduced, MDM2 becomes an

Table 27.1. Some Cellular Oncogene Products That Foster Proliferation by Acting in Signal Transduction Pathways

Location in pathway	Oncogene products
Extracellular growth factor (ligand)	INT-2, SIS, HST
Receptor at cell membrane	ERBB, FMS, MAS
Membrane-associated G-protein	H-RAS, N-RAS, K-RAS
Membrane/cytoplasmic protein tyrosine kinase	ABL/BCR, SRC, CRK
Cytoplasmic protein serine kinase	MOS, RAF, ERBA
Nuclear transcription factor	FOS, MYC

oncoprotein. An isochromosome 3q is seen in a number of sarcomas. This duplication of the *ATR* (ataxia telangiectasia and rad-3-related) gene in 3q also blocks p53 and Myo-D function (Smith et al., 1998). Acute myelogenous leukemia (AML) is frequently associated with a t(15;17) translocation. This is oncogenic because it fuses the retinoic acid receptor α (*RARA*) gene with the promyelocytic leukemia (*PML*) locus to form a chimeric receptor that activates histone deacetylase and blocks cell differentiation (Lin et al., 1998).

Some oncogenes enhance carcinogenesis by delaying programmed cell death (*apoptosis*). The t(14;18)(q32;q21) translocation present in 85% of follicular lymphomas fuses the *BCL2* gene at 18q21 with an active immunoglobulin heavy chain (*Ig*) locus and leads to overproduction of the *BCL2* gene product, a mitochondrial inner membrane protein that blocks apoptosis by preventing the creation of reactive oxygen species. Some oncogene proteins, such as Myb and Ras, induce *BCL2* expression (Adams and Corey, 1998). A second event is necessary for malignant transformation; this is usually the activation of *MYC*. The BCR-ABL fusion proteins 210 and 190 inhibit apoptosis by inducing *BCL2* expression, thus allowing more time for other oncogenic processes to act on the susceptible population of cells and produce chronic myelogenous leukemia (CML) and acute lymphatic leukemia (ALL), respectively (Sánchez-García and Grütz, 1995).

Reciprocal Translocations and Oncogene Activation

The *MYC* oncogene is consistently activated by a chromosomal translocation in the highly malignant Burkitt lymphoma. The most common chromosome finding in malignant cells of patients with this disease is t(8;14)(q24;q32.3) (Fig. 27.1). The *MYC* gene has been mapped to 8q24, and the immunoglobulin heavy chain (*IgH*) locus to 14q32.3. The new location of the *MYC* gene places it next to and just 3′ to (downstream of) the promoter of the broken *IgH* gene; this up-regulates *MYC* expression in lymphoid cells in which the immunoglobulin genes are expressed. In a minority of cases, the end of 8q is translocated to 2p11 or to 22q11, placing *MYC* next to the strong promoter of the gene coding for the Ig kappa or lambda light chain, respectively. The *MYC* oncogene is also activated in the same manner in other lymphomas (Fig. 27.2 a and b).

The first chromosome aberration consistently seen in a malignancy was the Philadelphia chromosome, found in 1960 by Nowell and Hungerford in CML;

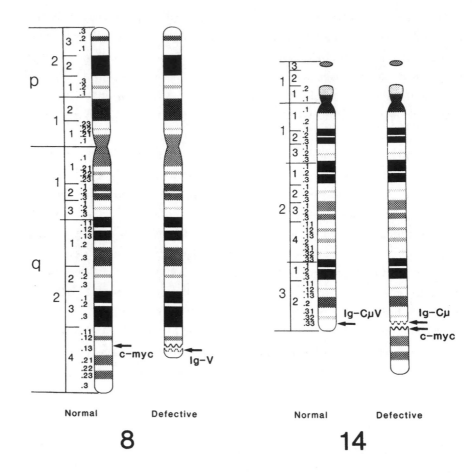

Figure 27.1. Locations (arrows) of c-*MYC* oncogene and heavy-chain immunoglobulin (Ig) variable (*V*) and constant (Cμ) genes on normal and rearranged chromosomes 8 and 14 in Burkitt lymphoma, represented at the 1200-G-band stage. Broken ends indicate breakpoints (Yunis JJ. The chromosomal basis of human neoplasia. Science 221:227–236, copyright 1983, by the AAAS).

Rowley (1973) showed this was a reciprocal translocation between 9q and 22q. With prophase banding, the breakpoints were defined as 9q34.1 and 22q11.2 (Fig. 27.1c). Molecular cytogenetic techniques have shown that the translocation activates the oncogene *ABL* at 9q34.1 by placing it next to the strong promoter of the *BCR* (breakpoint cluster region) gene at 22q11.2, which is strongly expressed in lymphocytes. The result is a novel fusion gene. Since the breakpoints are in introns of the two genes, there is no frame shift in the coding sequence of the exons, so a functional fusion protein is produced. Even when

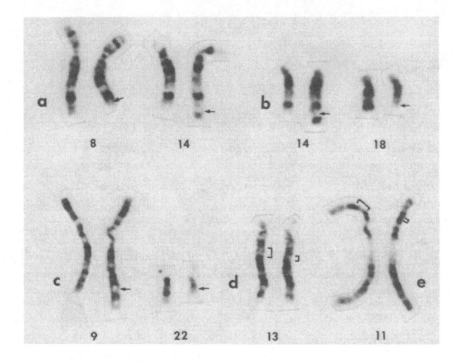

Figure 27.2. High-resolution (850- and 1200-band) G-banded chromosomes from (a) non-Burkitt lymphoma with t(8;14), (b) follicular cell lymphoma with t(14;18), (c) chronic myelogenous leukemia with t(9;22), (d) retinoblastoma with del(13)(q14), and (e) Wilms tumor with del(11)(p13). Arrows indicate breakpoints in the translocations, and brackets indicate deletions (Yunis JJ. The chromosomal basis of human neoplasia. Science 221:227–236, copyright 1983, by the AAAS).

the breakpoints are in the same band at a cytological level, they may be at different molecular locations. This accounts for the fact that some t(9;22)(q34;q11) translocations lead to CML and some to acute lymphoblastic leukemia (ALL); a BCR-ABL fusion protein 210 kDa in size leads to CML and one 190 kDa in size leads to ALL.

How Do Translocations Activate Cellular Oncogenes?

There are two general mechanisms by which translocations activate oncogenes. The first is to place the protooncogene under the control of a strong active pro-

Table 27.2. Oncogene Activation by V(D)J Recombinase-mediated Translocations

Translocation	Type of cancer	Activator, oncogene
t(8;14)(q24;q32)	BL, B-ALL	IgH, MYC
t(2;8)(p12;q24)	BL, B-ALL	IgL-κ, MYC
t(8;22)(q24;q11)	BL, B-ALL	IgL-λ, MYC
t(11;14)(q13;q32)	B-CLL	IgH, BCL1
t(14;18)(q32;q21)	FL	IgH, BCL2
t(8;14)(q24;q11)	T-ALL	TCR-α, MYC
t(7;19)(q35;p13)	ALL	TCR-β, LYL1
t(7;10)(q35;q24)	T-ALL	TCR-β, HOX11

BL, Burkitt lymphoma; B-ALL, B-cell acute lymphatic leukemia; B-CLL, B-cell chronic lymphatic leukemia; FL, follicular lymphoma; T-ALL, T-cell acute lymphatic leukemia

moter, as just described for the *MYC* gene. In hematopoietic malignancies, such as lymphomas, leukemias, and myelomas, one breakpoint is usually in an *IgH*, *Igκ*, *Igλ*, or T-cell receptor (*TCR*) gene. This suggests that the translocations in hematopoietic cells usually arise as a result of aberrant action of the enzymes normally involved in the DNA double-strand breakage and repair that lead to the V(D)J recombination responsible for antibody and histocompatibility receptor diversity. There are many examples of this; a few are shown in Table 27.2. This is supported by precise molecular evidence in many cases. An intriguing example is the t(4;14) translocation seen in some multiple myelomas. This fuses the *IgH* gene with the Wolf–Hirschhorn critical region gene *WHSC1*, which is normally expressed in early development (Stec et al., 1998).

Many of the translocations seen in specific types of lymphoid malignancies and virtually all solid tumors do *not* involve *Ig* or *TCR* genes. How do they arise? A precise molecular answer is not available for most of them. However, the *BCR-ABL* translocations seen in CML arise preferentially within the 5-kb breakpoint cluster region (*BCR*) by recombination involving Alu repeat elements (Jeffs et al., 1998). In most of these cases, and in many types of solid tumors, the specific translocation associated with the tumor leads to a novel fusion gene that

Table 27.3. Oncogene Activation or Fusion Neooncogene Formation by Translocations

Translocation	Type of cancer	Fusion genes
t(9;22)(q34;q11)	CML, B-ALL	BCR, ABL
t(9;12)(q34;p13)	AML	TEL, ABL
t(11;22)(q24;q12)	EWS, neuroblastoma	EWS, FL1
t(12;22)(q13;q12)	Melanoma	EWS, ATF1
t(2;13)(q37;q14)	Rhabdomyosarcoma	FKHR, PAX3
t(15;17)(q21;q11–q22)	AML, PML	PML, RARA
t(12;21)(p13;q22)	ALL	TEL, AML1

CML, chronic myelogenous leukemia; B-ALL, B cell acute lymphatic leukemia; AML, acute myelogenous leukemia; EWS, Ewing sarcoma; PML, promyelocytic leukemia

acts as a neooncogene. The *BCR-ABL* fusion gene seen in CML is the best-known example, but there are many others, some listed in Table 27.3.

Amplification and Oncogene Activation

Gene amplification is an important chromosomal mechanism of carcinogenesis (Chapter 25). Oncogenes are frequently activated or the steady-state level of their gene products increased by amplification (Table 27.4). Homogeneously stained regions (HSRs) and double minutes (DMs) have been observed in chromosomes of many primary tumors and cultured malignant cells. In colon carcinoma, the HSRs and DMs reflect the amplification of the *MYC* gene. The gene for the cell cycle regulator, cyclin D1, in the 11q13 region, is amplified in many types of cancer. This includes cancer of the esophagus, which is closely associated with environmental factors, such as alcohol consumption, tobacco smoking, and nitrosamine ingestion (Jiang et al., 1992). Tumor promoters, too, may act by promoting gene amplification. Growth of cells for several generations in the phorbol ester TPA (12-O-tetradecanoyl-phorbol-13-acetate), one of the most potent tumor promoters, enhances 3- to 16-fold the amplification of the *DHFR* (dihydrofolate reductase) gene under methotrexate selection and presumably has

Table 27.4. Oncogenes Amplified in Various Types of Cancer

Oncogene	Map position
ERBB1/EGFR	7p12.3–p12.1
CDK6	7q21–q22
MYC	8q24.12–q24.13
FGF4	11q13.3
INT2	11q13
HST1	11q13
Cyclin D1	11q13
Cyclin D2	12p13
CDK4	12q13
MDM2	12q13–q14
SAS	12q13–q14
HER2/NEU	17q12–q21
Cyclin E	19q12–q13

the same effect on various oncogenes. Cyclin D2, a pituitary gonadotrophic hormone (FSH)-responsive gene involved in gonadal cell proliferation, is amplified in many ovarian and testicular cancers (Sicinski et al., 1996). Amplification and overexpression of cyclin E leads to the formation of modified cyclin E/CDK2 kinase complexes that remain active throughout the cell cycle, bypassing the usual p16/RB repression (Gray-Bablin et al., 1996).

The cyclin-dependent kinase 6 (*CDK6*) gene, which maps to 7q21–q22, is amplified up to 50-fold in many gliomas, as shown by restriction landmark genome scanning by two-dimensional pulsed-field gel electrophoretic (PFGE) separation of very large *Not*I restriction enzyme fragments of DNA (Costello et al., 1997). Multiple genes in 12q13–q14 are often amplified in malignant gliomas. One is the cell cycle gene *CDK4*. Another is the cell cycle regulator gene *MDM2*, which is also amplified in other tumors, such as sarcomas (Reifenberger et al., 1994). The MDM2 protein binds the RB (retinoblastoma) protein, causing it to release the E2F transcription factor and thus stimulate cell division (Xiao et al., 1995). MDM2 also represses the transcription of the *TP53* gene and thus prevents activation of cell cycle checkpoints by DNA damage or spindle assembly defects (Levine, 1997).

Relaxation of Imprinting and Oncogene Activation

Hypomethylation of many genes is observed in cancer (Laird and Jaenisch, 1996). When this involves an imprinted gene, the imprint is relaxed (lost) and the gene may be expressed, leading to an increased dosage of its product. Such a gene may function as an oncogene. The *H19* and *IGF2* (insulin-like growth factor 2) genes are imprinted (Chapter 21), showing monoallelic expression of the paternal allele. Loss of imprinting of these genes has been observed in lung, renal, cervical, ovarian, testicular, and gastric cancers. It is often seen in benign tumors, suggesting that it is an early event in multistep tumorigenesis (Kim et al., 1998).

References

Adams JM, Cory S (1908) The Bcl-2 protein family: arbiters of cell survival. Science 281:1322–1326

Bishop JM (1983) Cellular oncogenes and retroviruses. Annu Rev Biochem 52: 301–354

Cobaleda C, Pérez-Lozada J, Sánchez-García I (1998) Chromosome abnormalities and tumor development: from genes to therapeutic mechanisms. BioEssays 20:922–930

Costello JF, Plass C, Arap W, et al. (1997) Cyclin-dependent kinase 6 (CDK6) amplification in human gliomas identified using two-dimensional separation of genomic DNA. Cancer Res 57:1250–1254

Gray-Bablin J, Zulvide J, Fox MP, et al. (1996) Cyclin E, a redundant cyclin in breast cancer. Proc Natl Acad Sci USA 93:15215–15220

Helm S, Mitelman F (1995) Cancer cytogenetics, 2nd ed, Wiley-Liss, New York

Jeffs AR, Benjes SM, Smith TL, et al. (1998) The BCR gene recombines preferentially with Alu elements in complex BCR-ABL translocations of chronic myeloid leukemia. Hum Mol Genet 7:767–776

Jiang W, Kahn SM, Tomita N, et al. (1992) Amplification and expression of the human cyclin D gene in esophageal cancer. Cancer Res 52:2980–2983

Kim HT, Choi BH, Niikawa N, et al. (1998) Frequent loss of imprinting of the H19 and IGF-II genes in ovarian tumors. Am J Med Genet 80:391–395

Laird P, Jaenisch R (1996) The role of DNA methylation in cancer genetics and epigenetics. Annu Rev Genet 30:441–464

Levine AJ (1997) p53, the cellular gatekeeper for growth and division. Cell 88:323–331

Lin RJ, Nagy L, Inoue S, et al. (1998). Role of the histone deacetylase complex in acute myelocytic leukemia. Nature 391:811–814

Mitelman F, Martens F, Johansson B (1997) A breakpoint map of recurrent chromosomal rearrangements in human neoplasia. Nat Genet 15:417–474

Reifenberger G, Reifenberger J, Ichimura K, et al. (1994) Amplification of multiple genes from chromosomal region 12q13–14 in human malignant gliomas: preliminary mapping of the amplicon shows preferential involvement of CDK4, SAS, and MDM2. Cancer Res 54:4299–4303

Rowley JD (1973) A new consistent chromosomal abnormality in chronic myelogenous leukemia identified by quinacrine fluorescence and Giemsa staining. Nature 243:290–293

Sánchez-García I, Grütz G (1995) Tumorigenic activity of the BCR-ABL oncogene is mediated by BCL2. Proc Natl Acad Sci USA 92:5287–5291

Sicinski P, Donaher JL, Geng Y, et al. (1996) Cyclin D2 is an FSH-responsive gene involved in gonadal cell proliferation and oncogenesis. Nature 384:470–474

Smith L, Liu SJ, Goodrich L, et al. (1998) Duplication of ATR inhibits MyoD, induces aneuploidy and eliminates radiation-induced G1 arrest. Nat Genet 19:39–46

Stec I, Wright TJ, Van Ommen G-JB, et al. (1998) WHSC1, a 90 kb SET domain-containing gene, expressed in early development and homologous to a Drosophila dysmorphy gene maps in the Wolf–Hirschhorn syndrome critical region and is fused to IgH in t(4;14) multiple myeloma. Hum Mol Genet 7:1071–1082

Xiao Z-X, Chen J, Levine AJ, et al. (1995) Interaction between the retinoblastoma protein and the oncoprotein MDM2. Nature 375:694–697

Yunis JJ (1983) The chromosomal basis of human neoplasia. Science 221:227–236

28

Chromosomes and Cancer: Inactivation of Tumor Suppressor Genes

Tumor/Nontumor Cell Hybrids: First Evidence for Tumor Suppressor Genes

Harris et al. (1969) showed that the fusion of normal and malignant cells produced hybrid cells that were generally nonmalignant. Rarely, however, a hybrid cell gave rise to malignant subclones. These always showed some loss of chromosomes derived from the normal parent, suggesting that one or more of these chromosomes carried a *tumor suppressor gene*. This has been confirmed by studies showing that the addition of a specific chromosome, chromosome segment, or single gene can suppress the tumorigenic capability of cancer cells. Thus, chromosome 11 suppresses the malignancy of Wilms renal tumor cells and of HeLa or other cervical cancer cells. The gene responsible has been mapped to 11q13, using deletions and translocations (Jesudasan et al., 1995). Another tumor suppressor locus has been mapped to 11p11.2–p12 using microcell hybridization and PCR analysis of a battery of genetic markers scattered along the chromosome (Fig. 28.1; Coleman et al., 1997). The normal function of a tumor suppressor gene is to prevent unregulated growth of cells. When both copies of such a gene are deleted or inactivated by mutation, the absence of the product allows uncontrolled growth to occur. An individual who has inherited a mutant tumor suppressor gene has a greatly increased risk of developing one or more types of cancer, because somatic mutation (deletion or inactivation) of only the single normal copy of the gene is necessary to abolish the tumor-suppressing function of the gene.

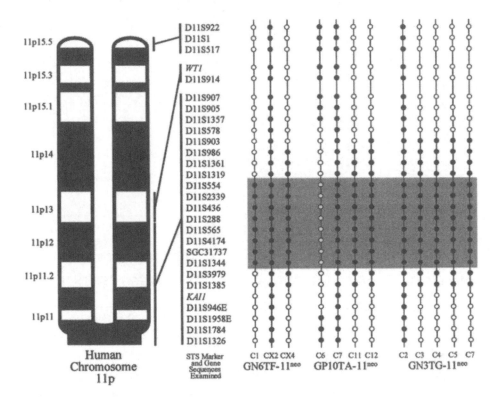

Figure 28.1. Summary of microsatellite PCR mapping in three microcell hybrid cell lines. Closed circles: retention of the marker; open circles: loss of the marker. The shaded area indicates the minimal liver tumor suppressor region (11p11.2–p14) identified using these three lines (courtesy of W. B. Coleman; based on data from three published studies and unpublished data).

Allele Loss and Loss of Heterozygosity

Loss of a chromosome segment carrying the wild-type allele from a cell heterozygous for a tumor suppressor gene leads to absence of its tumor-suppressing product. One mechanism for such loss of heterozygosity (LOH) is described in Chapter 24 and illustrated in Fig. 24.3. In normal cells, LOH occurs 10–100 times more frequently than point mutations (Tischfield, 1997). It is responsible for more than 90% of the renal cancers seen in von Hippel–Lindau syndrome (Gnarra et al., 1994) and more than 80% of familial breast and ovarian cancers (Neuhauser and Marshall, 1994). Studies have found that LOH of many specific chromosome regions is very common in many types of cancer. In addition, LOH analysis has shown consistent loss of the wild-type *BRCA1* allele in cancer cells from carriers of a *BRCA1* mutation, indicating that *BRCA1* is a tumor suppressor gene (Smith et al., 1992).

Retinoblastoma and the Two-Hit Model of Carcinogenesis

The most extensively studied of the cancers caused by inactivation of a tumor suppressor gene is the childhood tumor retinoblastoma, whose analysis led Knudsen to the two-hit model of carcinogenesis (Chapter 26). Retinoblastoma (RB) may occur as an isolated tumor, usually involving only one eye, or the tendency to it may be inherited, in which case there are multiple tumors, usually affecting both eyes. The genotype of a person with inherited retinoblastoma is either *RB/rb* or *RB/–*, where *rb* is a nonfunctional allele and—indicates a deletion of the locus. As Knudsen predicted, a somatic mutation or deletion of the remaining normal allele is required to produce malignant transformation of retinal cells. A small percentage of RBs arise in individuals with a constitutional heterozygous deletion of 13q14, historically the first evidence for the location of the gene (Chapter 1). Instances of LOH may be the result of chromosome loss (with or without duplication of the remaining chromosome), deletion, mitotic recombination, or gene conversion (Cavanee et al., 1983). Knowing its location made possible the positional cloning and sequencing of the gene, prediction of the amino acid sequence of the protein, isolation of the protein, and demonstration of the complete absence of RB mRNA or protein in RB cells.

In *RB/rb* or *RB/–* individuals, the risk of developing retinoblastoma is about 100,000 times higher than it is in the general population, and the risk for other types of cancer, especially osteosarcoma, is also increased. Many types of cancer are associated with inactivation of the gene, now called *RB1*, or its protein, pRB. Alternatively, many cancers instead show genetic alterations in proteins that regulate pRB, such as cyclin D, CDK4, or the p16 cell cycle inhibitor (Chapter 2). In over 90% of uterine cervical cancers, the RB protein is inactivated by binding to the E7 protein of the human papilloma virus (HPV), a major exogenous cause of cancer (Zur Hausen, 1996).

Mechanism of Tumor Suppression by a Functional *RB1* Gene

The RB protein acts as a negative regulator of growth and a tumor suppressor in two ways. It binds to the RNA polymerase I transcription factor upstream binding factor (UBF) and thus inhibits ribosomal RNA transcription (Cavanaugh et al., 1995). More important, pRB also arrests cell proliferation in the mid- to late G1 phase of the cell cycle by binding to E2F, a transcription factor that is essential for the G1/S transition and DNA replication (Chapter 2). The absence of a functional *RB1* gene thus leads to uncontrolled cell proliferation. Even when a normal *RB1* gene is present, pRB function may be impaired. For example, human papilloma virus E7 protein acts as an oncoprotein because it binds to pRB, causing it to release active E2F.

The two-hit model does not fully explain multistep carcinogenesis. Epstein-Barr virus infection early in life is a critical predisposing factor for both Burkitt lymphoma and nasopharyngeal carcinoma. Rearrangements involving chromosomes 1, 3, 9, 11, 12, and 17 are common in these tumors, but none of the usual oncogenes or tumor suppressor genes have yet been implicated. Loss of heterozygosity for loci on chromosome 3 led Stanbridge and his associates to use microcell-mediated transfer of single chromosomes or chromosome fragments into a nasopharyngeal carcinoma cell line. The malignant phenotype was suppressed by the addition of any chromosome segment containing 3p21.3, suggesting the presence of an unknown tumor suppressor in that region (Cheng et al., 1998). The role of the virus may be to destabilize the genome, leading ultimately to inactivation of one or more specific tumor suppressor genes.

The p53 Tumor Suppressor Gene, *TP53*

Many tumor suppressor genes, like *RB1*, are involved in normal cell cycle control of proliferation (Table 28.1). Perhaps the most important of these for carcinogenesis is the p53 gene, *TP53*. Mutations in *TP53* are extremely common in cancers (Chapter 26). The p53 protein is essential for the G1 checkpoint that arrests cells with DNA damage in G1 until DNA repair can be carried out. The p53 protein is also required for apoptosis, by which cells with irreparable DNA damage are destroyed. Mutations in *TP53* foster carcinogenesis by both these mechanisms. If cells enter S with damaged DNA, they can produce daughter cells with mutations and chromosome breaks. Failure of apoptosis permits the generation of cells with progressively more mutations, and selection acting on these can lead to more malignant cells. Without apoptosis to destroy the damaged cells, a more malignant cancer can be generated by treatment with genotoxic drugs or radiation (Griffiths et al., 1997).

Most human tumors show abnormalities of both *RB1* and *TP53*, indicating that disruption of both pathways is usually necessary for tumorigenesis. Bates et al.

Table 28.1. Cell Cycle Regulators That Act as Tumor Suppressors

Gene	Product	Location
INK4C	p18	1p32
VHL	VHL	3p25
CIP1	p21	6p21
CDKN2A	p16, p19	9p21
INK4B	p15	9p21
*KIP2**	p57	11p15
ATM	ATM	11q22
KIP1	p27	12p13
RB1	pRB	13q14
TP53	p53	17p13
BRCA1	BRCA1	17q21
INK4D	p19	19p13

*An imprinted gene (maternal allele expressed)

(1998) showed why this is so. Loss of pRB frees E2F, which activates the *ARF* gene product, p14. This binds to p53-MDM2 complexes and prevents p53 degradation. This fail-safe mechanism of tumor suppression is disrupted if p53 is also mutated.

Other Genes That Affect the Cell Cycle

In addition to *RB1* and *TP53*, several other genes act as tumor suppressors through their inhibitory effects on the cell cycle (Table 28.1). A missing chromosome or band has often been the first clue to the localization of such a gene, but LOH analysis is also helpful. It can even detect submicroscopic deletions, which are very frequent in cancers.

Cyclin-dependent kinase (CDK) inhibitors arrest the cell cycle and thus control cell proliferation. The tumor-suppressing functions of pRB and p53 are also bypassed by mutations of the *ARF1/NK4a* (*CDKN2A*) gene locus at 9p21. These are almost as common as mutations in *TP53* itself (Kamb et al., 1994). This single locus produces two different gene products, p16/INK4A and p19/ARF (Fig. 28.2), by using separate promoters and two different reading frames, so that the resultant proteins have no amino acid sequence similarity. The p16 product normally inhibits CDK4 and CDK6 and thus blocks phosphorylation of the RB protein, maintaining pRB in its growth suppressor state (Monzon et al., 1998). The p19 protein is localized mainly to the nucleolus, but it leaves the nucleolus when pMDM2 binds to p53 and forms a larger complex with them. This blocks the nuclear export and degradation of p53 (Zhang and Xiong, 1999). Thus, mutations in *CDKN2A* can lead to a reduction in both pRB and p53 (Pomerantz et al., 1998). Many tumors show deletions that remove both copies of the *CDKN2A* gene. The deletions may be megabase pairs long. Many types of tumor show allele loss, as shown by LOH for marker loci in this region.

Mutations in the *BRCA1* gene are found in the majority of families with both breast and ovarian cancers (Easton et al., 1993). This gene appears to act as a tumor suppressor by activating the *WAF1/C1P1* gene product, p21, which blocks cell cycle progression into S (Somasundaram et al., 1997; see also Chapter 2). *BRCA1* is also required for one type of repair of oxidative DNA damage (Gowen et al., 1998).

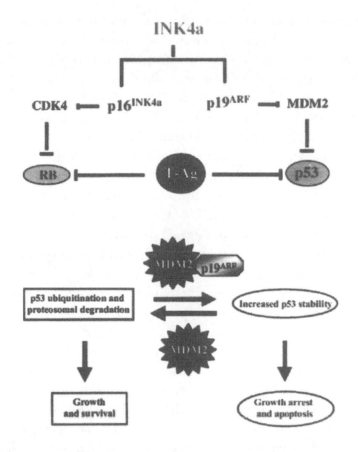

Figure 28.2. Two pathways affected by the two products of the *INK4A* gene, p16^{INK4A} and p19^{ARF}. (A) Functional relationship of the two products with p53 and pRB. (B) Proposed mechanism for the enhancement of p53 functions by p19^{ARF} (Pomerantz et al., copyright 1998, Cell Press).

Imprinted Tumor Suppressor Genes

Imprinted gene loci are those in which either the maternally or the paternally derived allele is inactivated, usually by methylation of the 5′ promoter or CpG island (Chapter 21). Cells therefore contain a single functional copy of each imprinted gene, and a *single* mutation is enough to abolish this function, in contrast to the *two independent* mutations necessary to abolish function of any non-imprinted autosomal gene. As a result, any imprinted gene with tumor suppressor activity may play a major role in some cancers. Breast tumors are one of the most frequent types of cancer. One reason may be that the breast tumor suppressor gene *HIC-1* (hypermethylated in cancer-1) is imprinted in breast ductal epithelium (although not in other cell types), with one allele hypermethylated and the other allele unmethylated. *HIC-1* maps to 17p13.3, a region noted for LOH in breast cancer. Loss of heterozygosity of distal 17p, with loss of the unmethylated (functional) allele, was observed in 22 of 26 breast cancers with only a hypermethylated *HIC-1* gene or genes (Fujii et al., 1998).

A second imprinted tumor suppressor gene is *H19* (Chapter 21). *H19* RNA, which is not translated into a protein product, has tumor suppressor activity (Hao et al., 1993). It is interesting that several other tumor suppressor genes, including *RB1*, *VHL*, *p16/MTS*, *p15/INK4A*, *E-cadherin*, and *MLH1*, are hypermethylated in many types of cancer, although they are not known to be imprinted (Baylin et al., 1998). Hypermethylation inactivates these genes just as effectively as mutation or deletion, and this is found quite frequently in cancer cells. Loss of heterozygosity of 9p21, a region that contains the p16 tumor suppressor gene, is one of the most frequent changes in cancer. Cancer cell lines in which this has *not* occurred show methylation of the 5' CpG island of the p16 gene, with complete transcriptional silencing that is reversible upon demethylation with 5-azacytidine. De novo methylation of this CpG island is seen in about 20% of primary neoplasms but not in normal tissues (Merlo et al., 1995).

Beckwith–Wiedemann syndrome (BWS) was described in Chapter 21. In BWS there is an increased risk of childhood tumors, including Wilms tumor, adrenal or liver carcinoma, and sarcoma. The few familial cases show autosomal dominant inheritance, with the disorder limited to those who inherited the mutation maternally. Linkage analysis in these families places the *BWS* gene at 11p15.5. Most cases are sporadic, with a normal karyotype, but a few involve a duplication, a translocation, or paternal uniparental disomy of chromosome 11, with the critical region 11p15.5. In normal kidney, only the maternal genes in this region are expressed, and the maternal 11p15 region tends to be lost from Wilms tumor cells. Microcell-mediated transfer of fragments containing 11p15 suppressed the tumorigenicity of Wilms tumor cells (Dowdy et al., 1991).

A candidate tumor suppressor gene in 11p15 is *KIP2*. Its protein product, p57, is an inhibitor of G1 cyclin-dependent kinases. Overexpression of p57 blocks the cell cycle in G1 (Chapter 2). In its absence, some cells will proliferate without control. *KIP2* is an imprinted gene: Only the maternal allele is expressed, the inactive paternal copy being fully methylated. There is reduced *KIP2* expression in Wilms tumor samples. The residual KIP2 protein presumably reflects the presence of a few normal cells in the tumor samples analyzed. Confirmation that *KIP2* is indeed a tumor suppressor gene came from the demonstration that BWS can be caused not only by disruption or deletion but by an inactivating mutation of the maternally derived allele (Hatada et al., 1997).

Genes That Suppress Oncogenes or Influence Transcription

Colon carcinomas frequently overexpress *MYC* RNA and protein *without* the amplification or rearrangement of the gene discussed in Chapters 25 and 27. This appears to be the result of loss or inactivation of a tumor suppressor gene on chromosome 5, because introducing a normal chromosome 5 reduces the level of *MYC* expression and suppresses the malignancy of such colon carcinoma cells (Rodriguez-Alfageme et al., 1992). MYC functions only in the form of a heterodimer with a protein called MAX. Four proteins, the MAD/MXI1 family, compete with MYC for binding to MAX and thus block MYC. The genes for the four MAD/MXI1 proteins map to loci often disturbed in cancer cells. For example, *MXI1* maps to 10q24–q26 (Table 28.2), a region that often shows translocations or deletions in major kinds of cancer. In one study, half of prostate cancers showed a submicroscopic loss of one *MXI1* allele (LOH) and mutational inactivation of the retained allele (Prochownik et al., 1996).

SNF5 (Table 28.2) is a tumor suppressor gene with quite a different mechanism of action: It is involved in chromatin remodeling that fosters transcription (Chapter 5). Deletions of 22q11.2 are common in malignant rhabdoid childhood cancers of kidney, brain, and soft tissues. Homozygous deletions were found in 6 out of 13 of these tumor cell lines, and the shortest region of overlap contained the *SNF5* gene. In the other seven lines, there was a frameshift or nonsense (inactivating) mutation of one *SNF5* allele and loss of the other allele, leaving no functional copy (Versteege et al., 1998).

Chromosome band 10q23 is frequently deleted in endometrial, prostate, and breast carcinomas and in glioblastomas. Mapping homozygous 10q23 deletions led to the isolation from 10q23.3 of a tumor suppressor gene called *PTEN* (Table 28.2). The PTEN protein is a phosphatase whose target is phosphatidylinositol-3,4,5-triphosphate (PIP3), a key component of an important growth control pathway that stimulates proliferation and blocks apoptosis (Hopkin, 1998). The

Table 28.2. Normal Functions of Some Tumor Suppressor Genes

Gene	Location	Product*	Function regulated
KISS1	1q32–q41	KISS1	Metastasis suppressor
MSH2	2p22	MSH2	DNA mismatch repair
PMS1	2q31–q33	PMS1	DNA mismatch repair
VHL	3p25	VHL	Transcript elongation
MLH1	3p21–p23	MLH1	DNA mismatch repair
APC	5q21	APC	Cell adhesion
PMS2	7p22	PMS2	DNA mismatch repair
PTEN	10q23.3	PTEN	Signal transduction
MXI1	10q24–q26	MXI1	Inhibits MYC
H19	11p15.5	RNA only	Expression of nearby genes
MEN1	11q13	Menin	Transcription activation
BRCA2	13q12	BRCA2	Double-strand break repair
NF1	17q12	Neurofibromin	G-protein signal transduction
BRCA1	17q21	BRCA1	Double-strand break repair
DCC	18q21	DCC	Cell interactions
DPC4	18q21	DPC4	Signal transduction
SNF5	22q11.2	SNF5	Chromatin remodeling
NF2	22q12	Merlin	Signaling at membrane

*The letter symbols for the protein products are sometimes preceded by a "p," as in pVHL

PTEN protein has sequence similarity to the cytoskeletal protein tensin, which binds actin filaments. Overexpression of *PTEN* inhibits cell migration, spreading, and focal adhesion, suggesting that it may act as a tumor suppressor by negatively regulating cell interactions with the extracellular matrix (Tamura et al., 1998).

Mutations in the *MEN1* gene are responsible for familial multiple endocrine neoplasia type 1, an autosomal dominant disorder associated with tumors of the parathyroid, anterior pituitary, pancreas, and other neuroendocrine tissues. *MEN1* maps to 11q13 (Table 28.2), and LOH of 11q13 is seen in MEN1 tumors. As a result, the tumor tissue of affected individuals has lost the wild-type (normal) *MEN1* allele present in their normal tissues. Menin, the product of *MEN1*, interacts with the transcription factor JunD, blocking JunD–activated transcription (Agarwal et al., 1999).

Genes That Affect Cell Adhesion

Most normal cells are anchorage dependent, requiring attachment to an extracellular substrate for cell division. The reason for this is that anchorage is essential for the activation of cyclin E/CDK2 kinase in late G1 that initiates the G1/S transition (Chapter 2). In contrast, the kinase is always active in cancer cells, which are anchorage independent (Fang et al., 1996). This may indicate that normal cells produce a suppressor of cyclin E unless they are anchored to the extracellular matrix and that when inactivation or deletion silences the suppressor, cells can divide even if not anchored. Several of the cadherin genes have been mapped to regions that exhibit cancer-related LOH (Kremmidotis et al., 1998). E-cadherin is the main adhesion molecule in epithelia and is frequently absent in highly aggressive, invasive epithelial cancers. Restored expression of E-cadherin suppresses tumor invasiveness. *APC* (mutated in adenomatous polyps of the colon) may act as a tumor suppressor gene (Table 28.2) in two different ways. APC binds β-catenin, leading to its degradation. This blocks the activation of the Tcf4 transcription factor by β-catenin (Korinek et al., 1997). APC also interferes with spindle assembly by binding the EB1 protein that is normally found on microtubules (Su et al., 1995). Loss of APC function results in uncontrolled transcriptional activation of β-catenin/Tcf4 target genes.

Von Hippel–Lindau (VHL) syndrome is a rare autosomal dominant cancer predisposition disorder associated with both benign and malignant tumors, particularly of the kidney, pancreas, and brain. The *VHL* gene (Table 28.2) has been mapped to 3p25–p26 and identified (Latif et al., 1993). VHL patients are heterozygous for either a deletion or a point mutation of the gene, and renal carcinoma cells show allele loss on LOH analysis, indicating the absence of a functional copy (Crossey et al., 1994). The VHL protein (pVHL) acts in various ways. It is required for proper assembly of an extracellular fibronectin matrix. Since loss of this matrix is a feature of malignant cellular transformation, this may be the major tumor suppressor function of pVHL (Ohh et al., 1998). The region of pVHL that binds elongin B and elongin C is a hotspot for mutation in VHL kindreds, suggesting that the involvement of pVHL in the elongation of RNA transcripts may account for its role in tumor suppression, although the mechanism is unclear. It may be more important that the *CUL2* gene product, cullin, binds to the pVHL/elongin B/elongin C complex and is transported into the nucleus, where it plays a role in targeting cell cycle proteins for degradation. Failure to degrade these could lead to uncontrolled cell proliferation (Pause et al., 1997).

Metastasis Suppressor Genes

Microcell-mediated transfer of a normal chromosome 6 into malignant melanoma cells suppresses their ability to metastasize by at least 95% but does not affect their tumorigenicity, the ability to proliferate into a tumor at the site of injection of some of the cells into an immunosuppressed animal host. Lee et al. (1996) used subtractive DNA hybridization to detect any differences in the expression of mRNAs in the metastatic and nonmetastatic cells and isolated a novel gene, called *KISS1* (Table 28.2), from the nonmetastatic cells Transfection of a full length *KISS* cDNA into cancer cells could also suppress their metastasis. Surprisingly, FISH mapped *KISS1* to 1q32–q41, indicating that chromosome 6 carries a different metastasis suppressor gene, whose expression is required for *KISS1* expression.

DAP kinase mediates cell death (apoptosis). Some metastatic lung cancers have a reduced level of DAP kinase. The introduction of a functional *DAP* gene into cancer cells blocks their ability to metastasize and increases their sensitivity to apoptotic stimuli of the sort metastasizing cells are exposed to (Inbal et al., 1997).

References

Agarwal SK, Guru SC, Heppner C, et al. (1999) Menin interacts with the AP1 transcription factor JunD and represses JunD-activated transcription. Cell 96:143–152

Bates S, Phillips AC, Clark PA, et al. (1998) p14(*ARF*) links the tumor suppressors *RB* and *p53*. Nature 395:124–125

Baylin SB, Herman JG, Graff JR, et al. (1998) Alterations in DNA methylation: a fundamental aspect of neoplasia. Adv Cancer Res 72:141–196

Cavanaugh AH, Hempel WM, Taylor LJ, et al. (1995) Activity of RNA polymerase I transcription factor UBF blocked by *Rb* gene product. Nature 374:177–180

Cavenee WK, Dryja TP, Phillips RA, et al. (1983) Expression of recessive alleles by chromosomal mechanisms in retinoblastoma. Nature 305:779–784

Cheng Y, Poulos NE, Lung ML, et al. (1998) Functional evidence for a nasopharyngeal carcinoma tumor suppressor gene that maps at chromosome 3p21.3. Proc Natl Acad Sci USA 95:3042–3047

Coleman WB, Esch GL, Burchert KM, et al. (1997) Localization of a putative liver tumor suppressor locus to a 950-kb region of human 11p11.2–p12 using rat liver tumor microcell hybrid lines. Mol Carcinogenesis 19:267–272

Crossey PA, Richards FM, Foster K, et al. (1994) Identification of intragenic mutations in the von Hippel-Lindau disease tumour suppressor gene and correlation with disease phenotype. Hum Mol Genet 3:1303–1308

Dowdy SF, Fasching CL, Araujo D, et al. (1991) Suppression of tumorigenicity in Wilms tumor by the p15.5–p14 region of chromosome 11. Science 254:293–295

Easton DF, Bishop T, Ford D, et al. (1993) Genetic linkage analysis in familial breast and ovarian cancer: results from 214 families. Am J Hum Genet 52:678–701

Fang F, Orend G, Watanabe N, et al. (1996) Dependence of cyclin E-CDK2 kinase activity on cell anchorage. Science 271:499–502

Fujii H, Biel MA, Zhou W, et al. (1998) Methylation of the HIC-1 candidate tumor suppressor gene in human breast cancer. Oncogene 16:2159–2164

Gnarra JR, Tory K, Weng Y, et al. (1994) Mutation of the VHL tumor suppressor gene in renal carcinoma. Nat Genet 7:85–90

Gowen LC, Avrutskaya AV, Latour AM, et al. (1998) BRCA1 required for transcription-coupled repair of oxidative DNA damage. Science 281:1009–1012

Griffiths SD, Clarke AR, Healy LE, et al. (1997) Absence of p53 permits propagation of mutant cells following genotoxic damage. Oncogene 14:526–531

Hao Y, Crenshaw T, Moulton T, et al. (1993) Tumor suppressor activity of H19 RNA. Nature 365:764–767

Harris H, Miller OJ, Klein G, et al. (1969) Suppression of malignancy by cell fusion. Nature 223:363–368

Hatada I, Ohashi H, Fukushima Y, et al. (1997) An imprinted gene p57 (KIP2) is mutated in Beckwith–Wiedemann syndrome. Nat Genet 7:85– 90

Hopkin K (1998) A surprising function of the PTEN tumor suppressor. Science 282:1027–1030

Inbal B, Cohen O, Polak-Charcon S, et al. (1997) DAP kinase links the control of apoptosis to metastasis. Nature 390:180–184

Jesudasan RA, Rahman RA, Chandrashekharappa S (1995) Deletion and translocation of chromosome 11q13 sequences in cervical carcinoma lines. Am J Hum Genet 56:705–715

Kamb A, Gruis NA, Weaver-Feldhaus J, et al. (1994) A cell cycle regulator potentially involved in genesis of many tumor types. Science 264:436–440

Korinek V, Barker N, Morin PJ, et al. (1997) Constitutive activation by a β-catenin-Tcf complex in APC–/– colon carcinoma. Science 275:1784–1787

Kremmidotis G, Baker E, Crawford J, et al. (1998) Localization of human cadherin genes to chromosome regions exhibiting cancer-related loss of heterozygosity. Genomics 49:467–471

Latif F, Tory K, Gnarra J, et al. (1993) Identification of the von Hippel–Lindau disease tumor suppressor gene. Science 260:1317–1320

Lee J-H, Miele ME, Hicks DJ, et al. (1996) KISS-1, a novel human malignant melanoma metastasis-suppressor gene. J Natl Cancer Inst 88:1731–1737

Merlo A, Herman JG, Mao L, et al. (1995) 5'CpG island methylation is associated with transcriptional silencing of the tumor suppressor p16/CDKN2/MTS1 in human cancers. Nature Med 1:686–692

Monzon J, Liu L, Brill H, et al. (1998) CDKN2A mutations in multiple primary melanomas. N Engl J Med 338:879–887

Neuhauser SL, Marshall CJ (1994) Loss of heterozygosity in familial tumors from three BRCA1-linked families. Cancer Res 54:6069–6072

Ohh M, Yauch RL, Lonergan KM, et al. (1998) The von Hippel–Lindau tumor suppressor protein is required for proper assembly of an extracellular fibronectin matrix. Mol Cell 1:959–968

Pause A, Lee S, Worrell RA, et al. (1997) The von Hippel–Lindau tumor-suppressor gene product forms a stable complex with human CUL-2, a member of the Cdc53 family of proteins. Proc Natl Acad Sci USA 94:2156–2161

Pomerantz J, Schreiber-Agus N, Liégeois NJ, et al. (1998) The *INK4a* tumor suppressor gene product, p19^{Arf}, interacts with MDM2 and neutralizes MDM2's inhibition of p53. Cell 92:713–723

Prochownik EV, Grove LE, Deubler D, et al. (1996) Commonly occurring loss and mutation of the MXI1 gene in prostate cancer. Genes Chromosom Cancer 22:295–304

Rodriguez-Alfageme C, Stanbridge EJ, Astrin SM (1992) Suppression of deregulated c-MYC expression in human colon carcinoma cells by chromosome 5 transfer. Proc Natl Acad Sci USA 89:1482–1486

Smith SA, Easton DF, Evans DGR, et al. (1992) Allele losses in the region 17q12–q21 in familial breast and ovarian cancer non-randomly involves the wild-type chromosome. Nat Genet 2:128–131

Somasundaram K, Zhang H, Zeng Y-X, et al. (1997) Arrest of the cell cycle by the tumour-suppressor BRCA1 requires the CDK-inhibitor p21(WAF1/CIP1). Nature 389:187–190

Su LK, et al. (1995) APC binds to the novel protein EB1. Cancer Res 55:2972–2977

Tamura M, Gu J, Matsumoto K, et al. (1998) Inhibition of cell migration, spreading, and focal adhesions by tumor suppressor PTEN. Science 280:1614–1617

Tischfield JA (1997) Somatic genetics 1997. Loss of heterozygosity or: how I learned to stop worrying and love mitotic recombination. Am J Hum Genet 61:995–999

Versteege I, Sévenet N, Lange J, et al. (1998) Truncating mutations of hSNF5/INI1 in aggressive paediatric cancer. Nature 394:203–206

Zhang Y, Xiong Y (1999) Mutations in the *ARF* exon 2 disrupt its nucleolar localization and impair its ability to block nuclear export of MDM2 and p53. Mol Cell 3:579–591

Zur Hausen H (1996) Papillomavirus infections: a major cause of human cancers. Biochim Biophys Acta 1288:155–159

29

Mapping Human Chromosomes

A central problem in cytogenetics is mapping the location of each gene. This involves determining their linear order and the physical distance and frequency of meiotic recombination between all components of each chromosome, especially the genes. The rate of progress in mapping the human chromosomes has increased rapidly. As recently as 1990, only about 1% of our genes had been mapped, many of them not very accurately. In contrast, a recent paper describes a physical map of over 30,000 genes, estimated to be between one-third and one-half of the total human complement (Deloukas et al., 1998). It is no longer possible to maintain an up-to-date catalog in book form. The twelfth edition of *Mendelian Inheritance in Man* (McKusick, 1997) has mapping information on more than 4000 genes of known function. It is now accessible as an on-line database (http://www.ncbi.nlm.nih.gov/omim) known as OMIM. This is the only form in which the rapidly burgeoning body of information can be kept current.

Leading scientific and medical journals are increasingly filled with reports of the mapping, positional cloning, sequencing, and functional importance of new genes. For nearly 20 years, frequent international human gene mapping conferences helped coordinate the mapping efforts and the management of the growing databases. They continue to do so, but governmental research organizations in the United States and several other countries are playing an increasing role, and enormous databases are now available on the Internet. For example, complete searching and editing of human genomic data and documentation are accessible at http://www.gdb.org and the Généthon CEPH (Centre d'Etude de Polymorphisme Humaine) genotype database (for distances between markers on any chromosome) at http://www.cephb.fr.

Genetic Linkage Maps

A genetic linkage map is a pictorial summary of meiotic recombination frequencies for various loci, showing their linear order and distance apart. The distance between adjacent loci is expressed in centiMorgans (1 cM = 1% recombination). Note that a locus may be either a gene or an arbitrary fragment of DNA. Linkage analysis requires genetic variation, the presence at each locus of two or more alleles (distinguishable forms of the gene or DNA sequence). These can be used as genetic markers to show whether the alleles from one parent show independent assortment during meiosis or are linked and, if so, what the recombination frequencies are for each pair of linked loci. The lower their frequency of recombination, the closer the loci are to one another on the chromosome. Distances between more distant loci are not so accurate, because two or four crossover events between the loci would not result in the recombination of their alleles. As pointed out in Chapter 9, chiasma distributions and frequencies show individual and sex-specific differences, and so do recombination frequencies and the resultant linkage maps (Fig. 29.1).

The most exciting discovery of the molecular era for linkage analysis is the enormous amount of sequence variation in DNA throughout the genome. Restriction endonucleases cut DNA at precise sites (for example, *Eco*RI at GAATTC, *Hae*III at GGCC), and they often cut a DNA sequence from different copies of a chromosome into a different number of fragments, because a mutation has produced a nucleotide substitution at one of these sites

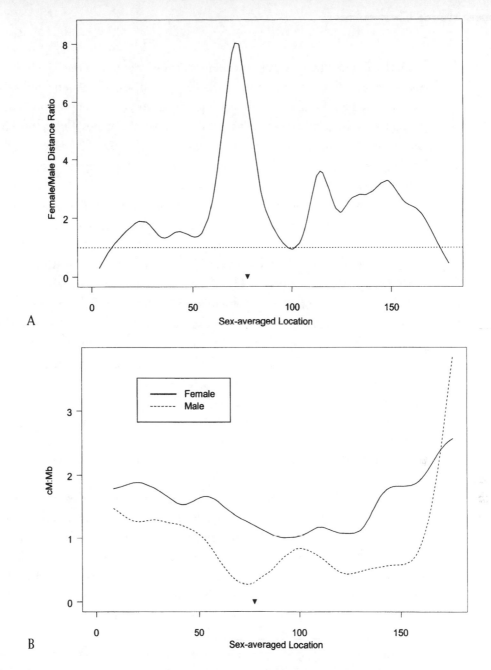

Figure 29.1. (A) Plot of the female : male genetic distance ratio against the sex-averaged genetic location (in cM, from 0 = 7pter to 182 = qter) along chromosome 7. Triangle: approximate location of the centromere; dashed line: equal male and female distance. The recombination rate in males exceeds that in females only near the telomeres. (B) Plot of sex-specific genetic distance : physical distance ratio (in cM : Mb) against sex-averaged genetic location. A dip in male genetic distance near the centromere corresponds to almost 20 Mb without male recombinations and accounts for the peak in this region in the female : male distance ratio. Comparable data for all the chromosomes can be obtained at http://www.marshmed.org/genetics (reproduced from Broman et al., Am J Hum Genet 63:861–869, copyright 1998, the American Society of Human Genetics, with permission of the University of Chicago Press).

(GAATTT or AGCC, say), blocking cutting. Such *restriction fragment length polymorphisms* (RFLPs) can serve as molecular alleles, or genetic markers, in linkage studies. RFLPs and similar molecular markers were instrumental in mapping a series of disease genes, for example, the genes for cystic fibrosis and familial adenomatous polyposis (Kerem et al., 1989; Kinzler et al., 1991), leading to their eventual isolation and identification by a process called *positional cloning*.

The enormous abundance of highly variable (polymorphic) microsatellites throughout the genome has made them the preferred genetic markers for linkage studies. Microsatellites are short, tandemly repetitive elements such as (CA)n, where n is usually 3–7. These are unstable because secondary structures formed within the repeat block processing of Okazaki fragments by flap endonuclease 1 (FEN-1), as described in Chapter 20. This leads to frequent increases or decreases in the number of repeats. Most individuals carry different numbers of repeats on an allelic pair at many of these sites, making microsatellites ideal genetic markers. A comprehensive human genetic map based on 5264 microsatellite markers was published in 1996 by Dib et al. Its use makes it possible to place any new gene on the genetic linkage map very quickly. An example is the mitotic protein kinase gene, *BUB1*, which was recently mapped to within 10 cM of a marker locus that is 120–129 cM from the end of the short arm (2pter) along the 227-cM-long chromosome 2. This corresponds to band 2q11–q21 (Pangilinan et al., 1997).

The availability of so many molecular markers and the ease of screening the total genome for them has made it possible to map rare recessive disease gene loci by a technique called *homozygosity mapping*. For this, one needs only a few individuals with a recessively inherited disease who have some normal sibs and consanguineous parents (first cousins, say). The affected individuals will be homozygous for the disease gene and also for the genetic markers that are very closely linked to it. For example, four subjects with ICF syndrome (Chapter 22) from three consanguineous matings were consistently homozygous only for markers in a 9-cM region of 20q11–q13, while all their normal sibs were heterozygous for these markers (Wijmenya et al., 1998). Microsatellite markers have also been used to integrate physical and genetic linkage maps (Deloukas et al., 1998; Mohrenweiser et al., 1998). This is important because linkage distances may differ markedly from physical distances as a result of localized differences in the frequency of meiotic crossing over.

Mapping Disease Genes: Family Studies with Genetic Markers

Family studies can show whether a gene lies on an autosome or on one of the sex chromosomes. An X-linked gene is never passed from father to son, whereas a Y-linked gene always is and is not passed from father to daughter. The first human gene locus assigned to a specific chromosome was that for red-green color blindness, which was assigned to the X chromosome in 1911. Now, more than 900 X-linked genes are known, based mainly on molecular cytogenetic methods (Deloukas et al., 1998). The Y chromosome has very few genes (Chapter 17).

For autosomal genes, linkage analysis may show whether two or more genes are on the same chromosome. In 1954, Mohr discovered the first autosomal linkage, between the Lutheran blood group gene locus and the ABO secretor gene locus. Such blood group systems were among the very few useful genetic markers available in those early years. Now there are thousands of genetic markers that can be used in mapping a disease gene or any other DNA sequence. In addition, the polymerase chain reaction (PCR) has made it possible to amplify DNA from single cells, including sperm cells (Li et al., 1988). Given the high frequency of DNA sequence variation throughout the genome, the frequency of recombination between any two sites can be measured directly by examination of gametes. In principle, one could produce a rather comprehensive genetic linkage map of a human genome from a single semen sample and DNA samples from both parents of the semen donor.

Assignment of Genes to Chromosomes: Synteny Groups

A gene or a group of linked genes (*linkage group*) can be assigned to a specific chromosome by family studies if a suitable marker chromosome is available that is readily distinguishable from its normal homologue or if linkage can be shown to a genetic marker already mapped to a chromosome. Typical chromosome markers consist of heterochromatin variants, fragile sites, or structurally rearranged chromosomes. The first gene assigned by means of a marker chromosome was the Duffy blood group locus, which segregated with a large heterochromatin variant of chromosome 1 (Donahue et al., 1968). One allele at the α-haptoglobin locus was inherited together with a fragile site on 16q in 30 family members and segregated from it in three. That is, the recombination frequency was 9%, indicating that the two loci were linked and about 9 cM apart on 16q (Magenis et al., 1970). Segregation of a pericentric inversion of chromosome 6 with one *haplotype* (a specific set of alleles at multiple closely linked loci) of the major histocompatibility complex localized the *MHC* locus to chromosome 6. Studies involving reciprocal translocations made possible the localization of this locus to 6p21 (Breuning et al., 1977).

For many years, the most important technique for assigning genes to specific chromosomes was somatic cell hybridization of normal human cells with immortal rodent cells. Human chromosomes tend to be lost from such human-rodent hybrid cells, leading to the formation of clones that differ in their human chromosome content. Any human genetic markers that are distinguishable from the homologous rodent markers can be mapped by analysis of a sufficient number of hybrid cells. Selection for or against a certain phenotype in the hybrid cells can also be used to aid the gene mapping process. The first human gene to be assigned in this way was the thymidine kinase (TK) gene, to chromosome 17 (Miller et al., 1971). The poliovirus receptor gene was mapped to chromosome 19 in a similar way, using hybrid cells that were quite sensitive to poliovirus (mouse cells are resistant) and had multiple human chromosomes; selection for resistance to the virus led to elimination of only human chromosome 19 (Miller et al., 1974).

The usual result of random chromosome elimination in human-rodent hybrid cells is the creation of clones that have different combinations of human chromosomes. To assign an unselected gene to its chromosome, one correlates the presence of the gene or gene product with the presence of a specific human chromosome. A panel of hybrid cell clones may be established in which each chromosome has a unique pattern of presence or absence among the clones. Assignments are made by testing each clone in the panel for the presence or absence of a particular gene product or DNA sequence. In principle, a panel of five clones might be sufficient to identify each of the 24 human chromosomes (see Table 29.1). Larger panels are used so that their redundancy can confirm each assignment and decrease the likelihood that unrecognized structural changes in some of the clones will lead to a false assignment. The summary of +/0 reactions with the members of a clone panel produces a binary pattern that uniquely identifies the human chromosome that is the site of the gene (Ruddle and Creagan, 1975). This technique can be used to determine whether two or

Table 29.1. This Panel of Five Hybrid clones, A–E, had been Selected for the Presence (+) or Absence (0) of the 24 Human Chromosomes Shown Across the Top. These Clones are Tested for the Presence or Absence of any Particular Human Gene Product or DNA Sequence. The +/0 Pattern, Emphasized in the Table by Shading, Allows the Product or Sequence to be Assigned to a Specific Chromosome, in this Case, Number 1

	1	2	3	4	5	6	7	8	9	10	11	12	13	14	15	16	17	18	19	20	21	22	X	Y
A	+	0	+	+	+	+	0	+	+	+	0	+	+	0	+	0	0	0	0	0	+	+	0	0
B	+	+	0	+	+	+	0	0	+	+	0	0	+	0	0	+	0	0	+	+	+	0	+	+
C	+	+	+	0	+	+	+	0	0	+	0	0	0	0	0	0	+	0	+	0	0	+	0	+
D	+	+	+	+	0	+	+	+	0	0	+	0	0	0	0	0	0	+	0	+	+	+	+	+
E	+	+	+	+	+	0	+	+	+	0	+	+	0	+	0	0	0	0	+	+	0	0	0	0

Source: From Ruddle and Creagan, 1975, p 422; modified and reproduced, with permission, from the *Annual Review of Genetics* Vol. 9, copyright 1975, by Annual Reviews, www.AnnualReviews.org

more genes are on the same chromosome but not to identify their map positions. Genes assigned to the same chromosome constitute a *synteny group*. Human chromosomes are so long that syntenic genes (for example, two genes on opposite arms of the chromosome) may not show linkage in family studies.

Physical Maps

A physical map shows the physical distances between loci on each chromosome. One of the first techniques used to determine the location and order of genes on a chromosome was the analysis of rearrangements. The gene for the X-linked recessive disorder Duchenne muscular dystrophy was mapped to Xp21 by showing that the X;autosome translocation breakpoint in an affected female was in this band (Boldrug et al., 1987). The retinoblastoma gene was mapped to 13q14 by noting its association with a deletion of this region (Chapter 15). Rearrangements have also been useful in mapping by cell hybridization. The occurrence of a particular phenotype in hybrid cells that have retained part of a chromosome restricts the location of the gene to the segment that is retained. For example, a translocation chromosome consisting of almost all of Xq attached to the distal end of 14q segregated with the X-linked genes for HGPRT (hypoxanthine-guanine phosphoribosyl-transferase), PGK (phosphoglycerate kinase), and G6PD (glucose-6-phosphate dehydrogenase) and the autosomal gene NP (nucleoside phosphorylase). The three X-linked genes could thus be assigned to Xq and the autosomal gene to 14q (Ricciuti and Ruddle, 1973). Using other X;autosome translocations, the order of the three genes was shown to be centromere-PGK-HGPRT-G6PD. By the use of overlapping deletions or translocations, the location of a gene can be narrowed to a smaller segment, the *smallest region of overlap*, or SRO.

Far more precise regional mapping can now be achieved by using radiation hybrid panels containing tiny fragments (some only a few hundred kb in size, others larger) produced by massive irradiation of human cells before hybridizing them with rodent cells (Chapter 23; Leach et al., 1995). A radiation hybrid physical map of the human genome has been produced (Gyapay et al., 1996). Such maps can be integrated with those produced by other methods, such as yeast artificial chromosome (YAC) contig maps (see below) (Bouzyk et al., 1997).

In situ hybridization with nonradioactive (especially fluorescent) labels is a rapid, reliable, and fairly precise method of mapping (Chapter 8). Thousands of genes or other DNA sequences have been mapped using this technique. *Microdissection* of a chromosome band, homogeneously staining region, double minute, or other region from a single metaphase spread, or a small number of spreads, provides a way to identify the genes in a specific region. The microdissected DNA is amplified by PCR (Erlich et al., 1989) and used for molecular studies to identify the genes that are present or as a probe for in situ hybridization to verify the origin of the DNA (Chapter 8). In both Mendelian and multifactorial diseases, the rare association of the disease with a translocation provides a direct approach to cloning the gene involved, by microdissecting and cloning the translocation breakpoint. Muir et al. (1995) used this approach with a t(1;11)(q42.2;q21) translocation that was associated with schizophrenia.

There are various kinds of physical maps (Fig. 29.2). The resolution possible with techniques that depend on chromosome banding, in situ hybridization, or microdissection is limited by the resolving power of the light microscope. Molecular methods have overcome this limitation. The ultimate limit of resolution in physical mapping is the nucleotide pair. DNA sequencing is the ultimate step in constructing a molecular map in which every gene and other feature of the genome is precisely positioned. Several intermediate stages have been important. The invention of pulsed-field gel electrophoresis (PFGE) provided a technique for separating and characterizing DNA fragments that were too large to

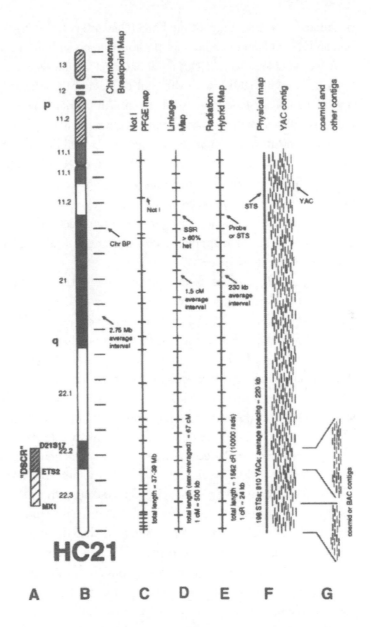

Figure 29.2. Maps for human chromosome 21 (HC21). (A) Down syndrome critical region. (B) Breakpoint map with 2.75-Mb average intervals. (C) Pulsed-field gel electrophoresis (PFGE) map showing *Not*I restriction enzyme cutting sites. (D) Linkage map with 1.5-cM average interval, 67 cM total length, and 1 cM = 500 kb. (E) Radiation hybrid map with 230-kb average interval, 1562 centivads (cR) total length, and 1 cR = 24 kb. (F) Physical map incorporating 196 STSs and 810 YACs, with average spacing 220 kb. (G) Sequence-ready contigs in bacterial clones (Antonarakis, 1998, used by permission, Academic Press).

be studied with the standard methods (Cantor et al., 1988). Large numbers of human DNA fragments hundreds of kilobase pairs in length can now be cloned in YACs (Schlessinger, 1990) and bacterial artificial chromosomes (BACs; Osoegawa et al., 1998), with the goal of placing these in a contiguous order, one overlapping the next to a small extent. An overlapping array of cloned fragments is called a *contig*. A contig map is a physical map showing the linear order and locations of overlapping DNA fragments cloned in YACs, BACs, or cosmids (Fig. 29.3). The construction of contigs has increased the speed at which geneticists can find genes of interest. Initially, contigs were short, perhaps just a few hundred kilobase pairs long. Now, many are megabase pairs long, and ultimately there will be one enormous contig representing each chromosome. Even short contigs are useful in mapping studies and identification of minute structural changes in various diseases.

By now, YAC contigs have been constructed for much of the genome, although there are still many gaps (Chumakov et al., 1995). Genes are not distributed at random (Chapter 6), and the gene-rich, highly GC-rich T-bands have about 12 times the gene density of the GC-poor G-bands. The YACs containing highly GC-rich DNA tend to be unstable, and a large fraction of the gene-rich regions is not yet represented in the existing YAC contigs (Zoudak et al., 1996). BACs with inserts over 200 kb long can now be produced (Osoegawa et al., 1998), and these may be more useful in studying GC-rich regions. The YACs from less GC-rich regions are reasonably stable. Perhaps the first YAC contig constructed included the entire 2.4-Mb Duchenne muscular dystrophy (*DMD*) gene (Monaco et al., 1992). A low-resolution (100 kb) YAC/STS map of a 15-Mb region in Xp21.3–p11.3 that includes the DMD gene has been constructed (Nagaraja et al., 1998). It contains only eight known genes, one per 2000 kb. Zoudak et al. (1996) estimated that the most gene-poor regions would have, on average, 1 gene per 150 kb. In contrast, a 220-kb region in Xq28, one of the most GC-rich bands, contains 13 known genes and 6 candidate genes, or one per 11–12 kb. This is close to the 1 per 9 kb they estimated for such bands.

The Human Genome Project

The Human Genome Project was originally based on the notion that the most efficient and comprehensive way to complete our genetic map was to sequence the entire genome: all 3.4 billion base pairs of it. Only in this way could one be

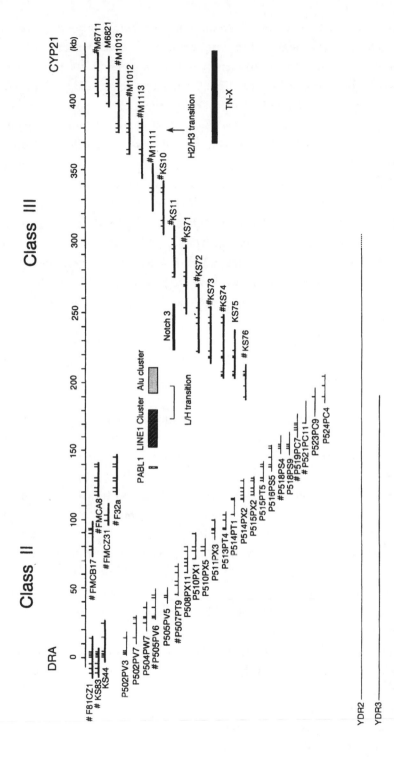

Figure 29.3. A 440-kb contig map of overlapping cosmid, λ phage, and YAC clones (horizontal lines) that cover the junction of *MHC* (major histocompatibility complex) classes II and III regions on chromosome 6. Vertical lines in clones: *Eco*RI sites. A pseudoautosomal boundary-like site (PABL) is near the L/H transition, the boundary between an AT-rich L isochore and a more GC-rich H isochore. A nearby LINE1 cluster is in the L isochore, and a nearby *Alu* cluster is in the H isochore. The site of an H2/H3 isochore transition is also shown (Fukugawa et al., 1995, with permission, Academic Press).

assured that every gene was identified and its neighborhood characterized. However, knowing the sequence is not enough: One must be able to recognize potential new genes, that is, to identify unknown genes by their sequence characteristics. Protein-coding genes have a series of open reading frames (*exons*) in which successive nucleotide triplets (*codons*) specify amino acids rather than calling for a stop in translation (*stop codon*). Most genes have multiple exons, separated by noncoding intervening sequences (*introns*), so how does one determine where one gene starts and where it ends? Fortunately, there are a number of sequence features that simplify this problem and make it possible to search for potential genes in a long, anonymous DNA sequence, but problems remain. A growing number of genes have been identified that do not encode a protein; their RNA transcript is the functional product. These genes, such as the *XIST* gene (Chapter 18), cannot be identified in a search for open reading frames. Another daunting problem in working from sequence data is that one still has to determine whether a potential gene indeed has a function and then characterize it.

An exciting alternative approach has been to start with expressed sequences, by isolating messenger RNA (mRNA) from many different types of cells and using reverse transcriptase to convert the mRNAs into a complementary DNA (cDNA) library for each type of cell. Initially, this approach was not very effective, because a few very abundant messages predominated in these libraries and mRNAs of low abundance were rarely seen. This problem has been overcome by eliminating most of the overabundant messages, thus producing normalized cDNA libraries representing virtually all of the mRNAs expressed in each type of cell. By sequencing a short segment of each cDNA (usually from the unique 3' untranslated region), one obtains an *expressed sequence tag* (EST) that serves as a molecular marker for a gene. Over 600,000 ESTs have been produced and are now in public databases (Marra et al., 1998). Many of them are redundant, representing the same gene, but as ESTs are produced from still more cell types, most genes that produce a protein product will be included in these libraries. There are an estimated 50,000–80,000 human genes (Fields et al., 1994; Antiqua and Bird, 1995).

The Human Genome Project was launched in 1990 by the United States Department of Energy and the National Institutes of Health and was soon joined by comparable efforts in other countries. Collins et al. (1998) reported that the Human Genome Project had successfully met all the goals of its current 5-year plan. A listing of these goals shows where we stand and how broad the goals really are; sequencing is only one element in this plan for maps of the human genome:

1. A genetic linkage map of each chromosome, with markers spaced an average of 1 cM apart;
2. A physical map of each chromosome, containing 52,000 STSs (sequence tagged sites);
3. DNA sequencing: 180-Mb human and 111-Mb model organisms;
4. Sequence technology: 90 Mb/year @ $0.50 per base;
5. Gene identification: 30,000 ESTs, mapped;
6. Model organisms: Fraction of genome sequenced: *E coli* (100%), *S cerevisiae* (100%), *C elegans* (80%), *Drosophila* (9%).

The projection for the future is that sequencing the more than 3-billion-bp human genome will be completed by the year 2003, 2 years ahead of schedule. The positions and sequence of most genes should be revealed by these coordinated mapping efforts.

References

Antiqua F, Bird A (1995) Number of CpG islands and genes in human and mouse. Proc Natl Acad Sci USA 90:11995–11999

Antonarakis SE (1998) Review: 10 years of *Genomics*, chromosome 21, and Down syndrome. Genomics 51:1–16

Boldrug SE, Ray PN, Gonzalez IL, et al. (1987) Molecular analysis of a constitutional X-autosome translocation in a female with muscular dystrophy. Science 237:1620–1624

Bouzyk M, Bryant SP, Evans C, et al. (1997) Integrated radiation hybrid and yeast artificial chromosome map of chromosome 9p. Eur J Hum Genet 5:299–307

Breuning MH, Berg-Loonen EM van den, Bernini LF, et al. (1977) Localization of HLA on the short arm of chromosome 6. Hum Genet 37:131–139

Broman KW, Murray JC, Sheffield VC, et al. (1998) Comprehensive human genetic maps: individual and sex-specific variation. Am J Hum Genet 63:861–869

Cantor CR, Smith CL, Matthew MK (1988) Pulsed-field gel electrophoresis of very large DNA molecules. Annu Rev Biophysics Biophys Chem 17:41–72

Chumakov IM, Rigault O, Le Gall I, et al. (1995) A YAC contig map of the human genome. Science 270:1945–1954

Collins FS, Patrinos A, Jordan E, et al. (1998) New goals for the U.S. human genome project: 1998–2003. Science 282:682–689

Deloukas P, Schuler GD, Gyapay G, et al. (1998) A physical map of 30,000 human genes. Science 282:744–746

Dib C, Fauré S, Fizames C, et al. (1996) A comprehensive genetic map of the human genome based on 5264 microsatellites. Nature 380:152–154

Donahue RP, Bias WB, Renwick JM, et al. (1968) Probable assignment of the Duffy blood group locus to chromosome 1 in man. Proc Natl Acad Sci USA 61:949–955

Erlich HA, Gibbs R, Kazazian HH Jr (eds) (1989) Current communications in molecular biology: polymerase chain reaction. Cold Spring Harbor Laboratory, Cold Spring Harbor, New York

Fields C, Adams MD, White O, et al. (1994) How many genes in the human genome? Nat Genet 7:345–346

Fukugawa T, Sugaya K, Matsumoto K-I, et al. (1995) A boundary of long-range G + C% mosaic domains in the human MHC locus: pseudoauto-somal boundary-like sequence exists near the boundary. Genomics 25: 184–191

Gyapay G, Schmitt K, Fizames C, et al. (1996) A radiation hybrid map of the human genome. Hum Mol Genet 5:339–346

Kerem B-S, Rommens JH, Buchanan JA, et al. (1989) Identification of the cystic fibrosis gene: genetic analysis. Science 245:1073–1080

Kinzler KW, Nilbert MC, Su LK, et al. (1991) Identification of FAP locus gene from chromosome 5q21. Science 253:661–665

Leach RJ, O'Connell P (1995) Mapping of mammalian genomes with radiation (Goss and Harris) hybrids. Adv Genet 33:63–99

Li H, Gyllensten UB, Xui X, et al. (1988) Amplification and analysis of DNA sequences in single human sperm and diploid cells. Nature 335:414–417

Magenis RE, Hecht F, Lovrien EW (1970) Heritable fragile site on chromosome 16: probable localization of haptoglobin locus in man. Science 170:85–87

Marra MA, Hillier L, Waterston RH (1998) Expressed sequence tags—ESTablishing bridges between genomes. Trends Genet 14:4–7

McKusick VA (1997) Mendelian inheritance in man, 12th ed, Johns Hopkins, Baltimore

Miller DA, Miller OJ, Dev VG, et al. (1974) Human chromosome 19 carries a poliovirus receptor gene. Cell 1:167–173

Miller OJ, Allderdice PW, Miller DA, et al. (1971) Human thymidine kinase gene locus: assignment to chromosome 17 in a hybrid of man and mouse cells. Science 173:244–245

Mohrenweiser HW, Tsujimoto S, Gordon L, et al. (1998) Regions of sex-specific hypo- and hyper-recombination identified through integration of 180 genetic markers with the metric physical map of human chromosome 19. Genomics 47:153–162

Monaco AP, Walker AP, Millwood I, et al. (1992) A yeast artificial chromosome contig containing the complete Duchenne muscular dystrophy gene. Genomics 12:465–473

Muir WJ, Gosden CM, Brookes AJ, et al. (1995) Direct microdissection and microcloning of a translocation breakpoint region t(1;11)(q42.2;q21), associated with schizophrenia. Cytogenet Cell Genet 70:35–40

Nagaraja R, Jermak C, Trusynich M, et al. (1998) YAC/STS map of 15 Mb of Xq21.3–q11.3, at 100 kb resolution, with refined comparisons of genetic distances and DMD structure. Gene 215:259–267

Osoegawa K, Woon PY, Zhao B, et al. (1998) An improved approach for construction of bacterial artificial chromosome libraries. Genomics 52:1–8

Pangilinan F, Li Q, Weaver T, et al. (1997) Mammalian BUB1 protein kinases: map positions and in vivo expression. Genomics 46:379–388

Ricciuti FC, Ruddle FH (1973) Assignment of three gene loci (PGK, HGPRT, and G6PD) to the long arm of the human X chromosome by somatic cell genetics. Genetics 74:661–678

Ruddle FH, Creagan RP (1975) Parasexual approaches to the genetics of man. Annu Rev Genet 9:407–486

Schlessinger D (1990) Yeast artificial chromosomes: Tools for mapping and analysis of complex genomes. Trends Genet 6:248–258

Wijmenya C, Heuvel LPWJ van den, Strengman E, et al. (1998). Localization of the ICF syndrome to chromosome 20 by homozygosity mapping. Am J Hum Genet 63:803–809

Zoudak S, Clay O, Bernardi G (1996) The gene distribution of the human genome. Gene 174:95–102

30

Genome Plasticity and Chromosome Evolution

A ll life as we know it traces back to a single common ancestor. The diversity of living organisms could not have been achieved in the 3–4 billion years they have existed on earth without a high level of genome plasticity. Here we will explore some of the aspects of this plasticity that are most relevant to understanding the behavior of human chromosomes. The common origin of all life forms was strongly supported by Charles Darwin, whose theory of evolution by natural selection provided a powerful explanation for the enormous diversity of living organisms. Genetics and molecular biology have confirmed this unity, demonstrating that all organisms store their genetic information in DNA or RNA and use a virtually universal genetic code for translating this information into protein sequences. Genes, too, show a remarkable degree of conservatism. As an example, over 70 human genes are already known that can function in yeast, substituting for the corresponding defective yeast gene (National Center for Biotechnology Information, http://www.ncbi.nim.nih.gov/Bassett/cerevisiae/index.htmi). A recent study in *Drosophila* used saturation mutagenesis of a 67-kb region to identify 12 new expressed genes. Nearly all these genes had close relatives in the human and round worm (*Caenorhabditis elegans*) databases. Half were present in the yeast (*Sacchiromyces cerevisiae*) database, and a few were even present in the bacterial databases (Maleszka et al., 1998). This level of sequence conservation is remarkable, especially since warm-blooded birds and mammals have evolved a rather different genome organization, marked by increased heterogeneity in base composition, with highly GC-rich and GC-poor isochores and attendant changes in codon usage (Bernardi, 1995; see also Chapter 7).

Sequencing the genome of the round worm, *C. elegans*, is essentially complete, with only minor gaps (The *C. elegans* Sequencing Consortium, 1998). The 97 million–bp genome contains more than 19,000 genes, with considerably more sequence similarity of the inferred proteins to human proteins than to those of yeast or bacteria. Analyses of amino acid sequences of proteins and nucleotide sequences of DNA are now widely used to clarify the phylogenetic relationships between different species and larger taxonomic groups. Ribosomal RNA gene sequence comparisons favor a complex, partially parasitic origin of eukaryotes (including humans) from eubacteria, archebacteria, and an unknown third source. Comparisons of the complete genomic sequences of 12 bacterial species and several yeast (eukaryotic) species has confirmed this and pointed out probable features of this third type of ancestor. Since neither eubacterian nor archebacteria have gene sequences resembling the cytoskeletal protein genes of eukaryotes, it is likely that the third life form from which eukaryotes also arose already had a cytoskeleton (Doolittle, 1998).

Genome Plasticity

The evolution of eukaryotes, and particularly multicellular metazoans, from simpler life forms has involved marked increases in genome size through processes that include gene duplication, transposition, insertion, exon shuffling, and genomic rearrangement. Gene duplication leads to an additional copy of a gene, usually as a direct tandem repeat. Unequal crossing over can lead to a third copy, and so on. For example, the *CYP2D6* gene exists as 1–13 tandem copies in different ethnic groups (Lundquist et al., 1999). Subsequent mutation may inactivate one or more of the copies or may lead to slightly altered function, so that a multigene family is created. Surprisingly, about half of all gene duplications appear to lead to functional divergence rather than inactivation of the extra copy (Nadeau and Sankoff, 1997). Both outcomes are possible within a single duplicated gene cluster. For example, the interferon (*IFN*) gene family at 9p21–p22 consists of 15 functional genes and 11 nonfunctional pseudogenes (Strissel et al., 1998). Gene amplification may also contribute to evolution, producing multiple copies of a gene that may remain at the same locus or be widely distributed throughout the genome. Thus, the keratinocyte growth factor multigene family emerged suddenly about 5–8 million years ago in the lineage that produced humans, chimpanzees, and gorillas. It arose by 16-fold amplification of part of a 3-exon gene containing exons 2 and 3 (Kelley et al., 1992).

Genome organization fosters rapid evolution in several ways. One is the behavior of introns. Most genes are made up of multiple short coding sequences (exons), averaging 120 bp long, separated by longer, noncoding sequences (introns). Gilbert has long maintained that exons can shuffle from one gene to another as a result of a chromosomal mechanism, ectopic recombination between introns, and that this has permitted the rapid evolution of eukaryotes. Another mechanism of potentially major importance involves L1 LINE (LINE1) elements. These retrotranspose at rather high frequency (they carry their own reverse transcriptase, which favors this), and can insert into an intron of a gene, carrying with them whatever sequence is at their 3' flank. This enables them to mobilize non-L1 sequences, including exons, and add them to an existing gene, where they will function as an additional exon (Moran et al., 1999). Each exon is usually, but not always, a functional domain, and exon shuffling can lead to the production of a protein with additional functional domains (e.g., DNA binding, kinase activity, hormone binding). Gilbert's group has estimated that all proteins that now exist on earth arose from no more than 7000, and possibly as few as 1000, exons, with exon shuffling greatly speeding up the evolution of proteins with new functions (Dorit et al., 1990). Many proteins, especially those in multigene families, are multifunctional, possibly as a result of exon shuffling. For example, the roles the imprinted *IGF2* (insulin-like growth factor 2) gene product plays in glycoprotein transport, growth and development, and tumor suppression involve three different functional domains (Desouza et al., 1997).

The presence of repeated sequences, especially those only a short distance from one another on a chromosome, greatly increases the risk of illegitimate ectopic pairing and recombination between sequences that are almost identical, that is, homologous recombination between nonsyntenic regions (Fig. 14.2). Such unequal crossing over can disrupt a functional gene; create a new gene; or

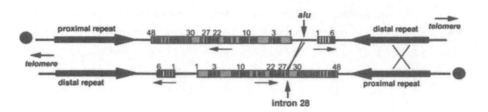

Figure 30.1. Rearrangement involving mispaired inverted repeats. Some of the exons (black) of the large *FLN1* gene and the small *emerin* gene are numbered for orientation. Arrows indicate direction of transcription. Black circles: centromeres. The inversion was mediated by unequal (ectopic) recombination between one *Alu* element between the two genes and one within intron 28 of *FLN1* (Small et al., Nat Genet 16:97, 1997).

lead to deletion, duplication, or rearrangement, as described in earlier chapters. It is worth emphasizing that the presence of inverted repeats can lead to inversion polymorphisms. Emerin is a lamin-associated nuclear membrane protein that is defective in the Emery-Dreifuss type of muscular dystrophy. The 48-kb region containing the *filamin* and *emerin* genes in Xq28 is flanked by 11-kb inverted repeats that recombine so often that the two repeats are virtually identical and the *filamin/emerin* region is in an inverted state in 33% of normal females (Fig. 30.1). Naturally, this affects recombination frequencies and the linkage map. How many such polymorphisms remain to be discovered throughout the human genome?

Repeated sequences are found in all eukaryotes. Mechanisms for reducing the frequency of illegitimate crossing over between them developed, perhaps hundreds of millions of years ago. The *MSH3* gene reduces microsatellite instability by detecting DNA mismatches in organisms as diverse as humans and yeast. Cells lacking a functional *MSH3* gene show microsatellite instability and hypermutability of a test gene, the *DHFR* locus. Two other genes involved in mismatch repair are *MSH2* and *MSH6*. The heterodimer complexes MSH2–MSH3 and MSH2–MSH6 detect mismatches. Introduction of a normal chromosome 5 (carrying the *MSH3* gene) into a uterine endometrial cancer cell line with defective *MSH3* and *MSH6* genes partially restores microsatellite stability (Drummond et al., 1997). A mechanism has also evolved for silencing (blocking) the transcription of long interspersed elements (LINEs) and short interspersed elements (SINEs): DNA methylation. The methylated DNA binds a specific protein, MCP2, which targets histone deacetylase to the DNA and converts it to heterochromatin, which is not transcribed (Chapter 5). This blocks the mobility of retrotransposons. In many cancers, there is widespread hypomethylation of DNA, and this is associated with greater movement of transposable elements (Yoder et al., 1997). Some interspecific hybrids show genome-wide under-methylation of DNA and associated activation of retrotransposons and chromosome remodeling; this could play a role in karyotypic evolution (O'Neill et al., 1998).

Evolution of the Autosomes

High-resolution chromosome banding and special FISH chromosome painting techniques have shown that very few chromosome rearrangements have occurred since the divergence of humans and great apes from a common ancestor some 5–8 million years ago. The only translocations are the fusion of two acrocentrics in the ancestor to produce human chromosome 2 and a t(5;17) translocation in the gorilla lineage (Wienberg et al., 1990). A few examples of activation of neocentromeres have been observed (Dutrillaux, 1979). Inversions account for more than 90% of the rearrangements. FISH analysis with YAC clones has led to the identification of five of the six evolutionary pericentric inversion breakpoints in human chromosomes 4, 9, and 12 (Nickerson and Nelson, 1998). It is interesting, though not surprising, that some of the inversions induced by radiation are the same as those that have been fixed (that is, become homozygous) during evolution (Dutrillaux et al., 1986).

Chromosomal rearrangements that reproduce the ancestral state of a genome are called *reverse chromosomal mutations*, and quite a number of these have been identified. For example, a paracentric inv(7)(q11.2q22.1) chromosome has the same banding pattern as the corresponding gorilla chromosome (Haaf and Schmid, 1987). Chromosome 7 is especially prone to inversion, accounting for 20% of all published cases. This may reflect the presence of hotspots for rearrangement that may be conserved over long evolutionary periods. The human 22q11 region is such a hotspot, and so is the homologous region in the mouse (Puech et al., 1997).

Some of the human chromosomes have been more highly conserved than others. The X chromosome and autosomes 6, 8, 11, 12, 18, and 19 are the most highly conserved, being little changed from their homologues among Old World monkeys (Clemente et al., 1990). Chromosomes 1, 3, and 7 are the least conserved and the most likely to show changes after radiation. More surprising is the evidence coming from comparisons with more distantly related species. The cat (2n = 38) and human (2n = 46) karyotypes obviously differ but nevertheless appear to have changed very little from their ancestral karyotype. Thus, the cat karyotype can be constructed from human chromosomes by postulating only seven chromosome breaks and one inversion (Rettenberger et al., 1995). A more comprehensive analysis including additional FISH mapping shows that human chromosome 11 is apparently identical to its cat homologue (O'Brien et al., 1997).

The origin of human chromosome 21 has been traced back 50 million years. It was formed after the divergence of platyrrhines (New World monkeys) and catarrhines (Old World monkeys, apes, and humans) but before the divergence of the Cercopithecidae (Old World monkeys) from apes and humans. Humans and Pongidae (great apes) have an equivalent chromosome 21, but in other catarrhines, a variety of translocations have produced quite different chromosomes. Only humans and Pongidae produce Down syndrome or its equivalent. The longer catarrhine and platyrrhine chromosomes containing human chromosome 21 material may be lethal when trisomic (Richard and Dutrillaux, 1998).

High-resolution banding and FISH mapping are excellent tools for evolutionary comparisons, but they do not detect the gross DNA differences between humans and great apes that are revealed by comparative genomic hybridization (Chapter 8) and interspecific representational difference analysis (Toder et al., 1998). Furthermore, they become increasingly unreliable as the evolutionary distance between species increases. Finer discrimination and greater reliability can be achieved using mapped genes and looking for regions of conserved synteny in species with fairly dense genetic maps, such as human and mouse (Ehrlich et al., 1997).

Evolution of the X and Y Chromosomes: Dosage Compensation

The mammalian sex chromosomes probably evolved from a pair of autosomes. One of the pair gained the male-determining *SRY* gene from the previous male-determining (proto-Y) chromosome and became the new Y chromosome, while its unchanged homologue became the X. During subsequent evolution, three further changes took place. Crossing over and recombination between X and Y was suppressed, most of the Y-linked genes were lost or inactivated by mutation, and a dosage compensation mechanism arose to equalize the expression levels of X-linked genes in the two sexes. This involved inactivation of most of the genes on one of the two X chromosomes in females. Doubling the expression level of genes expressed only from the single active X chromosome in both males and females equalized the expression levels of X-linked and autosomal genes (Ohno, 1967; Graves et al., 1998).

Ohno (1967) pointed out that mammalian X inactivation had certain consequences regarding the movement of genes between X and autosomes. The movement of any gene that shows a dosage effect would be likely to produce deleterious effects because of the altered gene dosage. We would now stipulate "unless the gene is imprinted in its autosomal location." Thus, the gene content of the X chromosome would change very slowly and be fairly constant in all mammals. This is generally true, but with significant exceptions. In monotremes and marsupials, the X chromosome makes up only 3% of the haploid genome, while in placental mammals it is 5%. Comparative mapping studies have shown that the long arm and proximal short arm of the human X correspond to the prototherian (monotreme) and metatherian (marsupial) X, and thus presumably to the ancestral mammalian X chromosome. The rest of the short arm represents one or more autosomal regions that have been added more recently to the eutherian X. The fusion point of the ancestral rearrangement maps to human Xp11.23 (Wilcox et al., 1996). In time, most of these transferred genes became subject to X-inactivation and the remaining pseudoautosomal region became shorter (Fig. 30.2; Perry et al., 1998).

Interesting secondary adjustments have been made to permit the successful translocation of some autosomal genes to the X. For example, the pyruvate dehydrogenase E1α subunit (PDHA) gene is autosomal in marsupials but X-linked in placental (eutherian) mammals. A single copy of the gene is present in the marsupial genome, but eutherians have a second, testis-specific, intronless copy that arose by reverse transcriptase action and retrotransposition to an autosome, just as happened with the phosphoglycerate kinase (PGK) gene (Fitzgerald et al., 1996). Both these genes are essential. The inactivation of the single X chromosome in eutherian males during spermatogenesis inactivates the PDHA and PGK genes on the X, but activation of the autosomal copies in spermatocytes enables these cells to function normally.

The functionality of a gene can sometimes be restored by recombination between two nonfunctional alleles that have mutations in different parts of the gene. However, this cannot occur in the absence of recombination. The suppression of crossing over between X and Y should thus have led to to gradual loss of virtually all genes from the Y not essential for sex determination or male

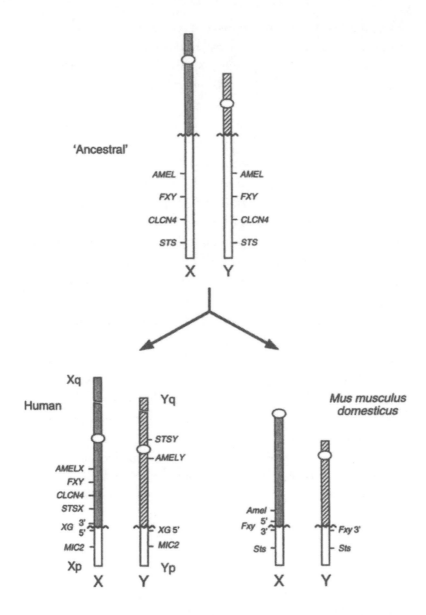

Figure 30.2. Model of the evolution of the human and mouse pseudoautosomal regions (PARs) on the X and Y chromosomes. The *XG* and *Fxy* genes span the pseudoautosomal boundaries (wavy lines) in humans and mice, respectively. X-unique (shaded) and Y-unique (hatched) regions are indicated (modified from Perry et al., 1998, Hum Mol Genet 7:299–305, with permission of Oxford University Press).

fertility. H. J. Muller pointed this out many years ago, and this phenomenon of inevitable gene degradation in the absence of recombination is referred to as *Muller's ratchet*. However, there is an interesting exception to this phenomenon: Some genes on the Y retain their function despite never crossing over with a homologue on the X. This includes several genes involved in male fertility.

The surprising feature of these genes involved in male fertility is that they are present in multiple copies on the Y. Since their deletion is associated with azoospermia, they are called *AZF* genes. *AZFb*, also known as *RBM*, was the first to be isolated. There are 20–50 *RBM* genes or pseudogenes scattered along both arms of the Y, with expression limited to the testis. The *AZFc* region on Yq contains the *DAZ* (deleted in azoospermia) multigene family, which is found only in Old World monkeys, apes, and humans. It appears to have arisen from the single-copy autosomal gene *DAZLA*, located at 3p24, which is expressed only in the germline of both testis and ovary and shares homology with *boule*, a gene whose mutation causes azoospermia in *Drosophila* (Agulnik et al., 1998). After transposition to the Y, *DAZ* underwent a complex series of duplications and rearrangements but remains as a cluster on Yq. This is deleted in 5–15% of infertile men with azoospermia, suggesting a functional role. Cooke and Elliott (1997) have suggested that the presence of multiple copies of these genes on the Y provides a means of forming normal copies from defective copies, by ectopic homologous recombination, thus enabling some Y-linked genes to escape Muller's ratchet.

In contrast to the relatively high level of conservation of the X chromosome, the mammalian Y chromosome shows remarkable diversity and rapid evolution among humans and higher primates, providing insights into human evolution and migration patterns. Most men worldwide have the same *DAZ* variants, supporting a recent (55,000–200,000 years ago) African origin of our species (Agulnik et al., 1998). The absence of significant variation in a 729-bp intron of the Y-linked *ZFY* gene in a worldwide sample of 38 males indicates either a very recent origin of our species, no more than about 270,000 years ago, or recurrent male population bottlenecks (Dorit et al., 1995). Comparative studies have shown that the present 2.6-Mb pseudoautosomal region on the short arm of X and Y chromosomes, PAR1, is only a remnant of a much larger pseudoautosomal region. For example, the ubiquitin activating enzyme (*UBE*) gene is present in the X-Y pairing segment in the platypus, a monotreme. It is not in that segment in marsupials or placental mammals, though it is still present on X and Y. The *UBE* gene is no longer on the Y in humans and other higher primates (Mitchell et al., 1998). Deterioration of genes on the Y chromosome is the driving force in this loss, as it is in sex chromosome evolution in general (Graves et al., 1998).

The human X and Y chromosomes have many blocks of similar DNA sequence in common. Some of these are located on Xq21 and Yp, regions that do not recombine during meiosis. This led to the idea that an inversion occurred during the evolution of the Y. Lambson et al. (1992) have reviewed the interchanges that occurred between X and Y chromosomes over the last 80 million years. A 4-Mb block on Xq is homologous to noncontiguous segments on Yp, suggesting that the segment was transposed from Xq to Y and later disrupted by a Y inversion. Schwartz et al. (1998) sequenced the breakpoints of this inversion and found LINE1 sequence elements, suggesting that ectopic recombination had occurred between two LINE1 elements, one inverted with respect to the other, in an otherwise nonhomologous region (see Fig. 30.1 for another example of this). Comparison with other primates indicated that the X-Y transposition occurred after the divergence of hominid and chimpanzee lineages. The 99.3% nucleotide identity between the X and Y chromosome segments places the transposition about 3–4 million years ago. Gläser et al. (1997) used multicolor FISH to map eight genes in Xp22 to prometaphase chromosomes. While the order was highly conserved on the X chromosome in all hominoid primates tested, there was a complex series of rearrangements affecting their order on the Y chromosome.

Evolution of Telomeric and Centromeric Regions

The human genome, like those of other mammals and of birds, has evolved a highly nonrandom arrangement of gene-rich and gene-poor isochores and bands (Bernardi, 1995). The most gene-rich isochores lie close to the telomeres of most chromosomes, regions noted for their high recombination frequencies. They are also preferred sites of origin of multigene families, such as the olfactory receptor family (Trask et al., 1998), and the sites where most minisatellite repeat elements are generated. In fact, these subtelomeric regions are sites of marked genomic plasticity and appear to serve as nurseries for generating new genes and enhancing genetic diversity.

The pericentromeric regions, with their abundant heterochromatin, are also regions of high genomic plasticity. Analysis of the sequences flanking the centromere of chromosome 10 found high levels of duplication, transposition, inversion, and deletion. Extrapolated to the whole 3400-Mb genome, perhaps 50 Mb of DNA in the regions closest to the centromeres is quite unstable (Jackson et al., 1999). These regions are preferred sites for integration or transposition of DNA sequences. Most of the transposed segments are truncated, nonexpressed pseudogenes, but some are complete genes. Eichler et al. (1997) described the duplication of a gene-rich cluster by transposition from Xq28 to the pericentromeric 16p11.1 region, a mechanism for genome evolution. They also observed transpositional duplication of a 9.7-kb segment of the adrenoleukodystrophy (ALD) gene from Xq28 into the pericentromeric regions of multiple chromosomes: 2p11, 10p11, 16p11, and 22q11. Their 92–96% sequence identity with the ALD gene indicates they probably arose only 5–10 million years ago, late in higher primate evolution.

Centromeres themselves appear to evolve rapidly. Most eukaryotic centromeres contain abundant repetitive sequences, but these vary markedly among species. Even within the human species, chromosome-specific variants of the centromeric α satellite are present on almost every chromosome (see Chapter 4). Neocentromeres also arise rather frequently, in evolutionary terms, by the sudden recruitment of totally different sequences. In humans, this is seen most frequently on the Y chromosome (Tyler-Smith et al., 1998).

References

Agulnik AI, Zharkikh A, Boettger-Tong H, et al. (1998) Evolution of the *DAZ* gene family suggests that Y-linked *DAZ* plays little, or a limited, role in spermatogenesis but underlines a recent African origin for human populations. Hum Mol Genet 7:1371–1377

Bernardi G (1995) The human genome: organization and evolutionary history. Annu Rev Genet 29:445–476

Clemente LC, Ponsa M, García M, et al. (1990) Evolution of the Similiformes and the phylogeny of human chromosomes. Hum Genet 84: 493–506

Cooke HJ, Elliott DJ (1997) RNA-binding proteins and human male fertility. Trends Genet 13:87–89

Desouza AT, Yamada T, Mills JJ, et al. (1997) Imprinted genes in liver carcinogenesis. FASEB J 11:60–67

Doolittle RF (1998) Microbial genomes opened up. Nature 392:339–342

Dorit RL, Akashi H, Gilbert W (1995) Absence of polymorphism at the ZFY locus in the human Y chromosome. Science 268:1183–1185

Dorit RL, Schoenbach L, Gilbert W (1990) How big is the universe of exons? Science 250:1377–1382

Drummond, JT, Genschel J, Wolf E, et al. (1997) *DHFF/MSH3* amplification in methotrexate—resistant cells alters the hMutSα/Hmutsβ ratio and reduces the efficiency of base—base mismatch repair. Proc Natl Acad Sci USA 94:10144–10149

Dutrillaux B (1979) Chromosomal evolution in primates: tentative phylogeny for Microcebus murinus (Prosimian) to man. Hum Genet 48:251–314

Dutrillaux B, Couturier J, Sabatier L, et al. (1986) Inversions in evolution of man and closely related species. Ann Génét 29:195–202

Ehrlich J, Sankoff D, Nadeau JH (1997) Synteny conservation and chromosome rearrangements during mammalian evolution. Genetics 147:289–296

Eichler EE, Budarf ML, Rocchi M, et al. (1997) Interchromosomal duplications of the adrenoleukodystrophy locus: a phenomenon of pericentromeric plasticity. Hum Mol Genet 6:991–1002

Fitzgerald J, Wilcox SA, Graves JAM (1996) A eutherian X-linked gene, PDHA1, is autosomal in marsupials: a model for the evolution of a second, testis-specific variant in eutherian mammals. Genomics 18:636–642

Gläser B, Grützner F, Taylor K, et al. (1997) Comparative mapping of Xp22 genes in hominoids—evolutionary linear instability of their Y homologues. Chrom Res 5:167–176

Graves JAM, Disteche CM, Toder R (1998) Gene dosage in the evolution and function of mammalian sex chromosomes. Cytogenet Cell Genet 80:94–103

Haaf T, Schmid M (1987) Paracentric inversion in human chromosome 7 as a graphic example of reverse chromosome mutation. Hum Evol 2:321–327

Jackson MS, Rocchi M, Thompson G, et al. (1999). Sequences flanking the centromere of human chromosome 10 are a complex patchwork of arm-specific sequences, stable duplications and unstable sequences with homologies to telomeric and other centromeric locations. Hum Mol Genet 8:205–215

Kelley MJ, Peck M, Seuanez HN, et al. (1992) Emergence of the keratinocyte growth factor multigene family during the great ape radiation. Proc Natl Acad Sci USA 89:9287–9291

Lambson B, Afara N, Mitchell M, et al. (1992) Evolution of DNA sequence homologies between the sex chromosomes in primate species. Genomics 14:1032–1040

Lundquist E, Johansson I, Ingelman-Sundberg M (1999) Genetic mechanisms for duplication and multiduplication of the CYP2D6 gene and methods for detection of duplicated CYP2D6 genes. Gene 226:327–338

Lupski JR (1998) Genomic disorders: structural features of the genome can lead to DNA rearrangements and human disease traits. Trends Genet 14:417–422

Maleszka R, Couet HG de, Miklos GLK (1998) Data transferability from model organisms to human beings: Insights from the functional genomics of the flightless region of Drosophila. Proc Natl Acad Sci USA 95: 3731–3736

Mitchell MJ, Wilcox SA, Watson JM, et al. (1998) The origin and loss of the ubiquitin activating enzyme gene on the mammalian Y chromosome. Hum Mol Genet 7:429–434

Moran JV, DeBerardinis RJ, Kazazian HH Jr (1999) Exon shuffling by L1 retro-transposition. Science 283:1530–1534

Nadeau JH, Sankoff D (1997) Comparable rates of gene loss and functional divergence after genome duplications early in vertebrate evolution. Genetics 147:1259–1266

Nickerson E, Nelson DL (1998) Molecular definition of pericentric inversion breakpoints occurring during the evolution of humans and chimpanzees. Genomics 50:368–372

O'Brien SJ, Wienberg J, Lyons LA (1997) Comparative genomics: lessons from cats. Trends Genet 13:393–399

Ohno S (1967) Sex chromosomes and sex-linked genes. Springer-verlag, New York.

O'Neill RJW, O'Neill MJ, Graves JAM (1998) Undermethylation associated with retroelement activation and chromosome remodelling in an interspecific mammalian hybrid. Nature 393:68–72

Perry J, Feather S, Smith A, et al. (1998) The human *FXY* gene is located within Xp22.3: implications for evolution of the mammalian X chromosome. Hum Mol Genet 7:299–305

Puech A, Saint-Jore B, Funke B, et al. (1997) Comparative mapping of the human 22q11 chromosomal region and the orthologous region in mice reveals complex changes in gene organization. Proc Natl Acad Sci USA 94: 14608–14613

Rettenberger G, Klett Ch, Zechner U, et al. (1995) ZOO-FISH analyses: cat and human karyotypes closely resemble the putative ancestral mammalian karyotype. Chrom Res 3:479–486

Richard F, Dutrillaux B (1998) Origin of human chromosome 21 and its consequences: a 50-million-year-old story. Chrom Res 6:263–268

Schwartz A, Chan DC, Brown LC, et al. (1998) Reconstructing hominid Y evolution: X-homologous block, created by X-Y transposition, was disrupted by Yp inversion through LINE-LINE recombination. Hum Mol Genet 7:1–11

Small K, Iber J, Warren ST (1997) Emerin deletion reveals a common X-chromosome inversion mediated by inverted repeats. Nat Genet 16:96–99

Strissel PL, Dann HA, Pomykala HM, et al. (1998) Scaffold-associated regions in the human type I interferon gene cluster on the short arm of chromosome 9. Genomics 47:217–229

The *C. elegans* Sequencing Consortium (1998) Genome sequence of the nematode *C. elegans*: a platform for investigating biology. Science 282:2012–2018

Toder R, Xia Y, Bausch E (1998) Interspecies comparative genome hybridization and interspecies representational difference analysis reveal gross DNA differences between humans and great apes. Chrom Res 6:487–494

Trask BJ, Friedman C, Martin-Gallardo A, et al. (1998) Members of the olfactory receptor gene family are contained in large blocks of DNA duplicated polymorphically near the ends of human chromosomes. Hum Mol Genet 7:13–26

Tyler-Smith C, Corish P, Burms E (1998) Neocentromeres, the Y chromosome and centromere evolution. Chrom Res 6:65–71

Wienberg J, Jauch A, Stanyon R, et al. (1990) Molecular cytotaxonomy of primates by chromosomal *in situ* suppression hybridization. Genomics 8:347–350

Wilcox SA, Watson JM, Spencer JA, et al. (1996) Comparative mapping identifies the fusion point of an ancient mammalian X-autosomal rearrangement. Genomics 35:66–70

Yoder JA, Walsh CP, Bestor TH (1997) Cytosine methylation and the ecology of intragenomic parasites. Trends Genet 13:335–340

31

The Future of
Human Cytogenetics

Continued improvements in cytogenetic technique have consistently
revealed chromosome aberrations not detectable by earlier methods. The
likeliest repository of additional aberrations in human populations remains the
same as before: spontaneous abortions, stillbirths, infants with multiple malformations, mentally retarded individuals, and cancers, examined using still newer
methods. For example, mental retardation affects up to 3% of the population.
Detectable chromosome imbalance accounts for a large percentage of severe
mental retardation (IQ <55) and perhaps 5–10% of milder retardation (IQ
55–70), but the cause is unknown in many cases. Flint et al. (1995) have taken
a novel approach, based on evidence that the subtelomeric regions are generally prone to recombination and that methods now exist for detecting unequal
recombinational events by probing the DNA of flow-sorted chromosomes with
polymorphic markers. They estimate that 6% of unexplained mental retardation

may be due to small chromosome changes in the subtelomeric regions alone. Better methods are also needed to identify paracentric insertions, which carry a 15% risk of duplications or deletions arising during meiotic segregation in carriers. Currently, as many as 90% of these insertions may be undetectable by present methods, based on their frequency in the two chromosomal regions where they can be most readily detected.

Unsolved Problems

Some of the most fundamental questions in human cytogenetics have yet to be answered. The role of telomere shortening in cell aging and carcinogenesis has become clear, but is telomere shortening also a major factor in other diseases of advancing age, such as non-insulin-dependent diabetes, atherosclerosis, and hypertension, and if so, how can its harmful effects be circumvented? What role do abnormalities of telomeres, centromeres, and the spindle play in nondisjunction? What are the causes of nondisjunction, and what is responsible for the profound maternal age effect?

How do trisomies, duplications, and deletions produce their phenotypic effects? That is, what are the critical genes responsible for these gene dosage effects? Cytogeneticists have approached the problem by trying to define the critical region for Down syndrome and for various duplication or deletion syndromes, and to use positional cloning to identify, in each critical region, specific genes with a major dosage effect. This could provide clues to the metabolic, signaling, or other pathways involved in the developmental errors. This exciting approach has already yielded insights into several deletion and imprinting syndromes and holds much promise. One should not forget that some genes act by suppressing other genes. The increased gene dosage associated with trisomy for a chromosome may exert its major effects by inhibiting genes elsewhere in the genome. An exciting example of this is the recent study of the synchrony of replication of the two alleles at four gene loci on three different autosomes (*MYC* on 8, *RB1* on 13, *TP53* and *HER2* on 17) in cultured amniotic fluid cells. In normal cells, the alleles at each of these four loci replicated synchronously, but in 21 trisomic cells they replicated asynchronously (Amiel et al., 1998). The presence of an earlier- and a later-replicating allele suggests that one allele has been inactivated. A reduction in gene dosage at these loci could be responsible for the increased cancer risk in Down syndrome. How many other genes are similarly affected in trisomy 21, and what features of Down syndrome are they responsible for? If trisomies or partial trisomies of other chromosomes also lead to replication asynchrony of particular genes at distant loci and reduced dosage of their products, our views on how chromosome imbalances lead to phenotypic abnormality could be profoundly changed.

What triggers late replication of both alleles of tissue-specific genes and of one allele of imprinted genes? What triggers X inactivation, and how do genes on the X chromosome or an attached autosomal segment escape inactivation? How are changes in the packaging of chromosomes brought about during cell differentiation, and how is this particular chromatin state maintained through mitosis? What are the evolutionary and mechanistic relationships between imprinting and X inactivation? The inactivation of most genes on one of the two X chromosomes in females equalizes the level of expression of these genes in males and females. A different dosage compensation mechanism is necessary to equalize the expression levels of genes on the single active X in both sexes and on the two expressed copies of most autosomal genes, but virtually nothing is known about this process. An attractive idea is that dosage compensation between X-linked and autosomal genes involves a doubling of the expression level of genes on the X (Ohno, 1967). Graves et al. (1998) have reviewed this topic and presented impressive evidence. The level of expression of the *Clc4* gene is twice as high in *Mus spretus*, in which it is on the X chromosome, as it is in *Mus musculus*, in which it is on an autosome. It is intriguing that dosage compensation between XY and XX in *Drosophila* is mediated the same way, suggesting that this very poorly understood autosome: X dosage compensation mechanism is far more highly conserved than anyone realized, and providing a lead for future research in humans.

Genome Organization

Some genes with closely related functions, which probably arose by duplication events, have maintained their contiguous locations throughout their very long evolutionary histories. A long-standing question in cytogenetics is whether the map position of any gene simply reflects its evolutionary history or whether there are functional constraints that maintain some tightly linked groups of genes. In general, the former seems to be the case, even for genes of similar function, such as cyclins or cyclin-dependent kinases (Chapter 2). The more than 80 genes encoding the structural proteins of the ribosome are widely scattered throughout the genome, even though they are coordinately regulated (Feo et al., 1992).

Table 31.1. Gene Clusters Maintained by Functional Constraints

Gene cluster	Symbol	Size (kb)	Location	Type of band
Hemoglobin α-globin	HBBA	70	16p13.3–pter	R (T)
Hemoglobin β-globin	HBBC	65	11p12	G
Homeobox A	HOXA	>100	7p15.3	R
Homeobox B	HOXB	>100	17q21.3	R
Homeobox C	HBAC	>60	12q13.3	R
Homeobox D	HOXD	>80	2q31	R
Major histocompatibility	MHC	>2000	6p21.3	R
Immunoglobulin κ light chain	IGK	160	2p11.2–p12	—
Immunoglobulin λ light chain	IGL	140	22q11.12	R
Immunoglobulin heavy chain	IGH	1200	14q32.33	R (T)
T-cell receptors α and δ	TCRA/D	140	14q11–q12	—
T-cell receptor β	TCRB	840	7q32–q33	—
T-cell receptor γ	TCRG	120	7p15	R
Odorant receptor family on 7		>100	7p22	R (T)
Odorant receptor family on 16		>100	16p13	R (T)
Odorant receptor family on 17		>100	17p13	R (T)
Imprinted cluster on 11			11p15.5	
Imprinted cluster on 15			15q11–q13	

However, there are several striking examples in which the position of a gene within a cluster is highly correlated with temporal and tissue-specific patterns of expression (Table 31.1). The best known may be the α- and β-globin gene clusters on chromosomes 16 and 11, respectively.

The genes at the 5′ end of each globin cluster are expressed earlier in development, and the genes at the 3′ end are expressed later. Each gene has its own promoter, but each cluster is under the control of a *locus control region* (LCR) that determines its chromatin structure, time of replication, and expression pattern. Tanimoto et al. (1999) used the powerful Cre-loxP recombination technique to invert either the human β-globin cluster of five genes or the LCR itself (Fig. 31.1). The ε-globin gene, normally at the 5′ end (nearest the LCR), is no longer expressed during the yolk-sac stage of erythropoiesis when it is separated from the LCR, and LCR activity itself is markedly diminished if the LCR is inverted. A similar mechanism may underlie the maintenance of small gene clusters in imprinted regions: Upstream regulatory elements must be maintained in addi-

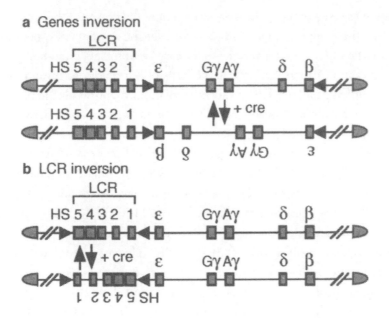

Figure 31.1. Gene manipulation: inversion of either the five-gene β-globin gene cluster (a) or the locus control region (LCR) (b), used to demonstrate the importance of order within each of these regions for gene function (reprinted with permission from Nature 398, p 345, Tanimoto et al., copyright 1999, Macmillan Magazines Limited).

tion to the genes themselves (Chapters 21 and 22). Homeobox (*HOX*) genes play critical roles in early embryogenesis. The four *HOX* gene clusters, each with about 8–11 contiguous *HOX* genes, show a very tight temporal and spatial correlation between gene order and function; that is, the genes are activated sequentially, from one end of each array to the other. Fine mapping of the four clusters places them all in R-bands (Apiou et al., 1996).

A different functional requirement has maintained the very large immunoglobulin heavy- and light-chain gene clusters and the T-cell receptor gene clusters. Here, site-specific breakage, elimination of much DNA, and V(D)J rejoining produce the exceptionally large number of antibody specificities and histocompatibility antigen cell surface receptor specificities so important for our resistance to infectious disease and cancer (Tonegawa, 1983). How many more constraints on genome organization will be found? Will their study reveal new genetic mechanisms that are as unexpected as imprinting?

A candidate for just such a discovery is the olfactory (odorant) receptor gene family, first identified in 1991 in the rat by Buck and Axel and then in 1992 in humans by several groups. There are several hundred copies of these genes. Nevertheless, there appears to be tight control over the entire family, because each olfactory neuron expresses very few olfactory receptor genes and only a single allele of each (Chapter 22). This is the basis of our ability to distinguish thousands of different odors. This complex regulation might involve imprinting the entire gene family in both oogenesis and spermatogenesis and activation of a single allele at a single locus in each olfactory neuron, but how is the activation of a different olfactory receptor gene in each cell achieved? An explanation would be at hand if all the family members were in a single long tandem array. However, most of the potentially functional copies are in tandem arrays hundreds of kilobase pairs long on chromosomes 7, 16, and 17. Additional smaller clusters are present at more than 20 locations throughout the genome, most of them in terminal bands (Rouquier et al., 1998). This may be related to another peculiar feature of this gene family, the presence of a single open reading frame (exon) in each olfactory receptor gene. No intron divides the exon, which is about 1 kb long. Intronless genes usually arise from genes with introns by reverse transcription of a processed mRNA and random reintegration into the genome. Such an origin might account for the widespread distribution of the gene family, but the precise regulation of these genes remains a mystery.

Genetic recombination is suppressed in a few regions of the genome. Genes in these regions are said to show *linkage disequilibrium*. One of these regions is on 17q21. It contains the *RNU2* array of 6–30 copies of the *U2* small nuclear RNA genes and the adjacent 175-kb region containing the tandemly duplicated *NBR1*, *LBRCA1* pseudogene, *NBR2*, and *BRCA1* genes. This 200 to 400-kb region shows complete suppression of crossing over, and only two major haplotypes (different allelic forms) of this region were observed in 275 Europeans and 34 Asians, indicating that crossover suppression in this region has persisted for at least the last 100,000 years (Liu and Barker, 1999). This region is rather close to the *HOXb* locus, but no functional relationship between the two clusters or their behavior is known.

Methods are needed to enable recognition of functionally important gene clusters. Close similarity in sequence (reflecting a common origin) will be readily identifiable, once the Human Genome Project is completed and the complete nucleotide sequence of every chromosome is known. However, genes of quite different sequences may remain closely linked because of shared regulatory controls, such as a promoter or LCR. Working out additional features of genome organization that delineate functional domains is a task for the future.

Mapping the human genome is still far from complete, and this is particularly so for locating and identifying all the transcribed sequences (genes). Yeast artificial chromosomes (YACs) may become a powerful tool for analyzing closely linked genes or gene clusters. All the human genes in a YAC are transcribed in yeast cells, which lack the tissue-specific regulatory systems that suppress the expression of most genes in human cells (Still et al., 1997). This technique provides a unique approach to identifying all the genes in a particular region and may be generally useful in completing our genetic map.

The highly nonrandom organization of the human genome presents the cytogeneticist with a bewildering array of unanswered questions. How did the 17-fold greater density of genes in the most GC-rich chromosome bands arise, how is it maintained, and what is its functional significance? Why are genes concentrated in subtelomeric locations, and is that concentration related to the high frequency of crossing over in these regions or to the fact that minisatellites almost always originate in these regions? The most gene-rich region yet defined is less than 50 kb long, in band 9q34.2. This region, called the *surfeit* locus, contains six housekeeping genes that have no sequence similarity or functional similarity. Five of the genes are separated from each other by tiny spacer sequences 97–302 bp long. Even the gene order is conserved in mammals and birds (Duhig et al., 1998), although there is no evidence that this particular gene cluster is maintained by functional constraints such as those involved in the globin, immunoglobulin, T-cell receptor, and homeobox genes. However, tighter clustering has been achieved by sharing CpG islands, so that four rather than six of these are sufficient. That is, there is a CpG island at the 5' end of each gene; the middle two CpG islands are each shared by two genes. This has the interesting consequence that the two genes sharing one CpG island are then transcribed in opposite directions (from opposite strands). Does such sharing have implications for gene regulation, and is there a preferred (nonrandom) orientation of transcribed strands along a chromosome?

Directions

The enormous successes of molecular cytogenetics in the last decade provide a clear indication of a number of developments to be expected in the future. Contigs based on ordered, overlapping cloned DNA fragments propagated in cosmids and bacterial and yeast artificial chromosomes will facilitate rapid mapping of new disease genes and their identification by positional cloning, aided by libraries of ESTs (expressed sequence tags) arrayed on DNA chips. Normalized cDNA libraries, containing nearly equal representation of all expressed sequences from sequential developmental stages, will speed understanding of the temporal and tissue specificity of many genes and perhaps throw light on the functional significance of chromosome organization. There will be a steady increase in the number of painting and other probes for specific chromosome segments, bands, and the breakpoint regions of specific rearrangements that cause cancer.

The introduction of human genes, DNA segments, or artificial chromosomes into transgenic or transgenomic mice may permit in vivo analysis of the effects of certain trisomies, duplications, and deletions on differentiation, providing needed animal models. Portions of various human chromosomes have been introduced into mouse embryonic stem cells by microcell-mediated cell fusion/chromosome transfer. Viable chimeric mice have been produced from these, and some of the transferred genes are expressed in their correct tissue-specific way in the adult tissues (Tomizuka et al., 1997). Methods are being developed for the conditional silencing of target genes. Unlike the constitutive silencing produced by standard gene targeting (gene knockout) techniques, conditional silencing methods will permit a gene to be turned off at a specific time and in a specific cell type (Porter, 1998). This could permit the study of otherwise lethal genes and those producing severe birth defects, as in various trisomies or deletions.

New Technology

Improvements in microscopes, development of *charge-coupled device* (CCD) cameras and computer-controlled digital image capture of very faint fluorescent signals, and use of computer interfaces that permit optical sectioning and three-dimensional reconstruction of nuclei have greatly expanded the usefulness of the cytogeneticist's classic instrument, the light microscope. This has been possible only because of equally impressive developments in the preparation of molecular probes, the DNA fragments essential for in situ hybridization. Continued developments in these areas are expected, such as an expansion of interphase and preimplantation cytogenetics.

More revolutionary advances are on the horizon, with some already coming into use. DNA chip technology is one of these. Using the techniques developed for making silicon microchips for computers, it is possible to prepare many copies of orderly arrays of tens of thousands of DNA fragments or oligonucleotides on a small nitrocellulose filter, glass slide, or silicon chip. The power of this approach is illustrated by the discovery of hundreds of cell cycle–regulated genes in yeast (Spellman et al., 1998). There are now 800 known, in contrast to 104 in 1997. Many, if not most, of these genes that are required only once per cell cycle will have human homologues that can now be rather quickly identified.

The same chip technology has been used to make microelectrophoresis channels that can be filled with a replaceable gel and used to type short tandem repeat genetic markers in just 30 seconds (Schmalzing et al., 1997). This could greatly speed the mapping of translocation breakpoints, the identification of new disease genes, and the understanding of multifactorial disorders like diabetes and hypertension. Chip arrays can be used to screen quickly and cheaply for specific rearrangements that cause cancer (Hacia et al., 1998). A microarray of 10,000 cDNAs could analyze the expression of thousands of genes in a single experiment, creating a need for new types of data management and analysis (Ermolaeva et al., 1998).

The Human Genome Project and its rapidly growing body of users are creating ever more massive databases. Their use has been greatly facilitated by ready Internet access. Frequent reference to these databases will become an essential part of keeping abreast of relevant advances, for both research and clinical cytogenetics laboratories. Bibliographic information can be obtained from the National Library of Medicine: PUBMED at http://www.ncbi.nlm.nih.gov. This is perhaps most easily accessed through the NCBI Entrez Browser at http://www.ncbi.nlm.nih.gov/Entrez/. The Online Mendelian Inheritance in Man (OMIM) Home Page, at http://www.ncbi.nlm.nih.gov/Omim, is a continuously updated catalogue of human genes and genetic disorders. The Genome Data Base (GDB) at http://www.gdb.org/ has molecular and mapping data.

Advances in cell culture will include discovery of mitogens for additional types of differentiated G0 cells, enabling them to enter mitosis. Fibroblast growth factor 4 (FGF4) has recently been shown to permit the development of permanent placental trophoblast stem cell lines that can differentiate into trophoblast subtypes (Tanaka et al., 1998). The introduction of genes for positive and negative selection of specific chromosomes or chromosome segments should make it possible to develop cell lines with specific deletions or monosomies. These could be used to study altered gene dosage or regulation, to show which aneuploidies are not viable at a cellular level, and to learn the reason. An immortal testicular Sertoli cell line has been produced in the mouse that supports the differentiation of meiotic and even postmeiotic cells from premeiotic germline cells (Rassoulzadegan et al., 1993). The ability to induce human germline cells to follow this meiotic pathway would provide a powerful impetus to the analysis of causes of nondisjunction and the mechanism of imprinting. Human pluripotent stem cell lines have recently been developed (Thomsen et al., 1998). Their potential use in basic research and cell or gene therapy will depend upon significant input from cytogeneticists.

Human artificial chromosomes have recently been created. Mitotically stable minichromosomes only 4–9 Mb long have been generated from the Y chromosome by telomere-directed chromosome breakage (Heller et al., 1996). Completely artificial chromosomes of comparable length have been produced from telomeric DNA, genomic DNA, and synthetic α-satellite arrays from chromosomes 17 and Y or chromosome 21; these are mitotically and cytogenetically stable (Harrington et al., 1997; Ikeno et al., 1998). Such artificial chromosomes may become useful for long-term correction of genetic defects, by serving as vehicles for the stable introduction of a functional copy of the missing gene. They could also be very useful research tools for investigating causes of nondisjunction in cell culture systems. A microchromosome that arose in a man with the CREST form of scleroderma was transmitted to a normal daughter, indicating meiotic as well as mitotic stability for such chromosomes (Haaf et al., 1992).

References

Amiel A, Avivi L, Gaber E, et al. (1998) Asynchronous replication of allelic loci in Down syndrome. Eur J Hum Genet 6:359–364

Apiou F, Flagiello D, Cillo C, et al. (1996) Fine mapping of human HOX gene clusters. Cytogenet Cell Genet 73:114–115

Duhig T, Ruhrberg C, Mor O, et al. (1998) The human *Surfeit* locus. Genomics 52:72–78

Ermolaeva O, Rastogi M, Pruitt KD, et al. (1998) Data management and analysis for gene expression arrays. Nat Genet 20:19–23

Feo S, Davies B, Fried M (1992) The mapping of seven intron-containing ribosomal protein genes shows they are unlinked in the human genome. Genomics 13:201–207

Flint J, Wilkie AOM, Buckle VJ, et al. (1995) The detection of subtelomeric chromosomal rearrangements in idiopathic mental retardation. Nat Genet 9:132–139

Graves JAM, Disteche CM, Toder R (1998) Gene dosage in the evolution and function of mammalian sex chromosomes. Cytogenet Cell Genet 80: 94–103

Haaf T, Sumner AT, Köhler J, et al. (1992) A microchromosome derived from chromosome 11 in a patient with the CREST syndrome of scleroderma. Cytogenet Cell Genet 60:12–17

Hacia JG, Makalowski W, Edgemon K, et al. (1998) Evolutionary sequence comparisons using high-density oligonucleotide arrays. Nat Genet 18:155–158

Harrington JJ, Bokkelen G Van, Mays RW, et al. (1997) Formation of de novo centromeres and construction of first-generation human artificial chromosomes. Nat Genet 15:345–355

Heller R, Brown KE, Burgtorf C, et al. (1996) Mini-chromosomes derived from the human Y chromosome by telomere directed chromosome breakage. Proc Natl Acad Sci USA 93:7125–7130

Ikeno M, Grimes B, Okazaki T, et al. (1998) Construction of YAC based mammalian artificial chromosomes. Nat Biotechnol 16:431–439

Liu X, Barker DF (1999) Evidence for effective suppression of recombination in the chromosome 17q21 segment spanning *RNU2–BRCA1*. Am J Hum Genet 64:1427–1439

Ohno S (1967) Sex chromosomes and sex-linked genes. Springer-Verlag, New York

Porter A (1998) Controlling your losses: conditional gene silencing in mammals. Trends Genet 14:73–79

Rassoulzadegan M, Paquis-Fluckinger V, Bertino B, et al. (1993) Transmeiotic differentiation of male germ cells in culture. Cell 75:997–1006

Rouquier S, Taviaux S, Trask BJ, et al. (1998) Distribution of olfactory receptor genes in the human genome. Nat Genet 18:243–250

Schmalzing D, Koutny L, Adoutian A, et al. (1997) DNA typing in thirty seconds with a microfabricated device. Proc Natl Acad Sci USA 94: 10273–10278

Spellman PT, Sherlock G, Zhang MO, et al. (1998) Comprehensive identification of cell cycle–regulated genes of the yeast Saccharomyces cerevisiae by microarray hybridization. Mol Biol Cell 9:3273–3297

Still IH, Vince P, Cowell JK (1997) Direct isolation of human transcribed sequences from yeast artificial chromosomes through the application of RNA fingerprinting. Proc Natl Acad Sci USA 94:10373–10378

Tanaka S, Kunath T, Hadjantonakis A-K, et al. (1998) Promotion of trophoblast stem cell proliferation by FGF4. Science 282:2072–2075

Tanimoto K, Liu Q, Bungert J, et al. (1999) Effects of altered gene order or orientation of the locus control region on human β-globin gene expression in mice. Nature 398:344–348

Thomsen JA, Itskowitz-Eldor J, Shapiro SS, et al. (1998) Embryonic stem cell lines derived from human blastocysts. Science 282:1145–1147

Tomizuka K, Yoshida H, Uejima H, et al. (1997) Functional expression and germline transmission of a human chromosome fragment in chimaeric mice. Nat Genet 16:133–143

Tonegawa S (1983) Somatic generation of antibody diversity. Nature 302:575–581

Index

Printed by Publishers' Graphics LLC